Electromagnetic Radiation:
Variational Methods, Waveguides and Accelerators

Kimball A. Milton   J. Schwinger

Electromagnetic Radiation: Variational Methods, Waveguides and Accelerators

 Springer

Kimball A. Milton

University of Oklahoma
Homer L. Dodge
Department of Physics and Astronomy
Norman, OK 73019, USA
E-mail: milton@nhn.ou.edu

Julian Schwinger

(1918-1994)

Library of Congress Control Number: 2005938671

ISBN-10   3-540-29304-3 Springer Berlin Heidelberg New York
ISBN-13   978-3-540-29304-0 Springer Berlin Heidelberg New York

Springer is a part of Springer Science+Business Media
springer.com
© Springer-Verlag Berlin Heidelberg 2006
Printed in The Netherlands

Typesetting: by the authors and techbooks using a Springer LaTeX macro package
Cover design: *design & production* GmbH, Heidelberg

Printed on acid-free paper   SPIN: 10907719   54/techbooks   5 4 3 2 1 0

We dedicate this book
to our wives, Margarita Baños-Milton and Clarice Schwinger.

Preface

Julian Schwinger was already the world's leading nuclear theorist when he joined the Radiation Laboratory at MIT in 1943, at the ripe age of 25. Just 2 years earlier he had joined the faculty at Purdue, after a postdoc with Oppenheimer in Berkeley, and graduate study at Columbia. An early semester at Wisconsin had confirmed his penchant to work at night, so as not to have to interact with Breit and Wigner there. He was to perfect his iconoclastic habits in his more than 2 years at the Rad Lab.[1]

Despite its deliberately misleading name, the Rad Lab was not involved in nuclear physics, which was imagined then by the educated public as a esoteric science without possible military application. Rather, the subject at hand was the perfection of radar, the beaming and reflection of microwaves which had already saved Britain from the German onslaught. Here was a technology which won the war, rather than one that prematurely ended it, at a still incalculable cost. It was partly for that reason that Schwinger joined this effort, rather than what might have appeared to be the more natural project for his awesome talents, the development of nuclear weapons at Los Alamos. He had got a bit of a taste of that at the "Metallurgical Laboratory" in Chicago, and did not much like it. Perhaps more important for his decision to go to and stay at MIT during the war was its less regimented and isolated environment. He could come into the lab at night, when everyone else was leaving, and leave in the morning, and security arrangements were minimal.

It was a fortunate decision. Schwinger accomplished a remarkable amount in 2 years, so much so that when he left for Harvard after the war was over, he brought an assistant along (Harold Levine) to help finish projects begun a mile away in Cambridge. Not only did he bring the theory of microwave cavities to a new level of perfection, but he found a way of expressing the results in a way that the engineers who would actually build the devices could understand, in terms of familiar circuit concepts of impedance and admittance. And he

---

[1] For a comprehensive treatment of Schwinger's life and work, see [1]. Selections of his writings appear in [2,3].

laid the groundwork for subsequent developments in nuclear and theoretical physics, including the perfection of variational methods and the effective range formulation of scattering.

The biggest "impedance matching" problem was that of Schwinger's hours, orthogonal to those of nearly everyone else. Communication was achieved by leaving notes on Schwinger's desk, remarkable solutions to which problems often appearing the very next day.[2] But this was too unsystematic. A compromise was worked out whereby Schwinger would come in at 4:00 p.m., and give a seminar on his work to the other members of the group. David Saxon, then a graduate student, took it on himself to type up the lectures. At first, Schwinger insisted on an infinite, nonconverging, series of corrections of these notes, but upon Uhlenbeck's insistence, he began to behave in a timely manner. Eventually, a small portion of these notes appeared as a slim volume entitled *Discontinuities in Waveguides* [5].

As the war wound down, Schwinger, like the other physicists, started thinking about applications of the newly developed technology to nuclear physics research. Thus Schwinger realized that microwaves could be used to accelerate charged particles, and invented what was dubbed the microtron. (Veksler is usually credited as author of the idea.) Everyone by then had realized that the cyclotron had been pushed to its limits by Lawrence, and schemes for circular accelerators, the betatron (for accelerating electrons by a changing magnetic field) and the synchrotron (in which microwave cavities accelerate electrons or protons, guided in a circular path by magnetic fields) were conceived by many people. There was the issue of whether electromagnetic radiation by such devices would provide a limit to the maximum energy to which an electron could be accelerated – Was the radiation coherent or not? Schwinger settled the issue, although it took years before his papers were properly published. His classical relativistic treatment of self-action was important for his later development of quantum electrodynamics. He gave a famous set of lectures on both accelerators and the concomitant radiation, as well as on waveguides, at Los Alamos on a visit there in 1945, where he and Feynman first met. Feynman, who was of the same age as Schwinger, was somewhat intimidated, because he felt that Schwinger had already accomplished so much more than he had.

The lab was supposed to publish a comprehensive series of volumes on the work accomplished during its existence, and Schwinger's closest collaborator and friend at the lab, Nathan Marcuvitz, was to be the editor of the *Waveguide*

---

[2] A noteworthy example of this was supplied by Mark Kac [4]. He had a query about a difficult evaluation of integrals of Bessel functions left on Schwinger's desk. Schwinger supplied a 40-page solution the following morning, which, unfortunately, did not agree with a limit known by Kac. Schwinger insisted he could not possibly have made an error, but after Kac had taught himself enough about Bessel functions he found the mistake: Schwinger had interpreted an indefinite integral in Watson's *Treatise on the Theory of Bessel Functions* as a definite one. Schwinger thereafter never lifted a formula from a book, but derived everything on the spot from first principles, a characteristic of his lectures throughout his career.

Handbook [6]. Marcuvitz kept insisting that Schwinger write up his work as *The Theory of Wave Guides*, which would complement Marcuvitz' practical handbook. Schwinger did labor mightily on the project for a time, and completed more than two long chapters before abandoning the enterprise. When he joined Harvard in February 1946, he taught a course on electromagnetic waves and waveguides at least twice. But the emerging problems of quantum electrodynamics caught his attention, and he never returned to classical electrodynamics while at Harvard. He did often recount how his experience with understanding radiation theory from his solution of synchrotron radiation led almost directly to his solution of quantum electrodynamics in terms of renormalization theory. In this, he had an advantage over Feynman, who insisted until quite late that vacuum polarization was not real, while Schwinger had demonstrated its reality already in 1939 in Berkeley [7].

It was not until some years after Schwinger moved to UCLA in 1971 that he seriously returned to classical electrodynamics.[3] It was probably my father-in-law Alfredo Baños Jr., who had been a part of the theory group at the Rad Lab, who in his capacity as Vice-Chairman of the Physics Department at the time suggested that Schwinger teach such a graduate course. I was Schwinger's postdoc then, and, with my colleagues, suggested that he turn those inspiring lectures into a book. The completion of that project took more than 20 years [9], and was only brought to fruition because of the efforts of the present author. In the meantime, Schwinger had undertaken a massive revision, on his own, on what was a completed, accepted manuscript, only to leave it unfinished in the mid-1980s.

These two instances of uncompleted book manuscripts are part of a larger pattern. In the early 1950s, he started to write a textbook on quantum mechanics/quantum field theory, part of which formed the basis for his famous lectures at Les Houches in 1955. The latter appeared in part only in 1970, as *Quantum Kinematics and Dynamics* [10], and only because Robert Kohler urged him to publish the notes and assisted in the process. Presumably this was envisaged at one time as part of a book on quantum field theory he had promised Addison-Wesley in 1955. At around the same time he agreed to write a long article on the "Quantum Theory of Wave Fields" for the *Handbuch der Physik*, but as Roy Glauber once told me, the real part of this volume was written by Källén, the imaginary part by Schwinger.

When he felt he really needed to set the record straight, Schwinger was able to complete a book project. He edited, with an introductory essay, a collection of papers called *Quantum Electrodynamics* [11] in 1956; and more substantially, when he had completed the initial development of source theory in the late 1960s, began writing what is now the three volumes of *Particles, Sources, and Fields* [12], because he felt that was the only way to spread his new gospel. But, in general, his excessive perfectionism may have rendered it

---

[3] That move to the West Coast also resulted in his first teaching of undergraduate courses since his first faculty job at Purdue. The resulting quantum lectures have been recently published by Springer [8].

nearly impossible to complete a textbook or monograph. This I have elsewhere termed tragic [1], because his lectures on a variety of topics have inspired generations of students, many of whom went on to become leaders in many fields. His reach could have been even wider had he had a less demanding view of what his written word should be like. But instead he typically polished and repolished his written prose until it bore little of the apparently spontaneous brilliance of his lectures (I say apparently, since his lectures were actually fully rehearsed and committed to memory), and then he would abandon the manuscript half-completed.

The current project was suggested by my editors, Alex Chao from SLAC and Chris Caron of Springer, although they had been anticipated a bit by the heroic effort of Miguel Furman at LBL who transcribed Schwinger's first fading synchrotron radiation manuscript into a form fit for publication in [2]. In spite of the antiquity of the material, they, and I, felt that there was much here that is still fresh and relevant. Since I had already made good use of the UCLA archives, it was easy to extract some more information from that rich source (28 boxes worth) of Schwinger material. I profusely thank Charlotte Brown, Curator of Special Collections, University Research Library, University of California at Los Angeles, for her invaluable help. The files from the Rad Lab now reside at the NE branch of the National Archives (NARA–Northeast Region), and I thank Joan Gearin, Archivist, for her help there. I thank the original publishers of the papers included in this volume, John Wiley and Sons, the American Physical Society, the American Institute of Physics, and Elsevier Science Publishers, for granting permission to reprint Schwinger's papers here. Special thanks go to the editor of Annals of Physics, Frank Wilczek, and the Senior Editorial Assistant for that journal, Eve Sullivan, for extraordinary assistance in making republication of the papers originally published there possible.[4] Throughout this project I have benefited from enthusiastic support from Schwinger's widow, Clarice. Most of all, I thank my wife, Margarita Baños-Milton, for her infinite patience as I continue to take on more projects than seems humanly achievable.

A brief remark about the assembly of this volume is called for. As indicated above, the heart of the present volume consists of those clearly typed and edited pages that were to make up the Rad Lab book. These manuscript pages were dated in the Winter 1945 and Spring 1946, before Schwinger left for Harvard. The bulk of Chaps. 6, 7, and 10 arise from this source. Some missing fragments were rescued from portions of hand-written manuscript. Chapter 8 seems to have lived through the years as a separate typescript entitled "Waveguides with Simple Cross Sections." Chapter 1 is based on another typed manuscript which may have been a somewhat later attempt to complete this book project. Chapter 15 is obviously based on "Radiation by Electrons in a Betatron," which is reprinted in Part II of this volume.[5] Most

---

[4,5] Refers to the hardcover edition which includes in addition the reprints of seminal papers by J. Schwinger on these topics.

of Chap. 11 is an extension of Chap. 25 of [9], the typescript of which was not discovered by me when I was writing that book. Chapters 2 and 12 were manuscripts intended for that same book. Other chapters are more or less based on various fragmentary materials, sometimes hard to decipher, found in the UCLA and Boston archives. For example, Chap. 16 is based on lectures Schwinger gave at the Rad Lab in Spring 1945, while the first part of Chap. 17 was a contract report submitted to the US Army Signal Corps in 1956. The many problems are based on those given many years later by Schwinger in his UCLA course in the early 1980s, as well as problems I have given in my recent courses at the University of Oklahoma. I have made every effort to put this material together as seamlessly as possible, but there is necessarily an unevenness to the level, a variation, to quote the Reader's Guide to [9], that "seems entirely appropriate." I hope the reader, be he student or experienced researcher, will find much of value in this volume.

Besides the subject matter, electromagnetic radiation theory, the reader will discover a second underlying theme, which formed the foundation of nearly all of Schwinger's work. That is the centrality of variational or action principles. We will see them in the first chapter, where they are used to derive conservation laws; in Chap. 4, where variational principles for harmonically varying Maxwell fields in media are deduced; in Chap. 10, where variational methods are used as an efficient calculational device for eigenvalues; in Chap. 16, where a variational principle is employed to calculate diffraction; and in the last chapter, where Schwinger's famous quantum action principle plays a central role in estimating quantum corrections. Indeed the entire enterprise is informed by the conceit that the proper formulation of any physical problem is in terms of a differential variational principle, and that such principles are not merely devices for determining equations of motion and symmetry principles, but they may be used directly as the most efficient calculational tool, because they automatically minimize errors.

I have, of course, tried to adopt uniform notations as much as possible, and adopt a consistent system of units. It is, as the recent example of the 3rd edition of David Jackson's *Electrodynamics* [13] demonstrates, impossible not to be somewhat schizophrenic about electrodynamics units. In the end, I decided to follow the path Schwinger followed in the first chapter which follows: For the microscopic theory, I use rationalized Heaviside–Lorentz units, which has the virtue that, for example, the electric and magnetic fields have the same units, and 4π does not appear in Maxwell's equations. However, when discussion is directed at practical devices, rationalized SI units are adopted. An Appendix concludes the text explaining the different systems, and how to convert easily from one to another.

St. Louis, Missouri, USA

February 2006

Kimball A. Milton

Contents

1

Maxwell's Equations

1.1 Microscopic Electrodynamics

Electromagnetic phenomena involving matter in bulk are approximately described by the Maxwell field equations, in SI units,[1]

$$\nabla \times \mathbf{H} = \frac{\partial}{\partial t}\mathbf{D} + \mathbf{J} , \quad \nabla \cdot \mathbf{D} = \rho , \tag{1.1a}$$

$$\nabla \times \mathbf{E} = -\frac{\partial}{\partial t}\mathbf{B} , \quad \nabla \cdot \mathbf{B} = 0 , \tag{1.1b}$$

together with constitutive equations of the medium which in their most common form are

$$\mathbf{D} = \varepsilon \mathbf{E} , \quad \mathbf{B} = \mu \mathbf{H} , \quad \mathbf{J} = \sigma \mathbf{E} . \tag{1.2}$$

This theory takes no cognizance of the atomic structure of matter, but rather regards matter as a continuous medium that is completely characterized by the three constants ε, μ, and σ. Here ε is the electric permittivity (or "dielectric constant"), μ is the magnetic permeability, and σ is the electric conductivity. The dependence of these material parameters on the nature of the substance, density, temperature, oscillation frequency, and so forth, is to be determined empirically. Opposed to this point of view, which we shall call macroscopic, is that initiated by Lorentz as an attempt to predict the properties of gross matter from the postulated behavior of atomic constituents. It is the twofold purpose of such a theory to deduce the Maxwell equations as an approximate consequence of more fundamental microscopic field equations and to relate the macroscopic parameters ε, μ, and σ to atomic properties. Although the macroscopic theory forms an entirely adequate basis for our work in this monograph, the qualitative information given by simple atomic models is of such value that we begin with an account of the microscopic theory.

---

[1] See the Appendix for a discussion of the different unit systems still commonly employed for electromagnetic phenomena.

1.1.1 Microscopic Charges

That attribute of matter which interacts with an electromagnetic field is electric charge. Charge is described by two quantities, the charge density $\rho(\mathbf{r}, t)$ and the current density $\mathbf{j}(\mathbf{r}, t)$. The charge density is defined by the statement that the total charge Q, within an arbitrary volume V at the time t, is represented by the volume integral [$(\mathrm{d}\mathbf{r}) = \mathrm{d}x\, \mathrm{d}y\, \mathrm{d}z$ is the element of volume]

$$Q = \int_V (\mathrm{d}\mathbf{r})\, \rho(\mathbf{r}, t) \ . \tag{1.3}$$

Of particular interest is the point charge distribution which is such that the total charge in any region including a fixed point $\mathbf{R}$ is equal to a constant q, independent of the size of the region, while the total charge in any region that does not include the point $\mathbf{R}$ vanishes. The charge density of the point distribution will be written

$$\rho(\mathbf{r}) = q\, \delta(\mathbf{r} - \mathbf{R}) \ , \tag{1.4}$$

with the δ function defined by the statements

$$\int_V (\mathrm{d}\mathbf{r})\, \delta(\mathbf{r} - \mathbf{R}) = \begin{cases} 1 \ , & \mathbf{R}\ \text{within}\ V \ , \\ 0 \ , & \mathbf{R}\ \text{not within}\ V \ . \end{cases} \tag{1.5}$$

It is a consequence of this definition that the δ function vanishes at every point save $\mathbf{R}$, and must there be sufficiently infinite to make its volume integral unity. No such function exists, of course, but it can be approximated with arbitrary precision. We need only consider, for example, the discontinuous function defined by

$$\delta_\epsilon(\mathbf{r} - \mathbf{R}) = \begin{cases} 0 \ , & |\mathbf{r} - \mathbf{R}| > \epsilon \ , \\ \frac{1}{\frac{4}{3}\pi\epsilon^3} \ , & |\mathbf{r} - \mathbf{R}| < \epsilon \ , \end{cases} \tag{1.6}$$

in the limit as $\epsilon \to 0$. Other possible representations are

$$\delta(\mathbf{r} - \mathbf{R}) = \lim_{\epsilon \to 0} \frac{1}{\pi^2} \frac{\epsilon}{(|\mathbf{r} - \mathbf{R}|^2 + \epsilon^2)^2} \ , \tag{1.7a}$$

$$\delta(\mathbf{r} - \mathbf{R}) = \lim_{\epsilon \to 0} \frac{1}{\epsilon^3} e^{-\pi|\mathbf{r} - \mathbf{R}|^2/\epsilon^2} \ . \tag{1.7b}$$

We shall not hesitate to treat the δ function as an ordinary, differentiable function.

The elementary constituents of matter, which for our purposes may be considered to be electrons and atomic nuclei, can ordinarily be treated as point charges, for their linear dimensions ($\sim 10^{-13}$ cm) are negligible in comparison with atomic distances ($\sim 10^{-8}$ cm). The charge density of a number of point charges with charges q_a located at the points $\mathbf{r}_a$, $a = 1, \ldots, n$, is

$$\rho(\mathbf{r}) = \sum_{a=1}^{n} q_a \delta(\mathbf{r} - \mathbf{r}_a) \, . \tag{1.8}$$

If the charges are in motion, the charge density will vary in time in consequence of the time dependence of $\mathbf{r}_a(t)$. The time derivative, for fixed $\mathbf{r}$, is

$$\frac{\partial}{\partial t}\rho(\mathbf{r}, t) = \sum_{a=1}^{n} q_a \mathbf{v}_a \cdot \nabla_{\mathbf{r}_a} \delta(\mathbf{r} - \mathbf{r}_a) = - \sum_{a=1}^{n} q_a \mathbf{v}_a \cdot \nabla_{\mathbf{r}} \delta(\mathbf{r} - \mathbf{r}_a) \, , \tag{1.9}$$

or

$$\frac{\partial}{\partial t}\rho(\mathbf{r}, t) + \nabla_{\mathbf{r}} \cdot \sum_{a=1}^{n} q_a \mathbf{v}_a \delta(\mathbf{r} - \mathbf{r}_a) = 0 \, , \tag{1.10}$$

where $\mathbf{v}_a = \frac{\mathrm{d}}{\mathrm{d}t}\mathbf{r}_a$ is the velocity of the ath point charge.

Charge in motion constitutes a current. The current density or charge flux vector $\mathbf{j}(\mathbf{r}, t)$ is defined by the equation

$$I = \int_S \mathrm{d}S \, \mathbf{n} \cdot \mathbf{j}(\mathbf{r}, t) \, , \tag{1.11}$$

where $I \, \mathrm{d}t$ is the net charge crossing an arbitrary surface S in the time interval $\mathrm{d}t$. Positive charge crossing the surface in the direction of the normal $\mathbf{n}$, or negative charge moving in the opposite direction, make a positive contribution to the total current I, while charges with the reversed motion from these are assigned negative weight factors in computing I. The total charge leaving an arbitrary region bounded by the closed surface S, in the time interval $\mathrm{d}t$, is

$$\mathrm{d}Q = \mathrm{d}t \oint_S \mathrm{d}S \, \mathbf{n} \cdot \mathbf{j}(\mathbf{r}, t) \, , \tag{1.12}$$

where $\mathbf{n}$ is the outward-drawn normal to the surface S. The fundamental property of charge, indeed its defining characteristic, is indestructibility. Thus the net amount of charge that flows across the surface S bounding V must equal the loss of charge within the volume. Hence

$$\oint_S \mathrm{d}S \, \mathbf{n} \cdot \mathbf{j}(\mathbf{r}, t) \equiv \int_V (\mathrm{d}\mathbf{r}) \, \nabla \cdot \mathbf{j}(\mathbf{r}, t) = -\frac{\partial}{\partial t} \int_V (\mathrm{d}\mathbf{r}) \, \rho(\mathbf{r}, t) \, , \tag{1.13}$$

in which we have also employed the divergence theorem relating surface and volume integrals. Since the statement must be valid for an arbitrary volume, we obtain as the conservation equation of electric charge

$$\nabla \cdot \mathbf{j}(\mathbf{r}, t) + \frac{\partial}{\partial t}\rho(\mathbf{r}, t) = 0 \, . \tag{1.14}$$

It will be noted that an equation of precisely this form has been obtained for an assembly of point charges in (1.10), with

$$\mathbf{j}(\mathbf{r}, t) = \sum_{a=1}^{n} q_a \mathbf{v}_a \delta(\mathbf{r} - \mathbf{r}_a) \,. \tag{1.15}$$

Thus, for a single point charge,

$$\mathbf{j} = \rho \mathbf{v} \,. \tag{1.16}$$

The elementary charged constituents of matter possess inertia. Associated with charges in motion, therefore, are the mechanical properties of kinetic energy, linear momentum, and angular momentum. The definitions of these quantities for a system of n particles with masses m_a, $a = 1, \ldots, n$, are, respectively,

$$E = \sum_{a=1}^{n} \frac{1}{2} m_a v_a^2 \,, \tag{1.17a}$$

$$\mathbf{p} = \sum_{a=1}^{n} m_a \mathbf{v}_a \,, \tag{1.17b}$$

$$\mathbf{L} = \sum_{a=1}^{n} m_a \mathbf{r}_a \times \mathbf{v}_a \,, \tag{1.17c}$$

provided all particle velocities are small in comparison with c, the velocity of light in vacuo. The more rigorous relativistic expressions are

$$E = \sum_{a=1}^{n} m_a c^2 \left(\frac{1}{\sqrt{1 - v_a^2/c^2}} - 1 \right) \,, \tag{1.18a}$$

$$\mathbf{p} = \sum_{a=1}^{n} \frac{m_a}{\sqrt{1 - v_a^2/c^2}} \mathbf{v}_a \,, \tag{1.18b}$$

$$\mathbf{L} = \sum_{a=1}^{n} \frac{m_a}{\sqrt{1 - v_a^2/c^2}} \mathbf{r}_a \times \mathbf{v}_a \,, \tag{1.18c}$$

but this refinement is rarely required in studies of atomic structure.

1.1.2 The Field Equations

The electromagnetic field is described by two vectors, the electric field intensity (or electric field strength) $\mathbf{e}(\mathbf{r}, t)$ and the magnetic field intensity (or magnetic induction) $\mathbf{b}(\mathbf{r}, t)$. [In this chapter, for pedagogical purposes, we will use lowercase letters to denote the microscopic fields, for which we will use (rationalized) Heaviside–Lorentz units. See the Appendix.] The equations defining these vectors in relation to each other and to the charge–current distribution are postulated to be

$$\boldsymbol{\nabla} \times \mathbf{b} = \frac{1}{c}\frac{\partial}{\partial t}\mathbf{e} + \frac{1}{c}\mathbf{j}, \quad \boldsymbol{\nabla} \cdot \mathbf{e} = \rho, \tag{1.19a}$$

$$\boldsymbol{\nabla} \times \mathbf{e} = -\frac{1}{c}\frac{\partial}{\partial t}\mathbf{b}, \quad \boldsymbol{\nabla} \cdot \mathbf{b} = 0, \tag{1.19b}$$

which are known variously as the microscopic field equations, or the Maxwell–Lorentz equations. Correspondence is established with the physical world by the further postulate that an electromagnetic field possesses the mechanical attributes of energy and momentum. These quantities are considered to be spatially distributed in the field, and it is therefore necessary to introduce not only measures of density, analogous to the charge density, but in addition measures of flux, analogous to the current density. We define

- energy density:

$$U = \frac{e^2 + b^2}{2}, \tag{1.20a}$$

- energy flux vector or the Poynting vector:

$$\mathbf{S} = c\,\mathbf{e} \times \mathbf{b}, \tag{1.20b}$$

- linear momentum density:

$$\mathbf{G} = \frac{1}{c}\mathbf{e} \times \mathbf{b}, \tag{1.20c}$$

- linear momentum flux dyadic or the stress dyadic:

$$\mathsf{T} = \mathsf{1}\frac{e^2 + b^2}{2} - \mathbf{ee} - \mathbf{bb}, \tag{1.20d}$$

The symbol 1 indicates the unit dyadic. The basis for these definitions are certain differential identities, valid in the absence of charge and current, which have the form of conservation equations, analogous to that for electric charge. It may be directly verified that ($\rho = 0$, $\mathbf{j} = \mathbf{0}$)

$$\frac{\partial}{\partial t}U + \boldsymbol{\nabla} \cdot \mathbf{S} = 0, \qquad \frac{\partial}{\partial t}\mathbf{G} + \boldsymbol{\nabla} \cdot \mathsf{T} = \mathbf{0}, \tag{1.21}$$

on employing the identities

$$\boldsymbol{\nabla} \cdot (\mathbf{A} \times \mathbf{B}) = (\boldsymbol{\nabla} \times \mathbf{A}) \cdot \mathbf{B} - (\boldsymbol{\nabla} \times \mathbf{B}) \cdot \mathbf{A}, \tag{1.22a}$$

$$(\boldsymbol{\nabla} \times \mathbf{A}) \times \mathbf{A} = -\mathbf{A} \times (\boldsymbol{\nabla} \times \mathbf{A}) = (\mathbf{A} \cdot \boldsymbol{\nabla})\mathbf{A} - \boldsymbol{\nabla}\frac{1}{2}A^2. \tag{1.22b}$$

The total energy,

$$E = \int (\mathbf{dr})\, U, \tag{1.23}$$

and the total linear momentum,

$$\mathbf{p} = \int (\mathrm{d}\mathbf{r})\,\mathbf{G}\ , \tag{1.24}$$

of an electromagnetic field confined to a finite region of space, are constant in time, for no energy or momentum flows through a surface enclosing the entire field. Energy and momentum, like charge, are recognized by the property of permanence.

The relation between the energy and momentum quantities expressed by

$$\mathbf{S} = c^2\mathbf{G} \tag{1.25}$$

is a consequence of, or at least is consistent with, the relativistic connection between energy and mass,

$$E = mc^2\ . \tag{1.26}$$

This may be seen from the remark that the momentum density can also be considered a mass flux vector, or alternatively, by the following considerations. On multiplying the energy conservation equation in (1.21) by $\mathbf{r}$ and rearranging terms, we obtain

$$\frac{\partial}{\partial t}\mathbf{r}\,U + \boldsymbol{\nabla}\cdot(\mathbf{S}\mathbf{r}) = \mathbf{S} = c^2\mathbf{G}\ , \tag{1.27}$$

which, on integration over a volume enclosing the entire field, yields

$$\mathbf{p} = \frac{\mathrm{d}}{\mathrm{d}t}\int (\mathrm{d}\mathbf{r})\,\mathbf{r}\,\frac{U}{c^2} = \frac{E}{c^2}\frac{\mathrm{d}\mathbf{R}}{\mathrm{d}t} = \frac{E}{c^2}\mathbf{V}\ , \tag{1.28}$$

where

$$\mathbf{R} = \frac{1}{E}\int (\mathrm{d}\mathbf{r})\,\mathbf{r}\,U \tag{1.29}$$

is the energy center of gravity of the field, which moves with velocity $\mathbf{V} = \mathrm{d}\mathbf{R}/\mathrm{d}t$. Here we have the conventional relation between momentum and velocity, with E/c^2 playing the role of the total mass of the electromagnetic field.

The velocity of the energy center of gravity, $\mathbf{V}$, which we shall term the group velocity of the field, is necessarily less in magnitude than the velocity of light. This is a result of the identity

$$(\mathbf{e}\times\mathbf{b})^2 = \left(\frac{e^2 + b^2}{2}\right)^2 - \left(\frac{e^2 - b^2}{2}\right)^2 - (\mathbf{e}\cdot\mathbf{b})^2\ , \tag{1.30}$$

and the consequent inequality

$$|\mathbf{e}\times\mathbf{b}| \le \frac{e^2 + b^2}{2}\ , \tag{1.31}$$

for from (1.24)

$$|\mathbf{p}| \le \frac{1}{c} \int (d\mathbf{r}) |\mathbf{e} \times \mathbf{b}| \le \frac{E}{c} , \qquad (1.32)$$

and therefore

$$|\mathbf{V}| \le c . \qquad (1.33)$$

Equality of $|\mathbf{V}|$ with c is obtained only when $\mathbf{e} \cdot \mathbf{b} = 0$, $e^2 = b^2$, and $\mathbf{e} \times \mathbf{b}$ has the same direction everywhere. That is, the electric and magnetic field intensities must be equal in magnitude, perpendicular to each other, and to a fixed direction in space, as is the case for an ideal plane wave. More generally, we call such a configuration a unidirectional light pulse, for which further properties are given in Problem 1.34.

Another velocity associated with the field can be defined in terms of the center of gravity of the momentum distribution. We proceed from the conservation of momentum equation in (1.21) written, for manipulatory convenience, in component form,

$$\frac{\partial}{\partial t}G_j + \sum_{i=1}^{3} \frac{\partial}{\partial x_i}T_{ij} = 0 , \quad j = 1, 2, 3 . \qquad (1.34)$$

On multiplying this equation by x_j, and summing with respect to the index j, we obtain

$$\frac{\partial}{\partial t}\sum_i x_i G_i + \sum_{i,j} \frac{\partial}{\partial x_i}(T_{ij}x_j) = \sum_i T_{ii} \equiv \mathrm{Tr}\,\mathsf{T} , \qquad (1.35)$$

(which introduces the concept of the trace of the dyadic T, $\mathrm{Tr}\,\mathsf{T}$) or, returning to vector notation,

$$\frac{\partial}{\partial t}(\mathbf{r} \cdot \mathbf{G}) + \boldsymbol{\nabla} \cdot (\mathsf{T} \cdot \mathbf{r}) = U , \qquad (1.36)$$

for (note that we do not use the summation convention over repeated indices here)

$$T_{ii} = U - (e_i^2 + b_i^2) , \qquad \mathrm{Tr}\,\mathsf{T} = U . \qquad (1.37)$$

The relation (1.36) thus established between the energy density and momentum quantities we shall call the virial theorem. On integration over the entire region occupied by the field, we find

$$E = \frac{d}{dt}\int (d\mathbf{r})\,\mathbf{r} \cdot \mathbf{G} \equiv \mathbf{W} \cdot \mathbf{p} , \qquad (1.38)$$

which defines a velocity $\mathbf{W}$, or at least its component parallel to $\mathbf{p}$, which we shall term the phase velocity of the field. Combining the two relations between the total energy and momentum, (1.28) and (1.38), we obtain

$$\mathbf{W} \cdot \mathbf{V} = c^2 , \qquad (1.39)$$

which implies that the magnitude of the phase velocity is never less than the speed of light.

A further conservation theorem, which is to be identified as that for angular momentum, can be deduced from the linear momentum conservation theorem. Multiplying the jth component of (1.34) by x_i and subtracting a similar equation with i and j interchanged, we obtain

$$\frac{\partial}{\partial t}(x_i G_j - x_j G_i) = \sum_k \frac{\partial}{\partial x_k}(T_{ki}x_j - T_{kj}x_i) + T_{ij} - T_{ji} , \qquad (1.40)$$

However, the stress dyadic is symmetrical,

$$T_{ij} = \delta_{ij}\frac{e^2 + b^2}{2} - e_i e_j - b_i b_j = T_{ji} , \qquad (1.41)$$

and therefore (in vector notation)

$$\frac{\partial}{\partial t}(\mathbf{r} \times \mathbf{G}) + \boldsymbol{\nabla} \cdot (-\mathsf{T} \times \mathbf{r}) = \mathbf{0} , \qquad (1.42)$$

which implies that the total angular momentum

$$\mathbf{L} = \int (\mathrm{d}\mathbf{r})\,\mathbf{r} \times \mathbf{G} \qquad (1.43)$$

of a field confined to a finite spatial volume is constant in time.

In the presence of electric charge, the energy and momentum of the electromagnetic field are no longer conserved. It is easily shown that

$$\frac{\partial}{\partial t}U + \boldsymbol{\nabla} \cdot \mathbf{S} = -\mathbf{j} \cdot \mathbf{e} , \qquad (1.44a)$$

$$\frac{\partial}{\partial t}\mathbf{G} + \boldsymbol{\nabla} \cdot \mathsf{T} = -(\rho\,\mathbf{e} + \frac{1}{c}\mathbf{j} \times \mathbf{b}) , \qquad (1.44b)$$

implying that electromagnetic energy is destroyed at the rate of $\mathbf{j} \cdot \mathbf{e}$ per unit volume, and that $\rho\,\mathbf{e} + \frac{1}{c}\mathbf{j} \times \mathbf{b}$ measures the rate of annihilation of linear electromagnetic momentum, per unit volume. In a region that includes only the ath elementary charge, electromagnetic energy and momentum disappear at a rate $q_a \mathbf{v}_a \cdot \mathbf{e}(\mathbf{r}_a)$, and $q_a \left(\mathbf{e}(\mathbf{r}_a) + \frac{1}{c}\mathbf{v}_a \times \mathbf{b}(\mathbf{r}_a)\right)$, respectively. If the indestructibility of energy and momentum is to be preserved, these expressions must equal the rate of increase of the energy and linear momentum of the ath elementary charge,

$$\frac{\mathrm{d}E_a}{\mathrm{d}t} = q_a \mathbf{v}_a \cdot \mathbf{e}(\mathbf{r}_a) , \qquad (1.45a)$$

$$\frac{\mathrm{d}\mathbf{p}_a}{\mathrm{d}t} = q_a \left(\mathbf{e}(\mathbf{r}_a) + \frac{1}{c}\mathbf{v}_a \times \mathbf{b}(\mathbf{r}_a)\right) = \mathbf{F}_a , \qquad (1.45b)$$

which determines the force, $\mathbf{F}_a$, exerted on the ath charge by the electromagnetic field, in terms of the rate of change of mechanical momentum $\mathbf{p}_a = m_a \mathbf{v}_a$. The consistency of the definitions adopted for field energy and momentum is verified by the observation that the rate of increase of the energy of the ath particle, in accord with mechanical principles, is equal to the rate at which the force $\mathbf{F}_a$ does work on the particle,

$$\frac{\mathrm{d}E_a}{\mathrm{d}t} = \mathbf{F}_a \cdot \mathbf{v}_a \,. \tag{1.46}$$

In a similar fashion, the rate of loss of electromagnetic angular momentum per unit volume $\mathbf{r} \times \left(\rho\mathbf{e} + \frac{1}{c}\mathbf{j} \times \mathbf{b}\right)$, when integrated over a region enclosing the ath charge, must equal the rate of increase of $\mathbf{L}_a$, the angular momentum of the particle;

$$\frac{\mathrm{d}\mathbf{L}_a}{\mathrm{d}t} = q_a\mathbf{r}_a \times \left(\mathbf{e}(\mathbf{r}_a) + \frac{1}{c}\mathbf{v}_a \times \mathbf{b}(\mathbf{r}_a)\right) = \mathbf{r}_a \times \mathbf{F}_a \,. \tag{1.47}$$

The identification of electromagnetic angular momentum is confirmed by this result, that the rate at which the angular momentum of the particle increases equals the moment of the force acting on it. For a further discussion of the local conservation of energy and momentum, see Problem 1.31.

1.2 Variational Principle

The equations of motion of the field and matter can be expressed in the compact form of a variational principle or Hamilton's principle. It is first convenient to introduce suitable coordinates for the field. These we shall choose as the vector potential $\mathbf{a}$ and the scalar potential ϕ, defined by

$$\mathbf{e} = -\frac{1}{c}\frac{\partial}{\partial t}\mathbf{a} - \boldsymbol{\nabla}\phi \,, \qquad \mathbf{b} = \boldsymbol{\nabla} \times \mathbf{a} \,, \tag{1.48}$$

which ensures that the second set of field equations (1.19b) is satisfied identically. The potentials are not uniquely determined by these equations; rather, the set of potentials

$$\mathbf{a}' = \mathbf{a} - \boldsymbol{\nabla}\psi \,, \qquad \phi' = \phi + \frac{1}{c}\frac{\partial}{\partial t}\psi \tag{1.49}$$

leads to the same field intensities as $\mathbf{a}$ and ϕ, for arbitrary ψ. Such a modification of the potentials is referred to as a gauge transformation, and those quantities which are unaltered by the transformation are called gauge invariant. The absence of a precise definition for the potentials will cause no difficulty provided that all physical quantities expressed in terms of the potentials are required to be gauge invariant.

A mechanical system is completely characterized by a Lagrangian L, which is such that $\int_{t_0}^{t_1} \mathrm{d}t \, L$ is an extremal for the actual motion of the system, in comparison with all neighboring states with prescribed values of the coordinates at times t_0 and t_1,

$$\delta \int_{t_0}^{t_1} dt\, L = 0\,. \tag{1.50}$$

We consider a general Lagrangian for the system of fields and matter which depends upon the positions and velocities of the particles, and the potentials and field quantities descriptive of the field. From the standpoint of the field, the Lagrangian is best regarded as the volume integral of a Lagrangian density $\mathcal{L}$. Thus, the effect of an arbitrary variation of the vector potential is expressed by

$$\delta_{\mathbf{a}} L = \int (d\mathbf{r}) \left(\frac{\partial \mathcal{L}}{\partial \mathbf{a}} \cdot \delta \mathbf{a} + \frac{\partial \mathcal{L}}{\partial \mathbf{b}} \cdot \boldsymbol{\nabla} \times \delta \mathbf{a} - \frac{1}{c}\frac{\partial \mathcal{L}}{\partial \mathbf{e}} \cdot \delta \dot{\mathbf{a}} \right)$$
$$= \int (d\mathbf{r}) \left(\frac{\delta L}{\delta \mathbf{a}} \cdot \delta \mathbf{a} + \frac{\delta L}{\delta \dot{\mathbf{a}}} \cdot \delta \dot{\mathbf{a}} \right)\,, \tag{1.51}$$

in which we have introduced the variational derivatives,

$$\frac{\delta L}{\delta \mathbf{a}} = \frac{\partial \mathcal{L}}{\partial \mathbf{a}} + \boldsymbol{\nabla} \times \frac{\partial \mathcal{L}}{\partial \mathbf{b}}\,, \tag{1.52a}$$

$$\frac{\delta L}{\delta \dot{\mathbf{a}}} = -\frac{1}{c}\frac{\partial \mathcal{L}}{\partial \mathbf{e}}\,, \tag{1.52b}$$

and discarded a surface integral by requiring that all variations vanish on the spatial boundary of the region, as well as at the initial and terminal times t_0 and t_1. In a similar fashion,

$$\delta_{\phi} L = \int (d\mathbf{r}) \left(\frac{\partial \mathcal{L}}{\partial \phi} \delta\phi - \frac{\partial \mathcal{L}}{\partial \mathbf{e}} \cdot \boldsymbol{\nabla}\delta\phi \right) = \int (d\mathbf{r}) \frac{\delta L}{\delta \phi} \delta\phi\,, \tag{1.53}$$

with

$$\frac{\delta L}{\delta \phi} = \frac{\partial \mathcal{L}}{\partial \phi} + \boldsymbol{\nabla} \cdot \left(\frac{\partial \mathcal{L}}{\partial \mathbf{e}} \right)\,, \tag{1.54}$$

provided the time derivative of the scalar potential is absent in the Lagrangian. These relations, (1.51) and (1.53), expressed in terms of variational derivatives, are formally analogous to the variation of a Lagrangian associated with a material particle's coordinates,

$$\delta_{\mathbf{r}_a} L = \frac{\partial L}{\partial \mathbf{r}_a} \cdot \delta \mathbf{r}_a + \frac{\partial L}{\partial \mathbf{v}_a} \cdot \frac{d}{dt}\delta \mathbf{r}_a\,. \tag{1.55}$$

Therefore, the condition expressing the stationary character of $\int_{t_0}^{t_1} dt\, L$ for variations of $\mathbf{r}_a$, subject to the vanishing of all variations at the termini,

$$\frac{d}{dt}\frac{\partial L}{\partial \mathbf{v}_a} = \frac{\partial L}{\partial \mathbf{r}_a}\,, \tag{1.56}$$

has a formally similar aspect for variations of $\mathbf{a}$ and ϕ,

$$\frac{\partial}{\partial t}\frac{\delta L}{\delta \mathbf{\dot{a}}} = \frac{\delta L}{\delta \mathbf{a}} , \qquad 0 = \frac{\delta L}{\delta \phi} . \tag{1.57}$$

Hence, the field equations deduced from a variational principle are

$$-\nabla \times \frac{\partial \mathcal{L}}{\partial \mathbf{b}} = \frac{1}{c}\frac{\partial}{\partial t}\frac{\partial \mathcal{L}}{\partial \mathbf{e}} + \frac{\partial \mathcal{L}}{\partial \mathbf{a}} , \qquad \nabla \cdot \frac{\partial \mathcal{L}}{\partial \mathbf{e}} = -\frac{\partial \mathcal{L}}{\partial \phi} , \tag{1.58}$$

which are identical with the Maxwell–Lorentz equations (1.19a) if

$$\mathcal{L} = \frac{e^2 - b^2}{2} - \rho\phi + \frac{1}{c}\mathbf{j} \cdot \mathbf{a} . \tag{1.59}$$

The Lagrangian thus consists of a part involving only the field quantities,

$$L_f = \int (d\mathbf{r}) \frac{e^2 - b^2}{2} , \tag{1.60}$$

a part containing the coordinates of both field and matter,

$$L_{fm} = -\int (d\mathbf{r}) \left(\rho\phi - \frac{1}{c}\mathbf{j} \cdot \mathbf{a} \right) = -\sum_a q_a \left(\phi(\mathbf{r}_a) - \frac{1}{c}\mathbf{v}_a \cdot \mathbf{a}(\mathbf{r}_a) \right) , \tag{1.61}$$

and a part involving only material quantities, which, as we shall verify, is for nonrelativistic particles

$$L_m = \sum_a \frac{1}{2} m_a v_a^2 . \tag{1.62}$$

(For the relativistic generalization, see Problem 1.32.) The Lagrangian form of the ath particle's equation of motion (1.56) is

$$\frac{d}{dt}\left(m_a \mathbf{v}_a + \frac{q_a}{c}\mathbf{a}(\mathbf{r}_a) \right) = -q_a \nabla_{\mathbf{r}_a}\left(\phi(\mathbf{r}_a) - \frac{1}{c}\mathbf{v}_a \cdot \mathbf{a}(\mathbf{r}_a) \right) , \tag{1.63}$$

where we see the appearance of the canonical momentum,

$$\boldsymbol{\pi}_a = m_a \mathbf{v}_a + \frac{q_a}{c}\mathbf{a}(\mathbf{r}_a) . \tag{1.64}$$

However,

$$\frac{d}{dt}\mathbf{a}(\mathbf{r}_a, t) = \frac{\partial}{\partial t}\mathbf{a} + \mathbf{v}_a \cdot \nabla\mathbf{a} = \frac{\partial}{\partial t}\mathbf{a} - \mathbf{v}_a \times \mathbf{b} + \nabla(\mathbf{v}_a \cdot \mathbf{a}) , \tag{1.65}$$

for in computing the time derivative, the implicit dependence of the particle's position on the time cannot be ignored. It is thus confirmed that the Lorentz force law (1.45b) holds,

$$\frac{d}{dt} m_a \mathbf{v}_a = q_a \left(\mathbf{e}(\mathbf{r}_a) + \frac{1}{c}\mathbf{v}_a \times \mathbf{b}(\mathbf{r}_a) \right) . \tag{1.66}$$

1.3 Conservation Theorems

The various conservation laws, those of charge, energy, linear momentum, and angular momentum, are consequences of the invariance of Hamilton's principle under certain transformations. These are, respectively, gauge transformations, temporal displacements, spatial translations, and spatial rotations. A gauge transformation (1.49) induces the variation

$$\delta\mathbf{a} = -\boldsymbol{\nabla}\psi\,, \qquad \delta\phi = \frac{1}{c}\frac{\partial}{\partial t}\psi\,, \tag{1.67a}$$

$$\delta\mathbf{e} = \delta\mathbf{b} = \mathbf{0}\,, \tag{1.67b}$$

whence

$$\delta L = \int (\mathrm{d}\mathbf{r})\left(\boldsymbol{\nabla}\cdot\frac{\partial\mathcal{L}}{\partial\mathbf{a}} - \frac{1}{c}\frac{\partial}{\partial t}\frac{\partial\mathcal{L}}{\partial\phi}\right)\psi + \frac{1}{c}\frac{\mathrm{d}}{\mathrm{d}t}\int (\mathrm{d}\mathbf{r})\frac{\partial\mathcal{L}}{\partial\phi}\,\psi\,, \tag{1.68}$$

from which we can infer from (1.59) that the local charge conservation equation,

$$\boldsymbol{\nabla}\cdot\left(c\frac{\partial\mathcal{L}}{\partial\mathbf{a}}\right) + \frac{\partial}{\partial t}\left(-\frac{\partial\mathcal{L}}{\partial\phi}\right) = \boldsymbol{\nabla}\cdot\mathbf{j} + \frac{\partial}{\partial t}\rho = 0\,, \tag{1.69}$$

must be a consequence of the field equations, for $\int_{t_0}^{t_1}\mathrm{d}t\, L$ is stationary with respect to arbitrary independent variations of $\mathbf{a}$ and ϕ.

The value of $\int_{t_0}^{t_1}\mathrm{d}t\, L$ is in no way affected by an alteration of the time origin,

$$\int_{t_0-\delta t}^{t_1-\delta t}\mathrm{d}t\, L(t + \delta t) - \int_{t_0}^{t_1}\mathrm{d}t\, L(t) = 0\,, \tag{1.70}$$

where δt is an arbitrary constant. We may conceive of the time displacement as a variation of the system's coordinates which consists in replacing the actual values at time t by the actual values which the system will assume at time $t + \delta t$. The statement of invariance with respect to the origin of time now reads

$$\int_{t_0}^{t_1}\mathrm{d}t\,\delta L = \delta t\,(L(t_1) - L(t_0)) = \delta t\int_{t_0}^{t_1}\mathrm{d}t\,\frac{\mathrm{d}L}{\mathrm{d}t}\,, \tag{1.71}$$

where δL is the consequence of the variations

$$\delta\mathbf{a} = \delta t\,\dot{\mathbf{a}}\,, \quad \delta\phi = \delta t\,\dot{\phi}\,, \quad \delta\mathbf{r}_a = \delta t\,\mathbf{v}_a\,, \tag{1.72a}$$

$$\delta L = \delta t\left[\int (\mathrm{d}\mathbf{r})\left(\frac{\delta L}{\delta\mathbf{a}}\cdot\dot{\mathbf{a}} + \frac{\delta L}{\delta\dot{\mathbf{a}}}\cdot\ddot{\mathbf{a}} + \frac{\delta L}{\delta\phi}\cdot\dot{\phi}\right) + \sum_a\left(\frac{\partial L}{\partial\mathbf{r}_a}\cdot\mathbf{v}_a + \frac{\partial L}{\partial\mathbf{v}_a}\cdot\dot{\mathbf{v}}_a\right)\right]\,. \tag{1.72b}$$

In writing this expression for δL various surface integrals have been discarded. This can no longer be justified by the statement that the variation vanishes at the surface of the integration region, for it is not possible to satisfy this condition with the limited type of variation that is being contemplated. Rather, it is assumed for simplicity that the volume integration encompasses the entire field. On rearranging the terms of δL and employing the Lagrangian equations of motion (1.56) and (1.57), we obtain

$$\delta L = \delta t \frac{d}{dt} \left(\int (\mathrm{dr}) \frac{\delta L}{\delta \dot{\mathbf{a}}} \cdot \dot{\mathbf{a}} + \sum_a \frac{\partial L}{\partial \mathbf{v}_a} \cdot \mathbf{v}_a \right) , \tag{1.73}$$

from which it follows from (1.71) that

$$E = \int (\mathrm{dr}) \frac{\delta L}{\delta \dot{\mathbf{a}}} \cdot \dot{\mathbf{a}} + \sum_a \frac{\partial L}{\partial \mathbf{v}_a} \cdot \mathbf{v}_a - L \tag{1.74}$$

is independent of time. It is easily verified from (1.59) that E is the total energy of the system,

$$E = \int (\mathrm{dr}) \frac{e^2 + b^2}{2} + \sum_a \frac{1}{2} m_a v_a^2 . \tag{1.75}$$

The Lagrangian is unaltered by an arbitrary translation of the position variable of integration, that is, if $\mathbf{r}$ is replaced by $\mathbf{r} + \delta \mathbf{r}$, with $\delta \mathbf{r}$ an arbitrary constant vector. The region of integration must be suitably modified, of course, but this need not be considered if the entire field is included, for the limits of integration are then effectively infinite. Under this substitution, the matter part of the Lagrangian, which corresponds to the Lagrange density $\mathcal{L}_m(\mathbf{r}) = L_m(\mathbf{r}) \delta(\mathbf{r} - \mathbf{r}_a)$, is replaced by $L_m(\mathbf{r} + \delta \mathbf{r}) \delta(\mathbf{r} + \delta \mathbf{r} - \mathbf{r}_a)$. Hence, viewed as the variation

$$\delta \mathbf{a} = (\delta \mathbf{r} \cdot \nabla) \mathbf{a} , \quad \delta \phi = (\delta \mathbf{r} \cdot \nabla) \phi , \quad \delta \mathbf{r}_a = -\delta \mathbf{r} , \tag{1.76}$$

the translation of the space coordinate system induces a variation of

$$\delta L = \int (\mathrm{dr}) \left[\frac{\delta L}{\delta \mathbf{a}} \cdot (\delta \mathbf{r} \cdot \nabla) \mathbf{a} + \frac{\delta L}{\delta \dot{\mathbf{a}}} \cdot (\delta \mathbf{r} \cdot \nabla) \dot{\mathbf{a}} + \frac{\delta L}{\delta \phi} (\delta \mathbf{r} \cdot \nabla) \phi \right]$$
$$- \sum_a \frac{\partial L}{\partial \mathbf{r}_a} \cdot \delta \mathbf{r} , \tag{1.77}$$

which must be zero. As a consequence of the Lagrangian equations of motion (1.56) and (1.57) and the relations

$$(\delta \mathbf{r} \cdot \nabla) \mathbf{a} = \nabla (\delta \mathbf{r} \cdot \mathbf{a}) + \mathbf{b} \times \delta \mathbf{r} , \tag{1.78a}$$

$$\nabla \cdot \left(\frac{\delta L}{\delta \dot{\mathbf{a}}} \right) = -\frac{1}{c} \nabla \cdot \left(\frac{\partial \mathcal{L}}{\partial \mathbf{e}} \right) = -\frac{1}{c} \rho , \tag{1.78b}$$

we obtain

$$\delta L = -\frac{d}{dt}\left[-\int (dr)\frac{\delta L}{\delta \dot{\mathbf{a}}}\times \mathbf{b} + \sum_a \left(\frac{\partial L}{\partial \mathbf{v}_a} - \frac{1}{c}q_a\mathbf{a}(\mathbf{r}_a)\right)\right]\cdot \delta \mathbf{r} = 0\,. \quad (1.79)$$

Therefore,

$$\mathbf{P} = -\int (dr)\frac{\delta L}{\delta \dot{\mathbf{a}}}\times \mathbf{b} + \sum_a \left(\frac{\partial L}{\partial \mathbf{v}_a} - \frac{1}{c}q_a\mathbf{a}(\mathbf{r}_a)\right)$$

$$= \int (dr)\frac{1}{c}\mathbf{e}\times \mathbf{b} + \sum_a m_a\mathbf{v}_a\,, \quad (1.80)$$

the total linear momentum of the system, must be constant in time.

Similar considerations are applicable to a rotation of the coordinate system. The infinitesimal rotation

$$\mathbf{r} \to \mathbf{r} + \boldsymbol{\epsilon}\times \mathbf{r} \quad (1.81)$$

induces the variation (because $\mathbf{a}$, like $\mathbf{r}_a$, is a vector)

$$\delta \mathbf{a} = (\boldsymbol{\epsilon}\times \mathbf{r}\cdot \boldsymbol{\nabla})\mathbf{a} - \boldsymbol{\epsilon}\times \mathbf{a}\,, \quad \delta \phi = (\boldsymbol{\epsilon}\times \mathbf{r}\cdot \boldsymbol{\nabla})\phi\,, \quad \delta \mathbf{r}_a = -\boldsymbol{\epsilon}\times \mathbf{r}_a\,, \quad (1.82)$$

which must leave the Lagrangian unaltered,

$$\delta L = \int (dr)\left\{\frac{\delta L}{\delta \mathbf{a}}\cdot [(\boldsymbol{\epsilon}\cdot \mathbf{r}\times \boldsymbol{\nabla})\mathbf{a} - \boldsymbol{\epsilon}\times \mathbf{a}] + \frac{\delta L}{\delta \dot{\mathbf{a}}}\cdot [(\boldsymbol{\epsilon}\cdot \mathbf{r}\times \boldsymbol{\nabla})\dot{\mathbf{a}} - \boldsymbol{\epsilon}\times \dot{\mathbf{a}}]\right.$$

$$\left. + \frac{\delta L}{\delta \phi}(\boldsymbol{\epsilon}\cdot \mathbf{r}\times \boldsymbol{\nabla})\phi\right\} - \sum_a \boldsymbol{\epsilon}\cdot \mathbf{r}_a\times \frac{\partial L}{\partial \mathbf{r}_a} - \sum_a \boldsymbol{\epsilon}\cdot \mathbf{v}_a\times \frac{\partial L}{\partial \mathbf{v}_a} = 0\,. \quad (1.83)$$

However, again using (1.78b),

$$\delta L = -\frac{d}{dt}\left[-\int (dr)\,\mathbf{r}\times \left(\frac{\delta L}{\delta \dot{\mathbf{a}}}\times \mathbf{b}\right) + \sum_a \mathbf{r}_a\times \left(\frac{\partial L}{\partial \mathbf{v}_a} - \frac{1}{c}q_a\mathbf{a}(\mathbf{r}_a)\right)\right]\cdot \boldsymbol{\epsilon}\,, \quad (1.84)$$

in consequence of the identity

$$(\boldsymbol{\epsilon}\cdot \mathbf{r}\times \boldsymbol{\nabla})\mathbf{a} - \boldsymbol{\epsilon}\times \mathbf{a} = \boldsymbol{\nabla}(\boldsymbol{\epsilon}\cdot \mathbf{r}\times \mathbf{a}) + \mathbf{b}\times (\boldsymbol{\epsilon}\times \mathbf{r})\,, \quad (1.85)$$

we conclude that

$$\mathbf{L} = \int (dr)\,\mathbf{r}\times \left(\frac{1}{c}\mathbf{e}\times \mathbf{b}\right) + \sum_a m_a\mathbf{r}_a\times \mathbf{v}_a\,, \quad (1.86)$$

the total angular momentum is unchanged in time.

1.4 Delta Function

Preparatory to determining the fields produced by given distributions of charge and current, it is useful to consider some properties of the δ function, and in particular, its connections with the Fourier integral theorem. A one-dimensional δ function is defined by the statements

$$\int_{x_0}^{x_1} dx\, \delta(x) = \begin{cases} 1, & x_1 > 0 > x_0, \\ 0, & x_1 > x_0 > 0, \text{ or } 0 > x_1 > x_0, \end{cases} \tag{1.87}$$

that is, the integral vanishes unless the domain of integration includes the origin, when the value assumed by the integral is unity. The function $\delta(x - x')$ has corresponding properties relative to the point x'. Particular examples of functions possessing these attributes in the limit are

$$\delta(x) = \lim_{\epsilon \to 0} \frac{1}{\pi} \frac{\epsilon}{x^2 + \epsilon^2}, \tag{1.88a}$$

$$\delta(x) = \lim_{\epsilon \to 0} \frac{1}{\epsilon} e^{-\pi x^2/\epsilon^2}. \tag{1.88b}$$

An integral representation for $\delta(x)$ can be constructed from the formulae

$$\frac{1}{\pi} \frac{\epsilon}{x^2 + \epsilon^2} = \frac{1}{2\pi} \int_{-\infty}^{\infty} dk\, e^{ikx} e^{-\epsilon|k|}, \tag{1.89a}$$

$$\frac{1}{\epsilon} e^{-\pi x^2/\epsilon^2} = \frac{1}{2\pi} \int dk\, e^{ikx} e^{-\epsilon^2 k^2/4\pi}. \tag{1.89b}$$

If we perform the limiting operation under the integral sign, either expression yields

$$\delta(x) = \frac{1}{2\pi} \int_{-\infty}^{\infty} dk\, e^{ikx} = \frac{1}{\pi} \int_0^{\infty} dk\, \cos kx. \tag{1.90}$$

The three-dimensional δ function already introduced, (1.5), is correctly represented by

$$\delta(\mathbf{r}) = \delta(x)\delta(y)\delta(z), \tag{1.91}$$

for $\delta(\mathbf{r})$ certainly vanishes unless x, y, and z are simultaneously zero, and the integral over any volume enclosing the origin is unity. More generally,

$$\delta(\mathbf{r} - \mathbf{r}') = \delta(x - x')\delta(y - y')\delta(z - z'). \tag{1.92}$$

The representation for $\delta(\mathbf{r})$, obtained by multiplying individual integrals (1.90) for the one-dimensional delta functions can be regarded as an integral extended over the entirety of the space associated with the vector $\mathbf{k}$,

$$\delta(\mathbf{r}) = \frac{1}{(2\pi)^3} \int (d\mathbf{k})\, e^{i\mathbf{k}\cdot\mathbf{r}}. \tag{1.93}$$

The functional representations mentioned previously, (1.7a) and (1.7b), are consequences of the formulae

$$\frac{1}{\pi^2} \frac{\epsilon}{(r^2 + \epsilon^2)^2} = \frac{1}{(2\pi)^3} \int (\mathbf{dk}) \, e^{i\mathbf{k}\cdot\mathbf{r}} e^{-\epsilon k} \, , \tag{1.94a}$$

$$\frac{1}{\epsilon^3} e^{-\pi r^2/\epsilon^2} = \frac{1}{(2\pi)^3} \int (\mathbf{dk}) \, e^{i\mathbf{k}\cdot\mathbf{r}} e^{-\epsilon^2 k^2/4\pi} \, . \tag{1.94b}$$

An arbitrary function of a coordinate x can be represented by a linear superposition of δ functions,

$$f(x) = \int_{-\infty}^{\infty} dx' \, \delta(x - x') f(x') \, , \tag{1.95}$$

for the entire contribution to the integral comes from the point $x' = x$. On employing the integral representation (1.90) for $\delta(x - x')$, we obtain

$$f(x) = \frac{1}{2\pi} \int_{-\infty}^{\infty} dk \, e^{ikx} \int_{-\infty}^{\infty} dx' \, e^{-ikx'} f(x') \, , \tag{1.96}$$

which states the possibility of constructing an arbitrary function from the elementary periodic function e^{ikx} – the Fourier integral theorem. The corresponding statements in three dimensions are

$$f(\mathbf{r}) = \int (\mathbf{dr'}) \, \delta(\mathbf{r} - \mathbf{r'}) f(\mathbf{r'})$$
$$= \frac{1}{(2\pi)^3} \int (\mathbf{dk}) \, e^{i\mathbf{k}\cdot\mathbf{r}} \int (\mathbf{dr'}) \, e^{-i\mathbf{k}\cdot\mathbf{r'}} f(\mathbf{r'}) \, , \tag{1.97}$$

while a function of space and time is represented by

$$f(\mathbf{r}, t) = \int (\mathbf{dr'}) \, dt' \, \delta(\mathbf{r} - \mathbf{r'}) \delta(t - t') f(\mathbf{r'}, t')$$
$$= \frac{1}{(2\pi)^4} \int (\mathbf{dk}) \, d\omega \, e^{i(\mathbf{k}\cdot\mathbf{r} - \omega t)} \int (\mathbf{dr'}) \, dt' \, e^{-i(\mathbf{k}\cdot\mathbf{r'} - \omega t')} f(\mathbf{r'}, t') \, .$$
$$\tag{1.98}$$

Thus, an arbitrary function $f(\mathbf{r}, t)$ can be synthesized by a proper superposition of the functions $\exp[i(\mathbf{k}\cdot\mathbf{r} - \omega t)]$, which are the mathematical descriptions of plane waves, harmonic disturbances propagating in the direction of the vector $\mathbf{k}$, with a space periodicity length or wavelength $\lambda = 2\pi/|\mathbf{k}|$, and a time periodicity or period $T = 2\pi/\omega$.

1.5 Radiation Fields

The treatment of an electrodynamic problem involves two preliminary stages; the evaluation of the fields produced by a given array of charges moving in a prescribed fashion, and the determination of the motion of a charge acted on by a given electromagnetic field. The correct solution of the problem is obtained when these two aspects of the situation are consistent, that is, when the charges move in such a way that the fields they generate produce precisely this state of motion. We turn to a discussion of the first stage, the calculation of the fields produced by a given distribution of charge and current.

The auxiliary quantities, the vector and scalar potentials, have been introduced in order to satisfy identically the second set of field equations (1.19b). Determining equations for the potentials are obtained on substituting the representations (1.48) for **e** and **b** in the first set of equations (1.19a), with the result

$$\left(\nabla^2 - \frac{1}{c^2}\frac{\partial^2}{\partial t^2}\right)\phi = -\frac{1}{c}\frac{\partial}{\partial t}\left(\nabla\cdot\mathbf{a} + \frac{1}{c}\frac{\partial}{\partial t}\phi\right) - \rho\,, \qquad (1.99a)$$

$$\left(\nabla^2 - \frac{1}{c^2}\frac{\partial^2}{\partial t^2}\right)\mathbf{a} = \nabla\left(\nabla\cdot\mathbf{a} + \frac{1}{c}\frac{\partial}{\partial t}\phi\right) - \frac{1}{c}\mathbf{j}\,. \qquad (1.99b)$$

It is always possible to impose the condition

$$\nabla\cdot\mathbf{a} + \frac{1}{c}\frac{\partial}{\partial t}\phi = 0\,, \qquad (1.100)$$

for if this quantity does not vanish, one can, by a suitable gauge transformation, introduce new potentials for which the condition is valid. Thus if $\mathbf{a}'$, ϕ' are obtained from $\mathbf{a}$ and ϕ by a gauge transformation associated with the function ψ, as in (1.49)

$$\nabla\cdot\mathbf{a}' + \frac{1}{c}\frac{\partial}{\partial t}\phi' = \nabla\cdot\mathbf{a} + \frac{1}{c}\frac{\partial}{\partial t}\phi - \left(\nabla^2 - \frac{1}{c^2}\frac{\partial^2}{\partial t^2}\right)\psi\,, \qquad (1.101)$$

and ψ can always be chosen to produce the desired result. With this restriction upon the potentials, which is referred to as the Lorenz condition,[2] the determining equations for the potentials become

$$\left(\nabla^2 - \frac{1}{c^2}\frac{\partial^2}{\partial t^2}\right)\phi = -\rho\,, \qquad (1.102a)$$

$$\left(\nabla^2 - \frac{1}{c^2}\frac{\partial^2}{\partial t^2}\right)\mathbf{a} = -\frac{1}{c}\mathbf{j}\,. \qquad (1.102b)$$

It should be noted that the potentials are still not unique, for a gauge transformation, with the scalar function ψ satisfying

$$\left(\nabla^2 - \frac{1}{c^2}\frac{\partial^2}{\partial t^2}\right)\psi = 0 \qquad (1.103)$$

is compatible with the Lorenz condition.

The charge and current densities, as prescribed functions of the space and time coordinates, can be represented in terms of plane waves as in (1.98). Thus,

---

[2] Often mistakenly attributed to H. A. Lorentz, the Lorenz condition actually originated with L.V. Lorenz.

$$\rho(\mathbf{r}, t) = \frac{1}{(2\pi)^4} \int (\mathrm{d}\mathbf{r}') \, \mathrm{d}t' \int (\mathrm{d}\mathbf{k}) \, \mathrm{d}\omega \, \mathrm{e}^{\mathrm{i}\mathbf{k}\cdot(\mathbf{r}-\mathbf{r}')-\mathrm{i}\omega(t-t')} \rho(\mathbf{r}', t') \,. \qquad (1.104)$$

The advantage of this Fourier integral representation is that a particular solution for the potentials can be constructed by inspection. For example, from (1.102a),

$$\phi(\mathbf{r}, t) = \frac{1}{(2\pi)^4} \int (\mathrm{d}\mathbf{k}) \, \mathrm{d}\omega \int (\mathrm{d}\mathbf{r}') \, \mathrm{d}t' \, \frac{\mathrm{e}^{\mathrm{i}\mathbf{k}\cdot(\mathbf{r}-\mathbf{r}')-\mathrm{i}\omega(t-t')}}{k^2 - \omega^2/c^2} \rho(\mathbf{r}', t') \,. \qquad (1.105)$$

As a first step in the simplification of this result, consider the Green's function

$$G(\mathbf{r}) = \int \frac{(\mathrm{d}\mathbf{k})}{(2\pi)^3} \frac{\mathrm{e}^{\mathrm{i}\mathbf{k}\cdot\mathbf{r}}}{k^2 - \omega^2/c^2} \,, \qquad (1.106)$$

which is a solution of the differential equation

$$\left(\nabla^2 + \frac{\omega^2}{c^2}\right) G(\mathbf{r}) = -\delta(\mathbf{r}) \,. \qquad (1.107)$$

Upon introducing polar coordinates in the $\mathbf{k}$ space, we obtain

$$G(\mathbf{r}) = \frac{1}{(2\pi)^3} \int \mathrm{d}\theta \, \sin\theta \, 2\pi \, k^2 \mathrm{d}k \frac{\mathrm{e}^{\mathrm{i}kr\cos\theta}}{k^2 - \omega^2/c^2} = \frac{1}{2\pi^2 r} \int_0^\infty k \, \mathrm{d}k \frac{\sin kr}{k^2 - \omega^2/c^2} \,, \qquad (1.108)$$

or, equivalently,

$$G(\mathbf{r}) = -\frac{\mathrm{i}}{8\pi^2 r} \int_{-\infty}^\infty \mathrm{d}k \left(\frac{\mathrm{e}^{\mathrm{i}kr}}{k - \omega/c} + \frac{\mathrm{e}^{\mathrm{i}kr}}{k + \omega/c}\right) \,. \qquad (1.109)$$

An essential complication can no longer be ignored; the integrand becomes infinite at $k = \pm\omega/c$. The difficulty can be avoided in a purely formal manner by supposing that $1/c$ has a small imaginary part which will be eventually be allowed to vanish. If the imaginary part of $1/c$ is positive,[3] the integrand, considered as a function of the complex variable k, has a simple pole at ω/c in the upper half plane, and a simple pole at $-\omega/c$ in the lower half plane. The path of integration along the real axis can be closed by an infinite semicircle drawn in the upper half plane without affecting the value of the integral, since r is positive. Within this closed contour the integrand is everywhere analytic save at the simple pole at $k = \omega/c$. Hence, by the theorem of residues,

$$G(\mathbf{r}) = \frac{\mathrm{e}^{\mathrm{i}\omega r/c}}{4\pi r} \,. \qquad (1.110)$$

If the imaginary part of $1/c$ is negative, the position of the poles is reflected in the real axis, and the pole at $k = -\omega/c$ lies in the upper half plane. For this situation,

---

[3] This is equivalent to distorting the k contour to avoid the poles by passing below the pole at $+\omega/c$, and above the pole at $-\omega/c$.

$$G(\mathbf{r}) = \frac{e^{-i\omega r/c}}{4\pi r} . \tag{1.111}$$

It can be directly verified that the two functions $e^{\pm i\omega r/c}/(4\pi r)$ are solutions of the differential equation (1.107) for $G(\mathbf{r})$. It must be shown that $(\nabla^2 + \omega^2/c^2)e^{\pm i\omega r/c}/(4\pi r)$ has the properties of $-\delta(\mathbf{r})$, which will be achieved on demonstrating that

$$\int (d\mathbf{r}) \left(\nabla^2 + \frac{\omega^2}{c^2} \right) \frac{e^{\pm i\omega r/c}}{4\pi r} = -1 \tag{1.112}$$

for any region of integration that includes the origin. It is sufficient to consider a sphere of arbitrary radius R. Thus, we are required to prove that

$$R^2 \frac{d}{dR} \left(\frac{e^{\pm i\omega R/c}}{R} \right) + \frac{\omega^2}{c^2} \int_0^R r \, dr \, e^{\pm i\omega r/c} = -1 , \tag{1.113}$$

which is easily checked. It is apparent, then, that the difficulty encountered by the Fourier integral method arises from the existence of two solutions for $G(\mathbf{r})$ and, in consequence, for the potentials. Which of these solutions to adopt can only be decided by additional physical considerations.

Tentatively choosing (1.110), we obtain from (1.105)

$$\phi(\mathbf{r}, t) = \frac{1}{2\pi} \int d\omega \, (d\mathbf{r}') \, dt' \, \frac{e^{i\omega|\mathbf{r}-\mathbf{r}'|/c - i\omega(t-t')}}{4\pi|\mathbf{r}-\mathbf{r}'|} \rho(\mathbf{r}', t') . \tag{1.114}$$

The integral with respect to ω is recognized as that of a delta function,

$$\phi(\mathbf{r}, t) = \int (d\mathbf{r}') \, dt' \, \frac{\delta\left(t' - t + |\mathbf{r} - \mathbf{r}'|/c\right)}{4\pi|\mathbf{r}-\mathbf{r}'|} \rho(\mathbf{r}', t') , \tag{1.115}$$

and if the integration with respect to t' is performed,

$$\phi(\mathbf{r}, t) = \int (d\mathbf{r}') \, \frac{\rho\left(\mathbf{r}', t - |\mathbf{r} - \mathbf{r}'|/c\right)}{4\pi|\mathbf{r}-\mathbf{r}'|} . \tag{1.116}$$

This result expresses the scalar potential at the point $\mathbf{r}$ and time t in terms of the charge density at other points of space and earlier times, the time interval being just that required to traverse the spatial separation at the speed c. The formula thus contains a concise description of the propagation of electromagnetic fields at the speed of light. Evidently, had the solution $e^{-i\omega r/c}/(4\pi r)$ been adopted for $G(\mathbf{r})$, the evaluation of the potential at a time t would have involved a knowledge of the charge density at later times. This possibility must be rejected, for it requires information which, by the nature of the physical world, is unavailable.[4] The corresponding solution of (1.102b) for the vector potential, in its several stages of development, is

---

[4] However, it is actually possible to use advanced Green's functions, with suitable boundary conditions, to describe classical physics (see [14]). This led Feynman to the discovery of the causal or Feynman propagator (see Problem 1.37).

$$\mathbf{a}(\mathbf{r},t) = \frac{1}{2\pi} \int d\omega \, (d\mathbf{r}') \, dt' \, \frac{e^{i\omega|\mathbf{r}-\mathbf{r}'|/c - i\dot{\omega}(t-t')}}{4\pi|\mathbf{r}-\mathbf{r}'|} \frac{1}{c} \mathbf{j}(\mathbf{r}',t')$$

$$= \int (d\mathbf{r}') \, dt' \, \frac{\delta\left(t'-t+|\mathbf{r}-\mathbf{r}'|/c\right)}{4\pi|\mathbf{r}-\mathbf{r}'|} \frac{1}{c} \mathbf{j}(\mathbf{r}',t')$$

$$= \frac{1}{c} \int (d\mathbf{r}') \, \frac{\mathbf{j}\left(\mathbf{r}',t-|\mathbf{r}-\mathbf{r}'|/c\right)}{4\pi|\mathbf{r}-\mathbf{r}'|} . \tag{1.117}$$

These solutions for the vector and scalar potentials, the so-called retarded potentials, satisfy the Lorenz condition. This is most easily demonstrated with the form the potentials assume before the integration with respect to the time t'. The quantity $\delta\left(t'-t+|\mathbf{r}-\mathbf{r}'|/c\right)/(4\pi|\mathbf{r}-\mathbf{r}'|)$ involves only the difference of time and space coordinates. Therefore derivatives with respect to t or $\mathbf{r}$ can be replaced by corresponding derivatives acting on t' and $\mathbf{r}'$, with a compensating sign change. Hence, with a suitable integration by parts,

$$\nabla \cdot \mathbf{a}(\mathbf{r},t) + \frac{1}{c} \frac{\partial}{\partial t} \phi(\mathbf{r},t) = \frac{1}{c} \int (d\mathbf{r}') \, dt' \, \frac{\delta\left(t'-t+|\mathbf{r}-\mathbf{r}'|/c\right)}{4\pi|\mathbf{r}-\mathbf{r}'|}$$

$$\times \left(\nabla' \cdot \mathbf{j}(\mathbf{r}',t') + \frac{\partial}{\partial t'} \rho(\mathbf{r}',t') \right) = 0 \,, \tag{1.118}$$

in consequence of the conservation of charge, (1.14).

As a particular example, consider a point charge moving in a prescribed fashion, that is, its position $\mathbf{r}(t)$ and velocity $\mathbf{v}(t)$ are given functions of time. The charge and current densities are, accordingly, represented by

$$\rho(\mathbf{r},t) = q \, \delta(\mathbf{r}-\mathbf{r}(t)) \,, \qquad \mathbf{j}(\mathbf{r},t) = q \, \mathbf{v}(t) \delta(\mathbf{r}-\mathbf{r}(t)) \,. \tag{1.119}$$

The most convenient form for the potentials is, again, that involving the delta function. On integrating over the space variable $\mathbf{r}'$, we obtain from (1.115) and (1.117)

$$\phi(\mathbf{r},t) = \frac{q}{4\pi} \int dt' \, \frac{\delta(t'-t+|\mathbf{r}-\mathbf{r}(t')|/c)}{|\mathbf{r}-\mathbf{r}(t')|} \,, \tag{1.120a}$$

$$\mathbf{a}(\mathbf{r},t) = \frac{q}{4\pi} \int dt' \, \frac{\mathbf{v}(t')}{c} \frac{\delta(t'-t+|\mathbf{r}-\mathbf{r}(t')|/c)}{|\mathbf{r}-\mathbf{r}(t')|} \,. \tag{1.120b}$$

The entire contribution to these integrals comes from the time τ defined by

$$t - \tau = \frac{|\mathbf{r}-\mathbf{r}(\tau)|}{c} \,, \tag{1.121}$$

which is evidently the time at which an electromagnetic field, moving at the speed c, must leave the position of the charge in order to reach the point of observation $\mathbf{r}$ at the time t. In performing the final integration with respect to t', one must be careful to observe that dt' is not the differential of the δ function's argument, and that therefore a change of variable is required. Thus calling

$$\eta(t') = t' - t + \frac{|\mathbf{r} - \mathbf{r}(t')|}{c} , \tag{1.122}$$

we obtain for the scalar potential

$$\phi(\mathbf{r}, t) = \frac{q}{4\pi} \left(\frac{\frac{dt'}{d\eta}}{|\mathbf{r} - \mathbf{r}(t')|} \right)_{t'=\tau} . \tag{1.123}$$

However,

$$\frac{d\eta}{dt'} = 1 - \frac{\mathbf{v}(t')}{c} \cdot \frac{\mathbf{r} - \mathbf{r}(t')}{|\mathbf{r} - \mathbf{r}(t')|} , \tag{1.124}$$

and therefore

$$\phi(\mathbf{r}, t) = \frac{q}{4\pi} \frac{1}{|\mathbf{r} - \mathbf{r}(\tau)| - \mathbf{v}(\tau) \cdot (\mathbf{r} - \mathbf{r}(\tau))/c} . \tag{1.125}$$

Similarly,

$$\mathbf{a}(\mathbf{r}, t) = \frac{q}{4\pi} \frac{\frac{1}{c}\mathbf{v}(\tau)}{|\mathbf{r} - \mathbf{r}(\tau)| - \mathbf{v}(\tau) \cdot (\mathbf{r} - \mathbf{r}(\tau))/c} = \frac{\mathbf{v}(\tau)}{c}\phi(\mathbf{r}, t) . \tag{1.126}$$

The direct evaluation of the fields from these potentials, the so-called Liénard–Wiechert potentials, is rather involved, for the retarded time τ is an implicit function of $\mathbf{r}$ and t. The calculation proceeds more easily by first deriving the fields from the δ function representation of the potentials and then performing the integration with respect to t'. However, no details will be given (see Problems 1.7 and 1.8).

1.5.1 Multipole Radiation

A problem of greater interest is that of a distribution of charge with a spatial extension sufficiently small so that the charge distribution changes only slightly in the time required for light to traverse it. Otherwise expressed, the largest frequency $\nu = \omega/2\pi$ that occurs in the time Fourier decomposition of the charge density must be such that $\nu a/c \ll 1$, where a is a length, representative of the system's linear dimensions. Equivalently, the corresponding wavelength $\lambda = c/\nu$ must be large in comparison with a. Molecular systems ($a \sim 10^{-8}$ cm) possess this property for optical and even for ultraviolet frequencies ($\lambda \sim 10^{-6}$ cm), and the condition $\lambda \gg a$ is more than adequately fulfilled for wavelengths in the microwave region ($\lambda \sim 1$ cm). Under these conditions, the difference in retarded time $t - |\mathbf{r} - \mathbf{r}'|/c$ between various parts of a molecule is of secondary importance, and to a first approximation all retarded times can be identified with that of some fixed point in the molecule, which we shall choose as the origin of coordinates. In a more precise treatment, the difference between $t - |\mathbf{r} - \mathbf{r}'|/c$ and $t - r/c$ can be taken into account by expansion of the charge and current densities, as follows:

$$\rho(\mathbf{r}',t-|\mathbf{r}-\mathbf{r}'|/c) = \rho(r',t-r/c)+(r-|\mathbf{r}-\mathbf{r}'|)\frac{1}{c}\frac{\partial}{\partial t}\rho(r',t-r/c)+\cdots . \quad (1.127)$$

However, we shall be concerned primarily with the field at a great distance from the center of the molecule (considered to be at rest). Rather than introduce our approximation in two steps, we proceed more directly by regarding r' as small in comparison with r wherever it occurs in the retarded potential expressions. It must not be forgotten that two approximations are thereby introduced, $r \gg a$ and $\lambda \gg a$. With these remarks, we insert the Taylor series expansion

$$\frac{\rho(\mathbf{r}',t-|\mathbf{r}-\mathbf{r}'|/c)}{|\mathbf{r}-\mathbf{r}'|} = \left(1-\mathbf{r}'\cdot\boldsymbol{\nabla}+\frac{1}{2}(\mathbf{r}'\cdot\boldsymbol{\nabla})^2 - \cdots\right)\frac{\rho(\mathbf{r}',t-r/c)}{r} \quad (1.128)$$

in the retarded scalar potential integral (1.116)

$$4\pi\phi(\mathbf{r},t) = \frac{\int(d\mathbf{r}')\,\rho(\mathbf{r}',t-r/c)}{r} - \boldsymbol{\nabla}\cdot\frac{\int(d\mathbf{r}')\,\mathbf{r}'\,\rho(\mathbf{r}',t-r/c)}{r}$$
$$+ \frac{1}{2}\boldsymbol{\nabla}\boldsymbol{\nabla}:\frac{\int(d\mathbf{r}')\,\mathbf{r}'\mathbf{r}'\rho(\mathbf{r}',t-r/c)}{r} - \cdots . \quad (1.129)$$

In terms of the total charge, electric dipole moment, and electric quadrupole moment dyadic,[5]

$$q = \int(d\mathbf{r})\,\rho(\mathbf{r},t)\,, \quad (1.131a)$$

$$\mathbf{d}(t) = \int(d\mathbf{r})\,\mathbf{r}\,\rho(\mathbf{r},t)\,, \quad (1.131b)$$

$$\mathsf{Q}(t) = \int(d\mathbf{r})\,\mathbf{r}\,\mathbf{r}\,\rho(\mathbf{r},t)\,, \quad (1.131c)$$

the first three terms of the expansion are

$$4\pi\phi(\mathbf{r},t) = \frac{q}{r} - \boldsymbol{\nabla}\cdot\frac{\mathbf{d}(t-r/c)}{r} + \frac{1}{2}\boldsymbol{\nabla}\boldsymbol{\nabla}:\frac{\mathsf{Q}(t-r/c)}{r}\,. \quad (1.132)$$

The notation A:B for dyadics designates the scalar product,

$$\mathsf{A}:\mathsf{B} = \sum_{i,j} A_{ij}B_{ji}\,. \quad (1.133)$$

In a similar fashion, the expansion of the vector potential (1.117) is

---

[5] Usually, the electric quadrupole dyadic is defined by

$$\mathsf{Q} = 3\int(d\mathbf{r})\left(\mathbf{r}\mathbf{r} - \frac{1}{3}r^2\mathbf{1}\right)\rho \quad (1.130)$$

so that $\mathrm{Tr}\,\mathsf{Q} = 0$ [see (2.48)].

$$4\pi\mathbf{a}(\mathbf{r},t) = \frac{1}{c}\frac{\int(d\mathbf{r}')\,\mathbf{j}(\mathbf{r}',t-r/c)}{r} - \boldsymbol{\nabla}\cdot\frac{1}{c}\frac{\int(d\mathbf{r}')\,\mathbf{r}'\,\mathbf{j}(\mathbf{r}',t-r/c)}{r} + \cdots . \quad (1.134)$$

Two terms suffice to give the same degree of approximation as the first three terms in the scalar potential expansion. The integrals can be re-expressed in convenient form with the aid of the conservation equation,

$$\boldsymbol{\nabla}\cdot\mathbf{j} + \frac{\partial}{\partial t}\rho = 0 . \quad (1.135)$$

Multiplying (1.135) by $\mathbf{r}$ and rearranging the terms, we obtain

$$\frac{\partial}{\partial t}\mathbf{r}\rho + \boldsymbol{\nabla}\cdot(\mathbf{j}\mathbf{r}) = \mathbf{j} . \quad (1.136)$$

The process of volume integration, extended over the entire region occupied by the charge distribution, yields

$$\int(d\mathbf{r})\,\mathbf{j}(\mathbf{r},t) = \frac{d}{dt}\mathbf{d}(t) . \quad (1.137)$$

Corresponding operations with $\mathbf{r}$ replaced by the dyadic $\mathbf{rr}$ give, successively,

$$\frac{\partial}{\partial t}\mathbf{rr}\,\rho + \boldsymbol{\nabla}\cdot(\mathbf{j}\mathbf{rr}) = \mathbf{rj}+\mathbf{jr} , \quad (1.138a)$$

$$\int(d\mathbf{r})\,(\mathbf{rj}(\mathbf{r},t)+\mathbf{j}(\mathbf{r},t)\mathbf{r}) = \frac{d}{dt}Q(t) . \quad (1.138b)$$

Now,

$$\mathbf{rj} = \frac{\mathbf{rj}+\mathbf{jr}}{2} + \frac{\mathbf{rj}-\mathbf{jr}}{2} = \frac{\mathbf{rj}+\mathbf{jr}}{2} - \frac{1}{2}\mathbf{1}\times(\mathbf{r}\times\mathbf{j}) , \quad (1.139)$$

and therefore

$$4\pi\mathbf{a}(\mathbf{r},t) = \frac{1}{c}\frac{\partial}{\partial t}\frac{\mathbf{d}(t-r/c)}{r} + \boldsymbol{\nabla}\times\frac{\mathbf{m}(t-r/c)}{r} - \frac{1}{2c}\frac{\partial}{\partial t}\boldsymbol{\nabla}\cdot\frac{Q(t-r/c)}{r} , \quad (1.140)$$

where

$$\mathbf{m}(t) = \frac{1}{2c}\int(d\mathbf{r})\,\mathbf{r}\times\mathbf{j}(\mathbf{r},t) \quad (1.141)$$

is the magnetic dipole moment of the system.

For a neutral molecule, $q = 0$, and the dominant term in the scalar potential expansion (1.132) is that associated with the electric dipole moment. The quadrupole moment contribution is smaller by a factor of the same magnitude as the larger of the two ratios a/λ, a/r, and will be discarded. The electric dipole moment term predominates in the vector potential expansion (1.140) save for static or quasistatic phenomena when the magnetic dipole moment effect may assume importance. The quadrupole moment term will also be discarded here. Thus, under the conditions contemplated, the potentials can be expressed in terms of two vectors, the electric and magnetic Hertz vectors,

$$\boldsymbol{\Pi}_e(\mathbf{r}, t) = \frac{1}{4\pi r} \mathbf{d}(t - r/c) , \tag{1.142a}$$

$$\boldsymbol{\Pi}_m(\mathbf{r}, t) = \frac{1}{4\pi r} \mathbf{m}(t - r/c) , \tag{1.142b}$$

by

$$\phi(\mathbf{r}, t) = -\boldsymbol{\nabla} \cdot \boldsymbol{\Pi}_e(\mathbf{r}, t) , \tag{1.143a}$$

$$\mathbf{a}(\mathbf{r}, t) = \frac{1}{c} \frac{\partial}{\partial t} \boldsymbol{\Pi}_e(\mathbf{r}, t) + \boldsymbol{\nabla} \times \boldsymbol{\Pi}_m(\mathbf{r}, t) . \tag{1.143b}$$

The consistency of the approximations for the vector and scalar potentials is verified on noting that these expressions satisfy the Lorenz condition (this statement also applies to the discarded quadrupole moment terms). The electric and magnetic field intensities are given by

$$\mathbf{e} = \boldsymbol{\nabla}\boldsymbol{\nabla} \cdot \boldsymbol{\Pi}_e - \frac{1}{c^2} \frac{\partial^2}{\partial t^2} \boldsymbol{\Pi}_e - \frac{1}{c} \frac{\partial}{\partial t} \boldsymbol{\nabla} \times \boldsymbol{\Pi}_m , \tag{1.144a}$$

$$\mathbf{b} = \boldsymbol{\nabla} \times (\boldsymbol{\nabla} \times \boldsymbol{\Pi}_m) + \frac{1}{c} \frac{\partial}{\partial t} \boldsymbol{\nabla} \times \boldsymbol{\Pi}_e$$

$$= \boldsymbol{\nabla}\boldsymbol{\nabla} \cdot \boldsymbol{\Pi}_m - \nabla^2 \boldsymbol{\Pi}_m + \frac{1}{c} \frac{\partial}{\partial t} \boldsymbol{\nabla} \times \boldsymbol{\Pi}_e . \tag{1.144b}$$

The Hertz vectors can be considered as the retarded solutions of the differential equations, because $-\nabla^2 1/r = 4\pi\delta(\mathbf{r})$,

$$\left(\nabla^2 - \frac{1}{c^2} \frac{\partial^2}{\partial t^2}\right) \boldsymbol{\Pi}_e(\mathbf{r}, t) = -\mathbf{d}(t)\delta(\mathbf{r}) \equiv -\mathbf{d}(\mathbf{r}, t) , \tag{1.145a}$$

$$\left(\nabla^2 - \frac{1}{c^2} \frac{\partial^2}{\partial t^2}\right) \boldsymbol{\Pi}_m(\mathbf{r}, t) = -\mathbf{m}(t)\delta(\mathbf{r}) \equiv -\mathbf{m}(\mathbf{r}, t) . \tag{1.145b}$$

The fields associated with the Hertz vectors can be regarded as produced by point distributions of charge and current. Since

$$\boldsymbol{\nabla} \times \mathbf{b} - \frac{1}{c} \frac{\partial}{\partial t} \mathbf{e} = -\frac{1}{c} \frac{\partial}{\partial t} \left(\nabla^2 - \frac{1}{c^2} \frac{\partial^2}{\partial t^2}\right) \boldsymbol{\Pi}_e - \boldsymbol{\nabla} \times \left(\nabla^2 - \frac{1}{c^2} \frac{\partial^2}{\partial t^2}\right) \boldsymbol{\Pi}_m$$

$$= \frac{1}{c} \frac{\partial}{\partial t} \mathbf{p}(\mathbf{r}, t) + \boldsymbol{\nabla} \times \mathbf{m}(\mathbf{r}, t) \tag{1.146}$$

and

$$\boldsymbol{\nabla} \cdot \mathbf{e} = \boldsymbol{\nabla} \cdot \left(\nabla^2 - \frac{1}{c^2} \frac{\partial^2}{\partial t^2}\right) \boldsymbol{\Pi}_e = -\boldsymbol{\nabla} \cdot \mathbf{d}(\mathbf{r}, t) , \tag{1.147}$$

the required distributions are

$$\rho_{\text{eff}} = -\boldsymbol{\nabla} \cdot \mathbf{d}(\mathbf{r}, t) , \tag{1.148a}$$

$$\mathbf{j}_{\text{eff}} = \frac{\partial}{\partial t} \mathbf{p}(\mathbf{r}, t) + c\boldsymbol{\nabla} \times \mathbf{m}(\mathbf{r}, t) . \tag{1.148b}$$

Notice that the dipole moments of this effective distribution are those of the original molecule,

$$\int (d\mathbf{r})\, \mathbf{r}\, \rho_{\text{eff}} = -\int (d\mathbf{r})\, \mathbf{r}\, \boldsymbol{\nabla} \cdot \mathbf{d}(\mathbf{r},t) = \int (d\mathbf{r})\, \mathbf{d}(\mathbf{r},t)$$
$$= \mathbf{d}(t)\,, \tag{1.149a}$$

$$\frac{1}{2c} \int (d\mathbf{r})\, \mathbf{r} \times \mathbf{j}_{\text{eff}} = \frac{1}{2} \int (d\mathbf{r})\, \mathbf{r} \times [\boldsymbol{\nabla} \times \mathbf{m}(\mathbf{r},t)] = \int (d\mathbf{r})\, \mathbf{m}(\mathbf{r},t)$$
$$= \mathbf{m}(t)\,. \tag{1.149b}$$

Use has been made of the two identities

$$\mathbf{r}\boldsymbol{\nabla} \cdot \mathbf{A} = \boldsymbol{\nabla} \cdot (\mathbf{A}\mathbf{r}) - \mathbf{A}\,, \tag{1.150a}$$
$$\mathbf{r} \times (\boldsymbol{\nabla} \times \mathbf{A}) = \boldsymbol{\nabla}(\mathbf{A} \cdot \mathbf{r}) - \boldsymbol{\nabla} \cdot (\mathbf{r}\mathbf{A}) + 2\mathbf{A}\,, \tag{1.150b}$$

and of the fact that

$$\int (d\mathbf{r})\, \mathbf{r}\, \delta(\mathbf{r}) = 0\,. \tag{1.151}$$

Therefore, the actual charge–current distribution in the molecule can be replaced by the effective point distribution without altering the values of the moments, or of the field at a sufficient distance from the molecule ($r \gg a$, $\lambda \gg a$).

Although the fields deduced from the effective distribution do not agree with the actual fields in the neighborhood of the molecule, nevertheless certain average properties of the fields are correctly represented. We shall show that the field intensities averaged over the volume contained in a sphere that includes the molecule, but is small in comparison with all wavelengths, is given correctly by the fields calculated from the effective distribution. It will follow, a fortiori, that the same property is maintained for any larger region of integration. In the immediate vicinity of the molecule, the potentials can be calculated, to a first approximation, by ignoring the finite propagation velocity of light,

$$\phi(\mathbf{r},t) = \int (d\mathbf{r}')\, \frac{\rho(\mathbf{r}',t)}{4\pi|\mathbf{r}-\mathbf{r}'|}\,, \tag{1.152a}$$

$$\mathbf{a}(\mathbf{r},t) = \frac{1}{c} \int (d\mathbf{r}')\, \frac{\mathbf{j}(\mathbf{r}',t)}{4\pi|\mathbf{r}-\mathbf{r}'|}\,. \tag{1.152b}$$

The fields are, correspondingly,

$$\mathbf{e}(\mathbf{r},t) = -\boldsymbol{\nabla} \int (d\mathbf{r}')\, \frac{\rho(\mathbf{r}',t)}{4\pi|\mathbf{r}-\mathbf{r}'|} = \int (d\mathbf{r}')\, \rho(\mathbf{r}',t)\boldsymbol{\nabla}'\frac{1}{4\pi|\mathbf{r}-\mathbf{r}'|}\,, \tag{1.153a}$$

$$\mathbf{b}(\mathbf{r},t) = \boldsymbol{\nabla} \times \frac{1}{c} \int (d\mathbf{r}')\, \frac{\mathbf{j}(\mathbf{r}',t)}{4\pi|\mathbf{r}-\mathbf{r}'|} = \frac{1}{c} \int (d\mathbf{r}')\, \mathbf{j}(\mathbf{r}',t) \times \boldsymbol{\nabla}'\frac{1}{4\pi|\mathbf{r}-\mathbf{r}'|}\,. \tag{1.153b}$$

The vector potential contribution to the electric field has been discarded in comparison with the electrostatic field. These fields are to be integrated over the volume V_S of a sphere which includes the entire molecule. The center of the sphere bears no necessary relation to the molecule. The integration for both fields requires an evaluation of

$$\int_{V_S} \frac{(\mathbf{dr})}{4\pi|\mathbf{r} - \mathbf{r}'|} \ , \tag{1.154}$$

extended over the sphere. On remarking that

$$\nabla'^2 \frac{1}{4\pi|\mathbf{r} - \mathbf{r}'|} = -\delta(\mathbf{r} - \mathbf{r}') \tag{1.155}$$

as the static limit of the differential equation satisfied by $G(\mathbf{r})$, (1.107), it is observed that

$$\nabla'^2 \int_{V_S} \frac{(\mathbf{dr})}{4\pi|\mathbf{r} - \mathbf{r}'|} = -1 \ , \tag{1.156}$$

provided that the point $\mathbf{r}'$ is within the sphere, as required by the assumption that the sphere encompasses the entire molecule. If the origin of coordinates is temporarily moved to the center of the sphere, it may be inferred that

$$\int_{V_S} \frac{(\mathbf{dr})}{4\pi|\mathbf{r} - \mathbf{r}'|} = -\frac{r'^2}{6} + \psi(\mathbf{r}') \ , \tag{1.157}$$

where $\psi(\mathbf{r}')$ is a solution of Laplace's equation,

$$\nabla'^2 \psi(\mathbf{r}') = 0 \ , \tag{1.158}$$

which, by symmetry, can depend only on the distance to the center of the sphere. Such a function must be a constant, for on integration of Laplace's equation over a sphere of radius r, and employing the divergence theorem, one obtains

$$4\pi r^2 \frac{\mathrm{d}}{\mathrm{d}r} \psi(r) = 0 \ , \tag{1.159}$$

which establishes the constancy of $\psi(\mathbf{r})$ within the sphere. Therefore,

$$\nabla' \int_{V_S} \frac{(\mathbf{dr})}{4\pi|\mathbf{r} - \mathbf{r}'|} = -\frac{1}{3}\mathbf{r}' \ , \tag{1.160}$$

which is independent of the radius of the sphere. It immediately follows from (1.153a) and (1.153b) that

$$\int_{V_S} (\mathbf{dr})\, \mathbf{e}(\mathbf{r}, t) = -\frac{1}{3}\int (\mathbf{dr}')\, \mathbf{r}'\, \rho(\mathbf{r}', t) \ , \tag{1.161a}$$

$$\int_{V_S} (\mathbf{dr})\, \mathbf{b}(\mathbf{r}, t) = \frac{1}{3c}\int (\mathbf{dr}')\, \mathbf{r}' \times \mathbf{j}(\mathbf{r}', t) \ . \tag{1.161b}$$

The vector $\mathbf{r}'$ is referred to the center of the sphere as origin. To return to the origin of coordinates established at the center of the molecule, $\mathbf{r}'$ must be replaced by $\mathbf{r}' + \mathbf{R}$, where $\mathbf{R}$ is the position vector of the center of the molecule relative to the center of the sphere. Finally, then, using (1.137),

$$\int_{V_S} (d\mathbf{r})\, \mathbf{e}(\mathbf{r}, t) = -\frac{1}{3}\mathbf{d}(t) \,, \tag{1.162a}$$

$$\int_{V_S} (d\mathbf{r})\, \mathbf{b}(\mathbf{r}, t) = \frac{2}{3}\mathbf{m}(t) + \frac{1}{3c}\mathbf{R} \times \dot{\mathbf{d}}(t) \,, \tag{1.162b}$$

provided that the molecule is electrically neutral. The essential result of this calculation is that the volume integrals depend only upon the moments of the system, not upon the detailed charge–current distribution. Now the effective point distribution, (1.148a) and (1.148b), predicts the correct values of the moments, and must therefore lead to the same integrated field intensities.

The explicit calculation of the fields derived by (1.144a) and (1.144b) from the electric Hertz vector (1.142a) gives

$$4\pi\mathbf{e} = \left(\frac{3\mathbf{r}\,\mathbf{r} \cdot \mathbf{d}}{r^5} - \frac{\mathbf{d}}{r^3}\right) + \frac{1}{c}\left(\frac{3\mathbf{r}\,\mathbf{r} \cdot \dot{\mathbf{d}}}{r^4} - \frac{\dot{\mathbf{d}}}{r^2}\right) + \frac{1}{c^2}\frac{\mathbf{r} \times (\mathbf{r} \times \ddot{\mathbf{d}})}{r^3} \,, \tag{1.163a}$$

$$4\pi\mathbf{b} = -\frac{1}{c}\frac{\mathbf{r} \times \dot{\mathbf{d}}}{r^3} - \frac{1}{c^2}\frac{\mathbf{r} \times \ddot{\mathbf{d}}}{r^2} \,. \tag{1.163b}$$

The electric dipole moment and its time derivative are to be evaluated at the retarded time $t - r/c$. The relative orders of magnitude of the three types of terms in the electric field are determined by the ratio r/λ. For $r/\lambda \ll 1$ (but $r/a \gg 1$), the electric field is essentially that of a static dipole. However, if $r/\lambda \gg 1$, the last term in both fields predominates, and

$$4\pi\mathbf{e} = \frac{1}{c^2}\frac{\mathbf{r} \times (\mathbf{r} \times \ddot{\mathbf{d}})}{r^3} = \frac{1}{c^2}\frac{\mathbf{n} \times (\mathbf{n} \times \ddot{\mathbf{d}})}{r} \,, \tag{1.164a}$$

$$4\pi\mathbf{b} = -\frac{1}{c^2}\frac{\mathbf{r} \times \ddot{\mathbf{d}}}{r^2} = -\frac{1}{c^2}\frac{\mathbf{n} \times \ddot{\mathbf{d}}}{r} \,, \tag{1.164b}$$

where

$$\mathbf{n} = \frac{\mathbf{r}}{r} \tag{1.165}$$

is a radial unit vector. Note that at these large distances, the electric and magnetic fields are transverse to the direction of observation and to each other, and equal in magnitude,

$$\mathbf{e} = \mathbf{b} \times \mathbf{n} \,, \qquad \mathbf{b} = \mathbf{n} \times \mathbf{e} \,. \tag{1.166}$$

Therefore, the energy flux vector

$$\mathbf{S} = c\,\mathbf{e} \times \mathbf{b} = \mathbf{n}\,c e^2 = \frac{\mathbf{n}}{4\pi r^2}\frac{(\mathbf{n} \times \ddot{\mathbf{d}})^2}{4\pi c^3} \tag{1.167}$$

is directed radially outward from the molecule and the net amount of energy which leaves a sphere of radius r per unit time is

$$P = \oint \mathrm{d}S\,\mathbf{n}\cdot\mathbf{S} = \int_0^\pi \frac{2\pi r^2 \sin\theta\,\mathrm{d}\theta}{4\pi r^2}\frac{(\ddot{\mathbf{d}})^2}{4\pi c^3}\sin^2\theta = \frac{2}{3c^3}\frac{1}{4\pi}(\ddot{\mathbf{d}})^2\;. \tag{1.168}$$

This expression is independent of the radius of the sphere, except insofar as the radius r determines the time of emission of the field under observation, and represents the rate at which the molecule loses energy by radiation. In the particular situation of a dipole moment that oscillates harmonically with a single frequency,

$$\mathbf{d}(t) = \mathbf{d}_0 \cos\omega t\;, \tag{1.169}$$

the rate of emission of energy is

$$P = \frac{2}{3c^3}\frac{\omega^4}{4\pi}(\mathbf{d}_0)^2 \cos^2\omega(t - r/c)\;, \tag{1.170}$$

which fluctuates about the average value

$$\overline{P} = \frac{\omega^4}{3c^3}\frac{1}{4\pi}(\mathbf{d}_0)^2\;. \tag{1.171}$$

The fields generated by a magnetic dipole moment can be obtained from the electric dipole fields by the substitutions[6]

$$\mathbf{d} \to \mathbf{m}\,, \quad \mathbf{e} \to \mathbf{b}\,, \quad \mathbf{b} \to -\mathbf{e}\,, \tag{1.172}$$

that is

$$4\pi\mathbf{b} = \left(\frac{3\mathbf{r}\,\mathbf{r}\cdot\mathbf{m}}{r^5} - \frac{\mathbf{m}}{r^3}\right) + \frac{1}{c}\left(\frac{3\mathbf{r}\,\mathbf{r}\cdot\dot{\mathbf{m}}}{r^4} - \frac{\dot{\mathbf{m}}}{r^2}\right) + \frac{1}{c^2}\frac{\mathbf{r}\times(\mathbf{r}\times\ddot{\mathbf{m}})}{r^3}\,, \tag{1.173a}$$

$$4\pi\mathbf{e} = \frac{1}{c}\frac{\mathbf{r}\times\dot{\mathbf{m}}}{r^3} + \frac{1}{c^2}\frac{\mathbf{r}\times\ddot{\mathbf{m}}}{r^2}\,, \tag{1.173b}$$

and correspondingly, the rate of radiation is

$$P = \frac{2}{3c^3}\frac{1}{4\pi}(\ddot{\mathbf{m}})^2\;. \tag{1.174}$$

---

[6] See (1.144a) and (1.144b), and the fact that the Hertz vectors satisfy the wave equation away from the origin, (1.145a) and (1.145b). This symmetry is an example of electromagnetic duality, which is further explored in Problems 1.23 and 1.25.

If oscillating electric and magnetic dipoles are present simultaneously, the total energy radiated per unit time is the sum of the individual radiation rates, for the outward energy flux is

$$\mathbf{n} \cdot \mathbf{S} = c e^2 = \frac{1}{4\pi r^2} \frac{1}{4\pi c^3} \left[(\mathbf{n} \times \ddot{\mathbf{d}})^2 + (\mathbf{n} \times \ddot{\mathbf{m}})^2 + 2\mathbf{n} \cdot (\ddot{\mathbf{d}} \times \ddot{\mathbf{m}}) \right] , \quad (1.175)$$

and the interference term disappears on integration over all directions of emission.

1.5.2 Work Done by Charges

It is instructive to calculate the rate of radiation by a system in a quite different manner, which involve evaluating the rate at which the charges in the molecule do work on the field and thus supply the energy which is dissipated in radiation. The precise statement of the consequence of energy conservation is obtained from (1.44a)

$$-\int_V (\mathrm{d}\mathbf{r}) \, \mathbf{e} \cdot \mathbf{j} = \frac{\mathrm{d}}{\mathrm{d}t} E + P , \quad (1.176)$$

in which the integration is extended over a region V encompassing the molecule, and from (1.20a)

$$E = \int_V (\mathrm{d}\mathbf{r}) \, \frac{e^2 + b^2}{2} \quad (1.177)$$

is the total electromagnetic energy associated with the molecule, while from (1.20b), the integral extended over the surface S bounding V

$$P = \oint_S \mathrm{d}S \, \mathbf{n} \cdot c\, \mathbf{e} \times \mathbf{b} \quad (1.178)$$

is the desired amount of energy leaving the system per unit time. This approach to the problem has the advantage of determining E and P simultaneously. In the evaluation of $\int (\mathrm{d}\mathbf{r}) \, \mathbf{e} \cdot \mathbf{j}$, we are concerned only with the fields within the region occupied by charge. The effect of retardation, or the finite speed of light, is slight and the difference between the charge density at the retarded time and at the local time can be expressed by a power series expansion, with $1/c$ regarded as a small parameter,

$$\rho(\mathbf{r}', t - |\mathbf{r} - \mathbf{r}'|/c) = \rho(\mathbf{r}', t) - \frac{|\mathbf{r} - \mathbf{r}'|}{c} \frac{\partial}{\partial t} \rho(\mathbf{r}', t)$$
$$+ \frac{|\mathbf{r} - \mathbf{r}'|^2}{2c^2} \frac{\partial^2}{\partial t^2} \rho(\mathbf{r}', t) - \frac{|\mathbf{r} - \mathbf{r}'|^3}{6c^3} \frac{\partial^3}{\partial t^3} \rho(\mathbf{r}', t) + \cdots .$$
$$(1.179)$$

Hence, from (1.116),

$$4\pi\phi(\mathbf{r}, t) = \int (d\mathbf{r}') \frac{\rho(\mathbf{r}', t)}{|\mathbf{r} - \mathbf{r}'|} + \frac{1}{2c^2} \frac{\partial^2}{\partial t^2} \int (d\mathbf{r}') |\mathbf{r} - \mathbf{r}'| \rho(\mathbf{r}', t)$$

$$- \frac{1}{6c^3} \frac{\partial^3}{\partial t^3} \int (d\mathbf{r}') |\mathbf{r} - \mathbf{r}'|^2 \rho(\mathbf{r}', t) + \cdots , \tag{1.180}$$

on employing charge conservation to discard the second term in the expansion. To the same order of approximation it is sufficient to write from (1.117)

$$4\pi\mathbf{a}(\mathbf{r}, t) = \frac{1}{c} \int (d\mathbf{r}') \frac{\mathbf{j}(\mathbf{r}', t)}{|\mathbf{r} - \mathbf{r}'|} - \frac{1}{c^2} \frac{d}{dt} \int (d\mathbf{r}') \mathbf{j}(\mathbf{r}', t)$$

$$= \frac{1}{c} \int (d\mathbf{r}') \frac{\mathbf{j}(\mathbf{r}', t)}{|\mathbf{r} - \mathbf{r}'|} - \frac{1}{c^2} \frac{d^2}{dt^2} \mathbf{d}(t) , \tag{1.181}$$

which uses (1.137), for we have consistently retained terms of the order $1/c^3$ in the electric field intensity, from (1.48)

$$4\pi\mathbf{e}(\mathbf{r}, t) = -\boldsymbol{\nabla} \int (d\mathbf{r}') \frac{\rho(\mathbf{r}', t)}{|\mathbf{r} - \mathbf{r}'|} - \frac{1}{2c^2} \frac{\partial^2}{\partial t^2} \int (d\mathbf{r}') \frac{\mathbf{r} - \mathbf{r}'}{|\mathbf{r} - \mathbf{r}'|} \rho(\mathbf{r}', t)$$

$$+ \frac{1}{3c^3} \frac{\partial^3}{\partial t^3} \int (d\mathbf{r}') (\mathbf{r} - \mathbf{r}') \rho(\mathbf{r}', t) - \frac{1}{c^2} \frac{\partial}{\partial t} \int (d\mathbf{r}') \frac{\mathbf{j}(\mathbf{r}', t)}{|\mathbf{r} - \mathbf{r}'|}$$

$$+ \frac{1}{c^3} \frac{d^3}{dt^3} \mathbf{d}(t) . \tag{1.182}$$

Now,

$$\frac{\partial}{\partial t} \int (d\mathbf{r}') (\mathbf{r} - \mathbf{r}') \rho(\mathbf{r}', t) = \mathbf{r} \frac{d}{dt} \int (d\mathbf{r}') \rho(\mathbf{r}', t) - \frac{d}{dt} \int (d\mathbf{r}') \mathbf{r}' \rho(\mathbf{r}', t)$$

$$= -\frac{d}{dt} \mathbf{d}(t) \tag{1.183}$$

and

$$\frac{\partial}{\partial t} \int (d\mathbf{r}') \frac{\mathbf{r} - \mathbf{r}'}{|\mathbf{r} - \mathbf{r}'|} \rho(\mathbf{r}', t) = -\int (d\mathbf{r}') \frac{\mathbf{r} - \mathbf{r}'}{|\mathbf{r} - \mathbf{r}'|} \boldsymbol{\nabla}' \cdot \mathbf{j}(\mathbf{r}', t) = -\int (d\mathbf{r}') \frac{\mathbf{j}(\mathbf{r}', t)}{|\mathbf{r} - \mathbf{r}'|}$$

$$+ \int (d\mathbf{r}') \frac{(\mathbf{r} - \mathbf{r}')(\mathbf{r} - \mathbf{r}') \cdot \mathbf{j}(\mathbf{r}', t)}{|\mathbf{r} - \mathbf{r}'|^3} , \tag{1.184}$$

whence

$$4\pi\mathbf{e}(\mathbf{r}, t) = -\boldsymbol{\nabla} \int (d\mathbf{r}') \frac{\rho(\mathbf{r}', t)}{|\mathbf{r} - \mathbf{r}'|} - \frac{1}{2c^2} \frac{\partial}{\partial t} \int (d\mathbf{r}') \left[\frac{\mathbf{j}(\mathbf{r}', t)}{|\mathbf{r} - \mathbf{r}'|} \right.$$

$$\left. + \frac{(\mathbf{r} - \mathbf{r}')(\mathbf{r} - \mathbf{r}') \cdot \mathbf{j}(\mathbf{r}', t)}{|\mathbf{r} - \mathbf{r}'|^3} \right] + \frac{2}{3c^3} \frac{d^3}{dt^3} \mathbf{d}(t) . \tag{1.185}$$

Therefore, in (1.176) we encounter

$$-4\pi \int (d\mathbf{r})\, \mathbf{j} \cdot \mathbf{e} = \frac{d}{dt}\left[\frac{1}{2} \int (d\mathbf{r})\, (d\mathbf{r}')\, \frac{\rho(\mathbf{r},t)\rho(\mathbf{r}',t)}{|\mathbf{r}-\mathbf{r}'|} \right.$$

$$+ \frac{1}{4c^2} \int (d\mathbf{r})\, (d\mathbf{r}') \left\{ \frac{\mathbf{j}(\mathbf{r},t)\cdot \mathbf{j}(\mathbf{r}',t)}{|\mathbf{r}-\mathbf{r}'|} \right.$$

$$+ \left. \left. \frac{(\mathbf{r}-\mathbf{r}')\cdot \mathbf{j}(\mathbf{r},t)\,(\mathbf{r}-\mathbf{r}')\cdot \mathbf{j}(\mathbf{r}',t)}{|\mathbf{r}-\mathbf{r}'|^3} \right\} \right] - \frac{2}{3c^3}\dot{\mathbf{d}}(t) \cdot \dddot{\mathbf{d}}(t)\,,$$

$$(1.186)$$

for

$$\int (d\mathbf{r})\, \mathbf{j}(\mathbf{r},t) \cdot \boldsymbol{\nabla} \int (d\mathbf{r}')\, \frac{\rho(\mathbf{r}',t)}{|\mathbf{r}-\mathbf{r}'|} = -\int (d\mathbf{r})\, (d\mathbf{r}')\, \boldsymbol{\nabla} \cdot \mathbf{j}(\mathbf{r},t)\, \frac{\rho(\mathbf{r}',t)}{|\mathbf{r}-\mathbf{r}'|}$$

$$= \frac{d}{dt}\frac{1}{2} \int (d\mathbf{r})\, (d\mathbf{r}')\, \frac{\rho(\mathbf{r},t)\rho(\mathbf{r}',t)}{|\mathbf{r}-\mathbf{r}'|}\,,$$

$$(1.187)$$

and we have used (1.137) again. As a last rearrangement,

$$-\frac{2}{3c^3}\dot{\mathbf{d}}\cdot\dddot{\mathbf{d}} = -\frac{d}{dt}\left(\frac{2}{3c^3}\dot{\mathbf{d}}\cdot\ddot{\mathbf{d}}\right) + \frac{2}{3c^3}(\ddot{\mathbf{d}})^2\,. \qquad (1.188)$$

The integral $-\int (d\mathbf{r})\,\mathbf{j}\cdot \mathbf{e}$ has thus been expressed in the desired form (1.176) as the time derivative of a quantity plus a positive definite expression which is to be identified with the rate of radiation,

$$P = \frac{2}{3c^3}\frac{1}{4\pi}(\ddot{\mathbf{d}})^2\,. \qquad (1.189)$$

Correct to terms of order $1/c^2$ the electromagnetic energy of the molecule is[7]

$$E = \frac{1}{2} \int (d\mathbf{r})\, (d\mathbf{r}')\, \frac{\rho(\mathbf{r},t)\rho(\mathbf{r}',t)}{4\pi|\mathbf{r}-\mathbf{r}'|}$$

$$+ \frac{1}{4c^2} \int (d\mathbf{r})\, (d\mathbf{r}') \left(\frac{\mathbf{j}(\mathbf{r},t)\cdot \mathbf{j}(\mathbf{r}',t)}{4\pi|\mathbf{r}-\mathbf{r}'|} + \frac{(\mathbf{r}-\mathbf{r}')\cdot \mathbf{j}(\mathbf{r},t)\,(\mathbf{r}-\mathbf{r}')\cdot \mathbf{j}(\mathbf{r}',t)}{4\pi|\mathbf{r}-\mathbf{r}'|^3} \right)\,.$$

$$(1.191)$$

Magnetic dipole radiation first appears in that approximation which retains terms of the order $1/c^5$, with the expected result (1.174).

---

[7] It is the presence of the third term in (1.191) that results in the attraction between like currents, described by the magnetostatic energy

$$E = -\frac{1}{2c^2} \int (d\mathbf{r})\, (d\mathbf{r}')\, \frac{\mathbf{J}(\mathbf{r})\cdot \mathbf{J}(\mathbf{r}')}{4\pi|\mathbf{r}-\mathbf{r}'|}\,. \qquad (1.190)$$

See [9], Chap. 33.

1.6 Macroscopic Fields

The electromagnetic fields within material bodies, composed of enormous numbers of individual particles, are extremely complicated functions of position and time, on an atomic scale. Even if the formidable task of constructing the fields in a given situation could be performed, the description thus obtained would be unnecessarily elaborate, for it would contain information that could not be verified by our gross, macroscopic measuring instruments which respond only to the effects of many elementary particles. A macroscopic measurement of the instantaneous value of the field at a point is, in reality, a measurement of an average field within a region containing many atoms and extending over an interval of time large in comparison with atomic periods. It is natural, then, to seek an approximate form of the theory, so devised that the quantities which are the object of calculation are such averaged fields from which microscopic inhomogeneities have been removed, rather than the actual fields themselves. Such a program can be carried out if a length L and a time interval T exist which are small in comparison with distances and times in which macroscopic properties change appreciably, but large compared with atomic distances and times. These conditions are adequately satisfied under ordinary circumstances, failing only for matter of very low density or periodic fields of extremely short wavelength. Any quantity exhibiting enormous microscopic fluctuations, such as a field intensity, can be replaced by a microscopically smoothed quantity possessing only macroscopic variations by an averaging process conducted over a temporal interval T and a spatial region of linear extension L. Thus

$$\overline{f}(\mathbf{r}, t) = \frac{1}{V} \int_{|\mathbf{r}'| < L/2} (d\mathbf{r}') \frac{1}{T} \int_{-T/2}^{T/2} dt' \, F(\mathbf{r} + \mathbf{r}', t + t') \tag{1.192}$$

defines a space–time average of the function $f(\mathbf{r}, t)$, extended through the time interval from $t - T/2$ to $t + T/2$, and over a spatial region of volume V which may be considered a sphere of diameter L drawn about the point $\mathbf{r}$. This averaging process has the important property expressed by

$$\overline{\boldsymbol{\nabla} f(\mathbf{r}, t)} = \boldsymbol{\nabla} \overline{f}(\mathbf{r}, t) \,, \qquad \overline{\frac{\partial}{\partial t} f(\mathbf{r}, t)} = \frac{\partial}{\partial t} \overline{f}(\mathbf{r}, t) \,, \tag{1.193}$$

providing that the averaging domains are identical for all points of space and time. Hence, any linear differential equation connecting field variables can be replaced by formally identical equations for the averaged fields. Thus, in terms of the averaged field intensities,

$$\overline{\mathbf{e}}(\mathbf{r}, t) = \sqrt{\epsilon_0}\, \mathbf{E}(\mathbf{r}, t) \,, \qquad \overline{\mathbf{b}}(\mathbf{r}, t) = \frac{1}{\sqrt{\mu_0}}\, \mathbf{B}(\mathbf{r}, t) \,, \tag{1.194}$$

the averaged Maxwell–Lorentz equations (1.19a) and (1.19b) read

$$\boldsymbol{\nabla} \times \frac{1}{\mu_0}\mathbf{B} = \frac{\partial}{\partial t}\epsilon_0\mathbf{E} + \sqrt{\epsilon_0}\,\bar{\mathbf{j}}\,, \quad \boldsymbol{\nabla} \cdot \epsilon_0\mathbf{E} = \sqrt{\epsilon_0}\,\bar{\rho}\,, \qquad (1.195a)$$

$$\boldsymbol{\nabla} \times \mathbf{E} = -\frac{\partial}{\partial t}\mathbf{B}\,, \quad \boldsymbol{\nabla} \cdot \mathbf{B} = 0\,, \qquad (1.195b)$$

in which we have introduced two new constants, ϵ_0 and μ_0, related by

$$\epsilon_0\mu_0 = \frac{1}{c^2}\,, \qquad (1.196)$$

in order to facilitate the eventual adoption of a convenient system of units (SI) for macroscopic applications. Such averaged equations will be meaningful to the extent that they are independent of the precise size of the space–time averaging regions, within certain limits. This will be true if the sources of the macroscopic fields $\mathbf{E}$ and $\mathbf{B}$, namely the averaged charge and current densities, can be expressed entirely in terms of the macroscopic field quantities and other large scale variables (temperature, density, etc.).

The actual charge distribution within a material medium arises not only from the charges within neutral atoms and molecules, which we shall call the bound charge, but also from relatively freely moving electrons (conduction electrons) and the charged atoms (ions) from which they have been removed.[8] The latter source of charge will be termed the free charge. We have already shown that the true bound charge–current distribution within a molecule can be replaced by an equivalent point distribution without affecting the values of integrated fields, or averaged fields, within a region large compared to the molecule. Hence, for the purpose of evaluating $\bar{\rho}$ and $\bar{\mathbf{j}}$, the actual charge–current distribution can be written as the sum of a free charge distribution and the equivalent point distributions for the neutral molecules (and the ions, save for their net charge), given in (1.148a) and (1.148b),

$$\rho(\mathbf{r}, t) = \rho_f(\mathbf{r}, t) - \boldsymbol{\nabla} \cdot \mathbf{d}(\mathbf{r}, t)\,, \qquad (1.197a)$$

$$\mathbf{j}(\mathbf{r}, t) = \mathbf{j}_f(\mathbf{r}, t) + \frac{\partial}{\partial t}\mathbf{d}(\mathbf{r}, t) + c\,\boldsymbol{\nabla} \times \mathbf{m}(\mathbf{r}, t)\,, \qquad (1.197b)$$

where, summed over the molecules,

$$\mathbf{d}(\mathbf{r}, t) = \sum_a \mathbf{d}_a(t)\,\delta(\mathbf{r} - \mathbf{r}_a)\,, \quad \mathbf{m}(\mathbf{r}, t) = \sum_a \mathbf{m}_a(t)\,\delta(\mathbf{r} - \mathbf{r}_a)\,. \qquad (1.198)$$

The averaged charge and current densities will be expressed, in the same form, in terms of averaged free charge and current densities and

$$\bar{\mathbf{d}}(\mathbf{r}, t) = n\,\bar{\mathbf{d}}\,, \qquad \bar{\mathbf{m}}(\mathbf{r}, t) = n\,\bar{\mathbf{m}} \qquad (1.199)$$

are the average dipole moments of a molecule within the smoothing region (in addition to the time average, a statistical average among the molecules is

---

[8] Holes in a semiconductor could also be contemplated.

implied), multiplied by the average density n of molecules at the macroscopic point in question. With the notation

$$\bar{\rho}_f(\mathbf{r}, t) = \frac{1}{\sqrt{\epsilon_0}}\, \rho(\mathbf{r}, t)\,, \quad \bar{\mathbf{j}}_f(\mathbf{r}, t) = \frac{1}{\sqrt{\epsilon_0}}\, \mathbf{J}(\mathbf{r}, t)\,, \tag{1.200a}$$

$$\bar{\mathbf{d}}(\mathbf{r}, t) = \frac{1}{\sqrt{\epsilon_0}}\, \mathbf{P}(\mathbf{r}, t)\,, \quad \overline{\mathbf{m}}(\mathbf{r}, t) = \sqrt{\mu_0}\, \mathbf{M}(\mathbf{r}, t) \tag{1.200b}$$

for the macroscopic quantities measuring the free charge and current densities, and the electric and magnetic intensities of polarization (dipole moment per unit volume),[9] the averaged charge and current densities are

$$\sqrt{\epsilon_0}\, \bar{\rho} = \rho - \boldsymbol{\nabla} \cdot \mathbf{P}\,, \tag{1.201a}$$

$$\sqrt{\epsilon_0}\, \bar{\mathbf{j}} = \mathbf{J} + \frac{\partial}{\partial t}\mathbf{P} + \boldsymbol{\nabla} \times \mathbf{M}\,. \tag{1.201b}$$

Therefore, the first set of the averaged microscopic field equations (1.195a) – the Maxwell equations – read

$$\boldsymbol{\nabla} \times \mathbf{H} = \frac{\partial}{\partial t}\mathbf{D} + \mathbf{J}\,, \quad \boldsymbol{\nabla} \cdot \mathbf{D} = \rho\,, \tag{1.202a}$$

where

$$\mathbf{H} = \frac{1}{\mu_0}\mathbf{B} - \mathbf{M}\,, \quad \mathbf{D} = \epsilon_0 \mathbf{E} + \mathbf{P}\,, \tag{1.202b}$$

while the second set (1.195b) are unchanged. Thus the starting equations (1.1a) and (1.1b) are recovered.

We also record the SI forms of the energy, energy flux vector, and momentum in vacuum ($\mathbf{M} = \mathbf{P} = \mathbf{0}$):

$$U = \frac{1}{2}(\varepsilon_0 E^2 + \mu_0 H^2)\,, \tag{1.203a}$$

$$\mathbf{S} = \mathbf{E} \times \mathbf{H}\,, \tag{1.203b}$$

$$\mathbf{G} = \mathbf{D} \times \mathbf{B}\,. \tag{1.203c}$$

The form of energy and momentum conservation in a medium is much more subtle, and will be treated subsequently.

1.7 Problems for Chap. 1

Note – In these problems, and in following chapters, we will use $\mathbf{E}$, $\mathbf{B}$, and $\mathbf{A}$ to denote the electric and magnetic fields, and the vector potential, both in macroscopic and microscopic situations, and we will use Heaviside–Lorentz

---

[9] $\mathbf{P}$ and $\mathbf{M}$ are also referred to as the electric polarization and the magnetization, respectively.

units for both, except in waveguide applications, which have a more engineering flavor. Again we remind the reader of the simple conversion factors necessary to pass between SI, Heaviside–Lorentz, and Gaussian units, described in the Appendix.

1. Verify the representations (1.7a) and (1.7b) for the three-dimensional delta function. Alternatively, derive them from the Fourier representations (1.94a) and (1.94b).
2. Establish the identities (1.22a) and (1.22b), and then prove the conservation statements (1.21) in empty space.
3. Prove the local statements of field energy and momentum nonconservation (1.44a) and (1.44b) from the inhomogeneous Maxwell equations (1.19a) and (1.19b).
4. Show that the energy is given by (1.75) by inserting (1.59) into (1.74); similarly, fill in the steps leading to (1.80) and (1.86).
5. Without reference to potentials, show that

$$-\square^2 \mathbf{E} = \left(-\boldsymbol{\nabla}\rho - \frac{1}{c^2}\frac{\partial}{\partial t}\mathbf{j} \right) , \tag{1.204a}$$

$$-\square^2 \mathbf{B} = \frac{1}{c}\boldsymbol{\nabla}\times\mathbf{j} . \tag{1.204b}$$

Here we have introduced the "d'Alembertian," or wave operator,

$$\square^2 = -\frac{\partial^2}{c^2\partial t^2} + \nabla^2 . \tag{1.205}$$

Use the retarded solution of these equations to arrive at the asymptotic radiation fields of a bounded current distribution. (Do not forget charge conservation, $\frac{\partial}{\partial t}\rho + \boldsymbol{\nabla}\cdot\mathbf{j} = 0$.)
6. Starting from the Liénard–Wiechert potentials (1.125) and (1.126), work out $\partial\tau/\partial t$ and $\boldsymbol{\nabla}\tau$ and so recognize that

$$\left\{ \begin{matrix} \phi \\ \mathbf{A} \end{matrix} \right\} (\mathbf{r},t) = \frac{q}{4\pi}\frac{1}{|\mathbf{r}-\mathbf{r}(\tau)|} \left\{ \begin{matrix} \partial\tau/\partial t \\ \frac{1}{c}\partial\mathbf{r}(\tau)/\partial t \end{matrix} \right\} . \tag{1.206}$$

Check that

$$(\boldsymbol{\nabla}\tau)^2 - \left(\frac{1}{c}\frac{\partial}{\partial t}\tau \right)^2 = 0 . \tag{1.207}$$

7. Work out the magnetic field of a moving point charge e by differentiating the δ-function form for the potentials, (1.120a) and (1.120b). Get the radiation field part by considering only the derivative of the δ function, and show that

$$\mathbf{B}(\mathbf{r},t) \sim -\frac{e}{4\pi c^2}\frac{1}{|\mathbf{r}-\mathbf{r}(\tau)|}\mathbf{n}\times\frac{d^2\mathbf{r}(\tau)}{dt^2} , \quad \text{where} \quad \mathbf{n} = \frac{\mathbf{r}-\mathbf{r}(\tau)}{|\mathbf{r}-\mathbf{r}(\tau)|} . \tag{1.208}$$

Note carefully that $d^2\mathbf{r}(\tau)/dt^2$ is *not* $d^2\mathbf{r}(\tau)/d\tau^2$.

8. What is the associated electric field as found by an analogous asymptotic computation from the potentials? Compare with (1.206).

9. Let $\mathbf{E}_c$ be the instantaneous Coulomb field due to the charge density ρ. Demonstrate that (1.204a) can be presented as

$$-\Box^2(\mathbf{E} - \mathbf{E}_c) = -\frac{1}{c^2}\frac{\partial}{\partial t}\left(\mathbf{j} + \frac{\partial}{\partial t}\mathbf{E}_c\right). \qquad (1.209)$$

Begin with the retarded solution of this equation and derive the expressions for the electromagnetic energy and radiation power of a small current distribution.

10. Use the radiation fields derived above to compute the energy flux at large distances, per unit solid angle, for a point particle of charge e in terms of the acceleration and velocity of the particle at the retarded time – the emission time. Convert this energy per unit detection time into energy per unit emission time to get the power radiated into a given solid angle,

$$\frac{dP}{d\Omega} = \frac{e^2}{(4\pi)^2 c^3}\left[\frac{\dot{\mathbf{v}}^2}{(1 - \mathbf{n}\cdot\mathbf{v}/c)^3} + 2\frac{\mathbf{n}\cdot\dot{\mathbf{v}}\frac{1}{c}\mathbf{v}\cdot\dot{\mathbf{v}}}{(1 - \mathbf{n}\cdot\mathbf{v}/c)^4}\right.$$
$$\left. - \left(1 - \frac{v^2}{c^2}\right)\frac{(\mathbf{n}\cdot\dot{\mathbf{v}})^2}{(1 - \mathbf{n}\cdot\mathbf{v}/c)^5}\right]. \qquad (1.210)$$

Show that this reduces to the formula for dipole radiation in the nonrelativistic limit, $v/c \ll 1$. For another expression of this result, see (3.111).

11. Integrate (1.210) over all directions to arrive at the expected result.

12. Use the result (1.210) to show, for the situation of linear acceleration, that is, when $\dot{\mathbf{v}}$ is in the same direction as $\mathbf{v}$, which makes an angle θ with respect to the direction of observation, that ($\beta = v/c$)

$$-\frac{d^2 E}{dt\, d\Omega}\bigg|_{\text{rad}} = \frac{e^2}{(4\pi)^2 c^3}\left(\frac{dv}{dt}\right)^2\frac{\sin^2\theta}{(1 - \beta\cos\theta)^5}. \qquad (1.211)$$

Integrate this over all solid angles to arrive at the energy loss rate for this circumstance.

13. Derive the dipole radiation formula for radiation emitted at a given frequency,

$$\frac{dE_{\text{rad}}}{d\omega} = \frac{2}{3\pi}\frac{1}{4\pi}\frac{1}{c^3}|\ddot{\mathbf{d}}(\omega)|^2. \qquad (1.212)$$

Apply this formula to an instantaneous collision of two particles, one with mass m_1 and charge e_1, the second with mass m_2 and charge e_2 in the center of mass frame (i.e., the total momentum is zero). Let the angle of scattering of either particle be θ. Ignoring radiation reaction, both particles have the same momentum magnitude p before and after the collision. What happens if $e_1/m_1 = e_2/m_2$? From the photon viewpoint, how does the assumption that the kinetic energy of the particles is not changed restrict the radiation frequencies to which your result can be applied?

14. Consider a particle undergoing an instantaneous reversal in direction, changing from velocity $\mathbf{v}$ to velocity $-\mathbf{v}$ in negligible time. Derive the following formula for the number of photons with energy $\hbar\omega$ emitted into a frequency interval $d\omega$ and into an element of solid angle $d\Omega$ making an angle θ with respect to the direction specified by $\mathbf{v}$: ($\alpha = e^2/4\pi\hbar c$)

$$\frac{d^2N}{d\Omega\,d\omega} = \frac{\alpha}{4\pi^2}\frac{1}{\omega}\left\{2\frac{1+\beta^2}{1-(\beta\cos\theta)^2}\right.$$
$$\left. - (1-\beta^2)\left[\frac{1}{(1-\beta\cos\theta)^2} + \frac{1}{(1+\beta\cos\theta)^2}\right]\right\}.\quad(1.213)$$

15. What is the result of integrating (1.213) over all angles, for any $\beta < 1$. Does your photon spectrum agree with the known result for $\beta \ll 1$? What does it become for $\beta \approx 1$? Can you understand this result by looking at the approximate form derived from (1.213) for $\beta \approx 1$, $\theta \ll 1$, $\pi - \theta \ll 1$?

16. Now suppose the charged particle stops on impact. Find the analog of the formula (1.213). Again, integrate it over all angles and look at the limits of $\beta \ll 1$ and $\beta \approx 1$. Are the last two results what you would have expected? Explain.

17. Point charges e and $-e$ are created at $\mathbf{r} = \mathbf{0}$, $t = 0$, and then move with constant velocities $\mathbf{v}$ and $-\mathbf{v}$, respectively. Derive the distribution in frequency and angle of the emitted radiation. Describe the angular distribution for $v/c \approx 1$. Repeat for one charge created at rest, the other with velocity $\mathbf{v}$.

18. Charge e is distributed uniformly over the surface of a sphere of radius a, which is rotating about an axis with constant angular velocity ω. Compute the power radiated, either by applying a general method or by considering electric and magnetic dipole radiation.

19. A free electron at rest acted on by a light wave, and also the radiation reaction force, is described by

$$m\dot{\mathbf{v}} = e\mathrm{Re}\,\mathbf{E}e^{-i\omega t} + \frac{1}{4\pi}\frac{2}{3}\frac{e^2}{c^3}\dddot{\mathbf{v}}.\quad(1.214)$$

Solve this equation to get the total scattering cross section, defined as the ratio of the total power removed from the incident field, P_{tot}, to the incident flux, $|\mathbf{S}|$,

$$\sigma_{\mathrm{tot}} = \frac{P_{\mathrm{tot}}}{|\mathbf{S}|}.\quad(1.215)$$

Express the cross section in terms of the so-called classical radius of the electron, $r_0 = e^2/4\pi mc^2$ and the reduced wavelength $\lambdabar = \lambda/(2\pi)$. What is the limiting form for $\lambdabar \ll r_0$?

20. Calculate the total cross section for the scattering of a plane wave by a dielectric sphere, assuming that the wavelength is large compared to the radius of the sphere.

21. Define a complex vector field by

$$\mathbf{F} = \mathbf{E} + i\mathbf{B}, \qquad \mathbf{F}^* = \mathbf{E} - i\mathbf{B}. \qquad (1.216)$$

Identify the scalar, vector, and dyadic, given by

$$\frac{1}{2}\mathbf{F}^* \cdot \mathbf{F}, \qquad \frac{1}{2i}\mathbf{F}^* \times \mathbf{F}, \quad \text{and} \quad \frac{1}{2}(\mathbf{F}\,\mathbf{F}^* + \mathbf{F}^*\,\mathbf{F}), \qquad (1.217)$$

respectively. What happens to these quantities if $\mathbf{F}$ is replaced by $e^{-i\phi}\mathbf{F}$, ϕ being a constant?

22. What magnetic field is described, almost everywhere, by the vector potential

$$\mathbf{A}(\mathbf{r}) = \mathbf{\nabla} \times \frac{g}{4\pi}\mathbf{n}\log(r - \mathbf{n}\cdot\mathbf{r}), \qquad (1.218)$$

where g is a constant and $\mathbf{n}$ is a unit vector?

23. Consider Maxwell's equations with both electric $(\rho_e, \mathbf{j}_e)$ and magnetic $(\rho_m, \mathbf{j}_m)$ charges. Show that these equations retain their form under the electromagnetic rotation (duality transformation) under which electric $(\mathcal{E})$ and magnetic $(\mathcal{M})$ quantities are redefined according to

$$\mathcal{E} \to \mathcal{E}\cos\phi + \mathcal{M}\sin\phi, \qquad \mathcal{M} \to \mathcal{M}\cos\phi - \mathcal{E}\sin\phi. \qquad (1.219)$$

Check that the generalized Lorentz force

$$\mathbf{F} = e\left(\mathbf{E} + \frac{\mathbf{v}}{c} \times \mathbf{B}\right) + g\left(\mathbf{B} - \frac{\mathbf{v}}{c} \times \mathbf{E}\right) \qquad (1.220)$$

also retains its form under this rotation. Can you give a two-dimensional geometrical interpretation of the latter fact? A unidirectional electromagnetic pulse [recall the discussion after (1.33)] is characterized by the relations

$$E^2 - B^2 = 0, \qquad \mathbf{E}\cdot\mathbf{B} = 0. \qquad (1.221)$$

How do these properties respond to the electromagnetic rotation? For general electromagnetic fields, how do U, $\mathbf{G}$, and T respond to electromagnetic rotations?

24. From the Maxwell equations with both electric and magnetic charges considered in the previous problem, derive second-order differential equations for $\mathbf{E}$ and for $\mathbf{B}$. Show that

$$\mathbf{E} = -\mathbf{\nabla}\phi_e - \frac{1}{c}\frac{\partial}{\partial t}\mathbf{A}_e - \mathbf{\nabla} \times \mathbf{A}_m, \qquad (1.222a)$$

$$\mathbf{B} = -\mathbf{\nabla}\phi_m - \frac{1}{c}\frac{\partial}{\partial t}\mathbf{A}_m + \mathbf{\nabla} \times \mathbf{A}_e, \qquad (1.222b)$$

and exhibit the differential equations for these potentials in the Lorenz gauge, and in the radiation gauge, where $\mathbf{\nabla}\cdot\mathbf{A} = 0$.

25. Solve for the above potentials in some gauge, and find the asymptotic radiation field. Now what is the relationship between $\mathbf{E}$ and $\mathbf{B}$? Construct the spectral–angular distribution of the radiated power. How do it change under the duality transformation (1.219)?

26. A point magnetic charge g is at rest, at the origin. A point electric charge e, carried by a particle of mass m is in motion about the electric charge. What is the Newton–Lorentz equation of motion? By taking the moment of this equation, verify that the conserved angular momentum is

$$\mathbf{J} = \mathbf{r} \times m\mathbf{v} - \frac{eg}{4\pi c}\frac{\mathbf{r}}{r} \ . \tag{1.223}$$

What follows if quantum ideas about angular momentum are applied to the radial component of $\mathbf{J}$?

27. Consider the relative motion of two particles with masses and electric and magnetic charges m_1, e_1, g_1, and m_2, e_2, and g_2, respectively. In deriving the equation of relative motion, which involves the reduced mass, remember that moving electric (magnetic) charges produce magnetic (electric) fields, but do not retain more than one factor of v_1/c or v_2/c. How do the combinations of e's and g's in this equation respond to the electromagnetic rotations of (1.219)? What is the conserved angular momentum?

28. A point magnetic charge g is located at the origin; a point electric charge e is located at the fixed point $\mathbf{R}$. What is the electromagnetic momentum density $\mathbf{G}$ at an arbitrary position $\mathbf{r}$? Write this vector as a curl. [This implies that $\mathbf{G}$ is divergenceless; why is that?] Now construct the total electromagnetic angular momentum as the integrated moment of $\mathbf{G}$, simplified by partial integration. You will recognize the remaining integral as the electric field at $\mathbf{R}$ produced by a charge density proportional to $1/|\mathbf{r}|$. Use spherical symmetry to solve the differential equation for the electric field (follow the known example of constant density). Compare your result with that of Problem 1.26.

29. Consider Maxwell's equations in vacuum with both electric and magnetic charges and currents, ρ_e, $\mathbf{j}_e$, ρ_m, and $\mathbf{j}_m$. Write the similar Maxwell equations satisfied by

$$\mathbf{E}' = \mathbf{E} - \mathbf{E}_s \ , \qquad \mathbf{B}' = \mathbf{B} - \mathbf{B}_s \ , \tag{1.224}$$

where $\mathbf{E}_s$ and $\mathbf{B}_s$ are the respective static fields at time t produced by the electric and magnetic charge densities at time t. That is,

$$\boldsymbol{\nabla} \cdot \mathbf{B}_s(\mathbf{r}, t) = \rho_m(\mathbf{r}, t) \ , \qquad \boldsymbol{\nabla} \times \mathbf{B}_s(\mathbf{r}, t) = 0 \ , \tag{1.225}$$

and so on. What is $\boldsymbol{\nabla} \cdot \mathbf{E}'$, $\boldsymbol{\nabla} \cdot \mathbf{B}'$? Then what can you say about $\mathbf{j}_e$, $\mathbf{j}_m$, the currents that appear in the Maxwell equations obeyed by $\mathbf{E}'$, $\mathbf{B}'$? Use that property to redefine $\mathbf{E}'$ so that you are left with the Maxwell equations without magnetic charge and current. Recognize that these fields can be constructed from a vector potential in the radiation gauge, and then exhibit $\mathbf{E}$ and $\mathbf{B}$.

30. An electron moves at speed $v \ll c$, in a circular orbit of radius r, about an infinitely massive proton. Compute the rate of radiation – the rate of energy loss – first, in terms of v and r, and then in terms of the electron energy E (recall the virial theorem). Integrate the resulting differential equation for E to find the time it takes an electron, initially of energy E_0, to fall into the nucleus of this classical hydrogen atom. State the collapse time in seconds when the initial energy is that of the first Bohr orbit. (Here is one reason for inventing quantum mechanics.)

31. Demonstrate that

$$U_{\text{charges}} = \sum_a \delta(\mathbf{r} - \mathbf{r}_a(t)) E_a(t) , \qquad (1.226a)$$

$$\mathbf{S}_{\text{charges}} = \sum_a \delta(\mathbf{r} - \mathbf{r}_a(t)) E_a(t) \mathbf{v}_a(t) , \qquad (1.226b)$$

obey

$$\frac{\partial}{\partial t} U_{\text{ch}} + \boldsymbol{\nabla} \cdot \mathbf{S}_{\text{ch}} = \mathbf{j} \cdot \mathbf{E} . \qquad (1.227)$$

How does this lead to a direct proof of local total energy conservation? Proceed similarly with

$$\mathbf{G}_{\text{ch}} = \sum_a \delta(\mathbf{r} - \mathbf{r}_a) m_a \mathbf{v}_a , \qquad (1.228a)$$

$$\mathbf{T}_{\text{ch}} = \sum_a \delta(\mathbf{r} - \mathbf{r}_a) m_a \mathbf{v}_a \mathbf{v}_a . \qquad (1.228b)$$

32. Use the relativistic Lagrangian

$$L = -m_0 c^2 \sqrt{1 - \mathbf{v}^2/c^2} - e\phi + \frac{e}{c} \mathbf{v} \cdot \mathbf{A} , \qquad \mathbf{v} = \frac{d\mathbf{r}}{dt} , \qquad (1.229)$$

to deduce the Einstein–Lorentz equation of motion.

33. The inference of the fundamental field equations discloses that imparting a small velocity $\delta \mathbf{v}$ to the system changes the fields by

$$\delta \mathbf{B} = \frac{\delta \mathbf{v}}{c} \times \mathbf{E} , \qquad \delta \mathbf{E} = -\frac{\delta \mathbf{v}}{c} \times \mathbf{B} . \qquad (1.230)$$

Show that Maxwell's equations, first without charge and current, retain their form if the meaning of the derivatives is also slightly altered:

$$\delta(\boldsymbol{\nabla}) = -\frac{\delta \mathbf{v}}{c} \frac{1}{c} \frac{\partial}{\partial t} , \qquad \delta\left(\frac{1}{c} \frac{\partial}{\partial t}\right) = -\frac{\delta \mathbf{v}}{c} \cdot \boldsymbol{\nabla} . \qquad (1.231)$$

Interpret this in terms of coordinate changes, $\delta \mathbf{r}$, δt. [Hint: $\boldsymbol{\nabla} t = 0$, $\partial \mathbf{r}/\partial t = 0$.] Now show that all this remains true in the presence of charges, provided

$$\delta \mathbf{j} = \delta \mathbf{v} \rho , \qquad \delta \rho = \frac{\delta \mathbf{v}}{c} \cdot \frac{1}{c} \mathbf{j} , \qquad (1.232)$$

the first of which is expected. This is a first suggestion of the Lorentz transformations of Einstein relativity, which will be explored further in Chap. 3.

34. Consider the stress dyadic T and the electromagnetic field of a unidirectional light pulse. Show that

$$\mathsf{T} \cdot \mathbf{E} = 0 , \quad \mathsf{T} \cdot \mathbf{B} = 0 , \quad \mathsf{T} \cdot \mathbf{E} \times \mathbf{B} = U \mathbf{E} \times \mathbf{B} . \tag{1.233}$$

Thus, in this situation, $\mathbf{E} \times \mathbf{B}$ is an eigenvector of T with the eigenvalue U, $\mathbf{E}$ and $\mathbf{B}$ are eigenvectors with the eigenvalue zero. Are these properties consistent with $\mathrm{Tr}\,\mathsf{T} = U$, (1.37)? What is the value of $\det \mathsf{T}$ for the light pulse field?

35. Prove that the last result is unique to the light pulse by demonstrating, for an arbitrary field, that

$$\det \mathsf{T} = -U[U^2 - (c\mathbf{G})^2] \le 0 . \tag{1.234}$$

[Hint: find the eigenvalues of T.] When does the equality sign hold? What is the value of $\mathrm{Tr}\,\mathsf{T}^2$?

36. Work out the three-dimensional Coulomb Green's function in the vacuum,

$$G(\mathbf{r}) = \int \frac{(d\mathbf{k})}{(2\pi)^3} \frac{e^{i\mathbf{k}\cdot\mathbf{r}}}{k^2} , \tag{1.235}$$

by writing

$$\frac{1}{k^2} = \int_0^\infty d\lambda\, e^{-\lambda k^2} , \tag{1.236}$$

and then performing first the three integrations over the rectangular coordinates of $\mathbf{k}$. Repeat this calculation in four dimensions. Use your result to verify explicitly that

$$\int_{-\infty}^\infty dx_4\, G(x_1, x_2, x_3, x_4) = G(\mathbf{r}) . \tag{1.237}$$

Make this understandable by considering the four-dimensional differential equation that $G(x_1, x_2, x_3, x_4)$ obeys.

37. Besides the advanced and retarded Green's functions considered in (1.106) et seq., another important Green's function is the causal or Feynman Green's function, defined by the $3 + 1$ dimensional Fourier integral

$$G_+(\mathbf{r} - \mathbf{r}', t - t') = \int \frac{(d\mathbf{k})}{(2\pi)^3} \frac{d\omega}{2\pi} \frac{e^{i[\mathbf{k}\cdot(\mathbf{r}-\mathbf{r}')-\omega(t-t')]}}{k^2 - \omega^2/c^2 - i\epsilon} . \tag{1.238}$$

Evaluate this as

$$G_+ = \frac{ic}{4\pi^2} \frac{1}{(\mathbf{r} - \mathbf{r}')^2 - c^2(t - t')^2 + i\epsilon} . \tag{1.239}$$

Write the analogous definition and form of $G_- = G_+^*$. Check that

$$\frac{1}{2}(G_+ + G_-) = \frac{1}{2}(G_{\text{ret}} + G_{\text{adv}}) , \tag{1.240}$$

where

$$G_{\text{ret,adv}} = \frac{\delta(t - t' \mp |\mathbf{r} - \mathbf{r}'|/c)}{4\pi|\mathbf{r} - \mathbf{r}'|} . \tag{1.241}$$

38. Solve the differential equation

$$(-\nabla^2 + \gamma^2)G(\mathbf{r}) = \delta(\mathbf{r}) , \tag{1.242}$$

by Fourier transformation followed by a contour integration.

2

Spherical Harmonics

Although spherical harmonics will not appear explicitly in the sequel, it seems impossible to write a textbook on the subject of electromagnetic theory without a brief account of some of their most fundamental features.

2.1 Connection to Bessel Functions

The ancient belief in the flatness of the Earth had a simple geometrical basis. A plane surface and a sphere of radius a are indistinguishable on or near the surface, within a region of linear dimensions that are small compared to a. There is a lesson here for us in electrostatics. It is elementary to solve the problem of the transition between the vacuum and a homogeneous dielectric medium, for both a plane and a spherical surface of contact. Under the geometrical restriction just cited for the Earth, the two solutions must be equivalent. And that implies a limiting relationship between the types of functions involved in the two situations.

Let us write, side by side, the electrostatic Green's functions referring to the plane surface, above the dielectric (see, e.g., (14.24) and (16.4) of [9]) $z, z' > 0$:

$$G(\mathbf{r}, \mathbf{r}') = \int_0^\infty dk \, J_0(kP) \left[e^{-k(z_> - z_<)} - \frac{\varepsilon - 1}{\varepsilon + 1} e^{-k(z + z')} \right], \qquad (2.1)$$

where P is the transverse distance between the two points,

$$P = |(\mathbf{r} - \mathbf{r}')_\perp| = \sqrt{\rho^2 + \rho'^2 - 2\rho\rho' \cos(\phi - \phi')}, \qquad (2.2)$$

in cylindrical coordinates, and that for the exterior of a dielectric sphere, $r, r' > a$, ((23.99) of [9])

$$G(\mathbf{r}, \mathbf{r}') = \sum_{l=0}^\infty P_l(\cos \Theta) \left[\frac{r_<^l}{r_>^{l+1}} - \frac{(\varepsilon - 1)l}{(\varepsilon + 1)l + 1} \frac{a^{2l+1}}{r^{l+1} r'^{l+1}} \right], \qquad (2.3)$$

where Θ is the angle between the two directions of $\mathbf{r}$ and $\mathbf{r}'$,

$$\cos\Theta = \cos\theta\cos\theta' + \sin\theta\sin\theta'\cos(\phi - \phi') , \qquad (2.4)$$

in terms of the spherical angles. In addition to the spherical coordinate system with its origin at the center of the sphere, we erect a rectangular coordinate system at the point on the spherical surface with $\theta = 0$, that is, at the north pole. The z-axis continues the line from the center of the sphere to that point, and the x–y plane is the tangential to the sphere. Under the restrictions

$$z \ll a , \quad \rho = (x^2 + y^2)^{1/2} \ll a , \qquad (2.5)$$

the distance from the center of the sphere to any point, with rectangular coordinates x, y, z, is

$$r = [(a + z)^2 + \rho^2]^{1/2} \approx a + z + \frac{\rho^2}{2a} \approx a + z , \qquad (2.6)$$

where $\rho^2/2a$ is omitted as being negligible compared to z.

Now note that

$$r_<^l \approx a^l \left(1 + \frac{z_<}{a}\right)^l \approx a^l e^{(l/a)z_<} , \qquad (2.7a)$$

and

$$r_>^{-l-1} \approx a^{-l-1} \left(1 + \frac{z_>}{a}\right)^{-l-1} \approx a^{-l-1} e^{-(l/a)z_>} , \qquad (2.7b)$$

based on the equivalence between $1 + t$ and $\exp t$, for small values of t. The consequences

$$\frac{r_<^l}{r_>^{-l-1}} \approx \frac{1}{a} e^{-(l/a)(z_> - z_<)} , \qquad (2.8a)$$

and

$$\frac{a^{2l+1}}{r^{l+1} r'^{l+1}} \approx \frac{1}{a} e^{-(l/a)(z+z')} , \qquad (2.8b)$$

display the asymptotic connection between the variables of (2.1) and (2.3),

$$\frac{l}{a} \to k . \qquad (2.9)$$

Indeed, the factor of $1/a$ appearing in (2.8a) and (2.8b) tells us that any particular value of l makes a negligible contribution in the limit of interest; l must range over a wide spectrum, corresponding to the continuous nature of k. The replacement of the l summation by the k integral is expressed by

$$\delta l = 1 : \quad \frac{\delta l}{a} \to dk . \qquad (2.10)$$

Then, if we use the geometrical relations,

$$\theta \approx \frac{\rho}{a}, \quad \theta' \approx \frac{\rho'}{a}, \quad \Theta \approx \frac{P}{a}, \tag{2.11}$$

we are led the asymptotic correspondence,

$$\Theta \ll 1, \, l \gg 1: \quad P_l(\cos\Theta) \to J_0(l\Theta). \tag{2.12}$$

Of course, had obtaining this result been the only objective, it would have sufficed to set $\varepsilon = 1$.

We can extend the connection (2.12) by comparing the addition theorems for the two types of functions ((16.69) of [9])

$$J_0(kP) = \sum_{m=-\infty}^{\infty} J_m(k\rho) J_m(k\rho') e^{im(\phi-\phi')}, \tag{2.13a}$$

and

$$P_l(\cos\Theta) = \frac{4\pi}{2l+1} \sum_{m=-l}^{l} Y_{lm}(\theta,\phi) Y_{lm}(\theta',\phi')^*$$

$$= \frac{1}{l+1/2} \sum_{m=-l}^{l} \Theta_{lm}(\theta) \Theta_{lm}(\theta') e^{im(\phi-\phi')}, \tag{2.13b}$$

where we have separated variables in defining

$$Y_{lm}(\theta,\phi) = \frac{1}{\sqrt{2\pi}} e^{im\phi} \Theta_{lm}(\theta), \tag{2.14}$$

$\Theta_{lm}(\theta)$ being essentially the associated Legendre function. It is clear that, within an algebraic sign (which we anticipate), one has

$$\theta \ll 1, \, l \gg 1: \quad (-1)^m \frac{1}{(l+1/2)^{1/2}} \Theta_{lm}(\theta) \to J_m(l\theta), \tag{2.15}$$

and (2.12) is recovered for $m = 0$,

$$\Theta_{l0}(\theta) = \sqrt{\frac{2l+1}{2}} P_l(\cos\theta). \tag{2.16}$$

· Reference to the initial terms in the small variable behavior of the two functions, for $m > 0$:

$$t \ll 1: \quad J_m(t) \sim \frac{\left(\frac{1}{2}t\right)^m}{m!}, \tag{2.17a}$$

$$\theta \ll 1: \quad \frac{1}{(l+1/2)^{1/2}} \Theta_{lm}(\theta) \sim \left[\frac{(l+m)!}{(l-m)!}\right]^{1/2} \frac{\left(-\frac{1}{2}\theta\right)^m}{m!}, \tag{2.17b}$$

suffices to show the need for the $(-1)^m$ factor in (2.15) (along with the remark that the relation between negative and positive m is the same for both functions).

The completeness statements in the forms

$$\int_0^\infty dk\, k\, J_0(k\rho) = \frac{1}{\rho}\delta(\rho) , \qquad (2.18a)$$

$$\sum_{l=0}^\infty \left(l + \frac{1}{2}\right) P_l(\cos\theta) = \frac{1}{\sin\theta}\delta(\theta) , \qquad (2.18b)$$

are also connected by this correspondence, as are the differential equations for $J_m(k\rho)$ and $\Theta_{lm}(\theta)$,

$$\left[\frac{1}{\rho}\frac{d}{d\rho}\left(\rho\frac{d}{d\rho}\right) - \frac{m^2}{\rho^2} + k^2\right] J_m(k\rho) = 0 , \qquad (2.19a)$$

$$\left[\frac{1}{\sin\theta}\frac{d}{d\theta}\left(\sin\theta\frac{d}{d\theta}\right) - \frac{m^2}{\sin^2\theta} + l(l+1)\right] \Theta_{lm}(\theta) = 0 . \qquad (2.19b)$$

Here is another way to arrive at the correspondence (2.15). First, we return to the initial definition of the spherical harmonics

$$\frac{(\boldsymbol{\nu}^* \cdot \frac{\mathbf{r}}{r})^l}{l!} = \sum_{m=-l}^l \psi_{lm}^* \left(\frac{4\pi}{2l+1}\right)^{1/2} Y_{lm}(\theta, \phi) , \qquad (2.20)$$

where $\boldsymbol{\nu}$ is a complex null vector constructed according to

$$\nu_x + i\nu_y = -\psi_+^2 , \quad \nu_x - i\nu_y = \psi_-^2 , \quad \nu_z = \psi_+\psi_- , \qquad (2.21)$$

from which the ψ_{lm} are constructed,

$$\psi_{lm} = \frac{\psi_+^{l+m}\psi_-^{l-m}}{[(l+m)!(l-m)!]^{1/2}} . \qquad (2.22)$$

If we insert into this

$$\psi_+ = i , \quad \psi_- = 1 : \quad \nu_x = 1 , \quad \nu_y = 0 , \quad \nu_z = i . \qquad (2.23)$$

Then (2.20), multiplied by $i^l l!$, becomes

$$(\cos\theta + i\sin\theta\cos\phi)^l = \sum_{m=-l}^l (-i)^m \frac{l!}{[(l+m)!(l-m)!]^{1/2}} \frac{\Theta_{lm}(\theta)e^{im\phi}}{(l+1/2)^{1/2}} , \qquad (2.24)$$

a Fourier series, from which we deduce

$$i^m \frac{l!}{[(l+m)!(l-m)!]^{1/2}} \frac{(-1)^m\Theta_{lm}(\theta)}{(l+1/2)^{1/2}} = \int_0^{2\pi} \frac{d\phi}{2\pi} e^{-im\phi}(\cos\theta + i\sin\theta\cos\phi)^l . \qquad (2.25)$$

We note the specialization to $m = 0$:

$$P_l(\cos\theta) = \int_0^{2\pi} \frac{d\phi}{2\pi} (\cos\theta + i\sin\theta\cos\phi)^l , \qquad (2.26)$$

which is known as Laplace's first integral representation. What is called Laplace's second integral representation, the result of the substitution $l \to -l - 1$, was pointed out by Jacobi in 1843 (see Problem 2.1).

Now let us recall that

$$i^m J_m(t) = \int_0^{2\pi} \frac{d\phi}{2\pi} e^{-im\phi} e^{it\cos\phi} , \qquad (2.27)$$

from which we derive, for any function f represented by a power series, that

$$f\left(\frac{d}{dt}\right) i^m J_m(t)\Big|_{t=0} = \int_0^{2\pi} \frac{d\phi}{2\pi} e^{-im\phi} f(i\cos\phi) . \qquad (2.28)$$

This is immediately applicable to (2.25) and yields

$$\frac{l!}{[(l+m)!(l-m)!]^{1/2}} (-1)^m \frac{\Theta_{lm}(\theta)}{(l+1/2)^{1/2}} = \left(\cos\theta + \sin\theta\frac{d}{dt}\right)^l J_m(t)\Big|_{t=0} , \qquad (2.29)$$

with the $m = 0$ specialization

$$P_l(\cos\theta) = \left(\cos\theta + \sin\theta\frac{d}{dt}\right)^l J_0(t)\Big|_{t=0} . \qquad (2.30)$$

An elementary example of the latter is

$$P_2(\cos\theta) = \left(\cos^2\theta + 2\cos\theta\sin\theta\frac{d}{dt} + \sin^2\theta\frac{d^2}{dt^2}\right)\left(1 - \frac{1}{4}t^2 + \cdots\right)\Big|_{t=0}$$

$$= \cos^2\theta - \frac{1}{2}\sin^2\theta = \frac{1}{2}(3\cos^2\theta - 1) . \qquad (2.31)$$

What we are looking for is realized by setting $\theta = s/l$, and, for fixed s, proceeding to the limit $l \to \infty$. First we observe, for example, that

$$\lim_{l\to\infty} \frac{(l!)^2}{(l+1)!(l-1)!} = \lim_{l\to\infty} \frac{l}{l+1} = 1 , \qquad (2.32)$$

and the outcome is the same for any given value of m. Accordingly, we get

$$\lim_{l\to\infty} (-1)^m \frac{1}{(l+1/2)^{1/2}} \Theta_{lm}\left(\frac{s}{l}\right) = \lim_{l\to\infty} \left(1 + \frac{s}{l}\frac{d}{dt}\right)^l J_m(t)\Big|_{t=0}$$

$$= e^{s\frac{d}{dt}} J_m(t)\Big|_{t=0} , \qquad (2.33)$$

or

$$\lim_{l\to\infty} (-1)^m \frac{1}{(l+1/2)^{1/2}} \Theta_{lm}\left(\frac{s}{l}\right) = J_m(s) , \qquad (2.34)$$

which is a precise version of (2.15).

2.2 Multipole Harmonics

The introduction of spherical harmonics may be motivated by an expansion of Coulomb's potential, as expressed symbolically by

$$r > r' : \qquad \frac{1}{|\mathbf{r} - \mathbf{r}'|} = \sum_{l=0}^{\infty} \frac{(-\mathbf{r}' \cdot \boldsymbol{\nabla})^l}{l!} \frac{1}{r}$$

$$= \sum_{l=0}^{\infty} \frac{(-1)^l}{l!} \{\mathbf{r}'\}^l \odot \{\boldsymbol{\nabla}\}^l \frac{1}{r} . \qquad (2.35)$$

The latter version introduces a notation (one can read $\odot$ as "scalar product all around") to express for arbitrary l what for $l = 2$, is the scalar dyadic product,

$$\mathbf{r}'\mathbf{r}' \odot \boldsymbol{\nabla}\boldsymbol{\nabla} = (\mathbf{r}' \cdot \boldsymbol{\nabla})^2 . \qquad (2.36)$$

[This notation is a slight generalization of that introduced in (1.133).] This is just the aspect of the Coulomb potential that is involved in expressing the potential $\phi(\mathbf{r})$ at points external to a bounded distribution of charge density $\rho(\mathbf{r})$,

$$\phi(\mathbf{r}) = \int (\mathrm{d}\mathbf{r}') \frac{1}{4\pi |\mathbf{r} - \mathbf{r}'|} \rho(\mathbf{r}') . \qquad (2.37)$$

It is desirable to choose the coordinate origin in the interior of the charge distribution. Then the introduction of the expansion (2.35) presents us with this related expansion of the potential:

$$\phi(\mathbf{r}) = \sum_{l=0}^{\infty} \frac{(-1)^l}{l!} \left[\int (\mathrm{d}\mathbf{r}')\{\mathbf{r}'\}^l \rho(\mathbf{r}') \right] \odot \{\boldsymbol{\nabla}\}^l \frac{1}{4\pi r} , \qquad (2.38)$$

which involves the successive moments of the charge distribution. The first of these is familiar:

$$l = 0 : \qquad \int (\mathrm{d}\mathbf{r}') \, \rho(\mathbf{r}') = e , \qquad (2.39)$$

the total charge, or electric monopole moment;

$$l = 1 : \qquad \int (\mathrm{d}\mathbf{r}') \, \mathbf{r}' \rho(\mathbf{r}') = \mathbf{d} , \qquad (2.40)$$

the electric dipole moment (1.131b). And the contribution to the potential of these first moments is exhibited in

$$4\pi\phi(\mathbf{r}) = \frac{e}{r} - \mathbf{d} \cdot \boldsymbol{\nabla}\frac{1}{r} + \cdots$$

$$= \frac{e}{r} + \mathbf{d} \cdot \frac{\mathbf{r}}{r^3} + \cdots , \qquad (2.41)$$

as is familiar [see (1.132)].

Before continuing, let us ask this question: Given a system for which the potential exterior to the charge distribution is adequately represented by the two terms displayed in (2.41), what fictitious charge distribution, '$\rho(\mathbf{r})$', would produce this potential if the latter were imagined to be valid everywhere? And what are the monopole and dipole moments of that fictitious distribution? The answer to the first question is immediate:

$$'\rho(\mathbf{r})' = -\frac{1}{4\pi}\nabla^2\left[\frac{e}{r} - \mathbf{d}\cdot\nabla\frac{1}{r}\right]$$
$$= e\delta(\mathbf{r}) - \mathbf{d}\cdot\nabla\delta(\mathbf{r}) \ . \tag{2.42}$$

Then the implied monopole moment is

$$\int (\mathrm{d}\mathbf{r})\,'\rho(\mathbf{r})' = \int (\mathrm{d}\mathbf{r})\,[e\delta(\mathbf{r}) - \nabla\cdot(\mathbf{d}\,\delta(\mathbf{r}))]$$
$$= e \ , \tag{2.43}$$

the surface integral resulting from the second, divergence term, being zero, and the dipole moment emerges as

$$\int (\mathrm{d}\mathbf{r})\,\mathbf{r}\,'\rho(\mathbf{r})' = \int (\mathrm{d}\mathbf{r})\,\mathbf{r}[e\delta(\mathbf{r}) - \nabla\cdot(\mathbf{d}\,\delta(\mathbf{r}))]$$
$$= \int (\mathrm{d}\mathbf{r})[-\nabla\cdot(\mathbf{d}\,\delta(\mathbf{r})\,\mathbf{r}) + \mathbf{d}\,\delta(\mathbf{r})]$$
$$= \mathbf{d} \ ; \tag{2.44}$$

they are the actual moments of the system. We shall see that this property, to which we referred in the previous chapter, extends to all the multipole moments we are in the process of developing.

A closely related remark is that the force and torque exerted on the fictitious charge distribution by an externally applied electric field are identical with those acting on the actual system in the presence of a sufficiently slowly varying field. Thus we have

$$\mathbf{F} = \int (\mathrm{d}\mathbf{r})\,'\rho(\mathbf{r})'\mathbf{E}(\mathbf{r})$$
$$= \int (\mathrm{d}\mathbf{r})[e\delta(\mathbf{r}) - \nabla\cdot(\mathbf{d}\,\delta(\mathbf{r}))]\mathbf{E}(\mathbf{r})$$
$$= e\mathbf{E} + \mathbf{d}\cdot\nabla\mathbf{E} \ , \tag{2.45}$$

where the final evaluation, referring to the origin, the reference point within the given charge distribution, can be recognized as the two leading terms in the force on the actual distribution; and

$$\tau = \int (\mathrm{d}\mathbf{r})\,\mathbf{r}\times\,'\rho(\mathbf{r})'\mathbf{E}(\mathbf{r})$$
$$= -\int (\mathrm{d}\mathbf{r})\nabla\cdot(\mathbf{d}\,\delta(\mathbf{r}))\,\mathbf{r}\times\mathbf{E}(\mathbf{r})$$
$$= \mathbf{d}\times\mathbf{E} \ , \tag{2.46}$$

again as expected.

The $l = 2$ term in (2.38) involves the symmetric dyadic

$$\nabla\nabla\frac{1}{r} = \frac{3\mathbf{r}\mathbf{r} - 1r^2}{r^5} , \tag{2.47}$$

which has $9 - 3 = 6$ components. There is, however, one relation among these components, stemming from Laplace's equation; the sum of the diagonal elements, three referring to the same components of the two vectors, vanishes. (This is, of course, the recognition that there are only five spherical harmonics for $l = 2$.) In consequence of the zero value of the diagonal sum – the so-called trace – any multiple of the unit dyadic in $\int(d\mathbf{r}')\mathbf{r}'\mathbf{r}'\rho(\mathbf{r}')$ occurring in (2.38) will make no contribution. That permits us to introduce the traceless dyadic structure

$$\mathbf{Q} = \int (d\mathbf{r}')(3\mathbf{r}'\mathbf{r}' - 1r'^2)\rho(\mathbf{r}') , \tag{2.48}$$

the quadrupole moment dyadic. Adding this contribution to the expansion of the potential gives us

$$4\pi\phi(\mathbf{r}) = \frac{e}{r} + \mathbf{d}\cdot\frac{\mathbf{r}}{r^3} + \frac{1}{6}\mathbf{Q}\odot\frac{3\mathbf{r}\mathbf{r} - 1r^2}{r^5} + \cdots . \tag{2.49}$$

For the following considerations it will be helpful to designate the reference point within the charge distribution, which we have been using as the origin of coordinates, by the vector $\mathbf{R}$. That only requires replacing $\mathbf{r}$ by $\mathbf{r} - \mathbf{R}$. Then the interaction energy of the charge distribution with a remote point charge e_1 that is situated at $\mathbf{r}_1$ appears as

$$\begin{aligned}E = e_1\phi(\mathbf{r}_1) = &\frac{ee_1}{4\pi|\mathbf{r}_1 - \mathbf{R}|} + \mathbf{d}\cdot\frac{e_1(\mathbf{r}_1 - \mathbf{R})}{4\pi|\mathbf{r}_1 - \mathbf{R}|^3}\\ &+ \frac{1}{6}\mathbf{Q}\odot\frac{e_1[3(\mathbf{r}_1 - \mathbf{R})(\mathbf{r}_1 - \mathbf{R}) - 1(\mathbf{r}_1 - \mathbf{R})^2]}{4\pi|(\mathbf{r}_1 - \mathbf{R})|^5} + \cdots ;\end{aligned} \tag{2.50}$$

or, in terms of the potential and electric field produced at $\mathbf{R}$ by the point charge,

$$\phi(\mathbf{R}) = \frac{e_1}{4\pi|\mathbf{R} - \mathbf{r}_1|} , \quad \mathbf{E}(\mathbf{R}) = e_1\frac{\mathbf{R} - \mathbf{r}_1}{4\pi|\mathbf{R} - \mathbf{r}_1|^3} , \tag{2.51}$$

we have

$$E = e\phi(\mathbf{R}) - \mathbf{d}\cdot\mathbf{E}(\mathbf{R}) - \frac{1}{6}\mathbf{Q}\odot\nabla_{\mathbf{R}}\mathbf{E}(\mathbf{R}) + \cdots . \tag{2.52}$$

Although we have used the field of a point charge in arriving at the interaction energy (2.52), it holds quite generally inasmuch as any electric potential or field is the superposition of the contributions of point charges. As an example, that part of the interaction energy between nonoverlapping charge distributions 1 and 2 that is attributed to their dipole moments is $(\mathbf{r} = \mathbf{r}_1 - \mathbf{r}_2)$

$$4\pi E_{\mathbf{dd}} = -\mathbf{d}_1 \cdot \left(-\boldsymbol{\nabla}_1 \mathbf{d}_2 \cdot \frac{\mathbf{r}}{r^3} \right)$$

$$= -\frac{3\mathbf{d}_1 \cdot \mathbf{r}\, \mathbf{d}_2 \cdot \mathbf{r} - \mathbf{d}_1 \cdot \mathbf{d}_2\, r^2}{r^5} \,. \tag{2.53}$$

Let us also note the symmetrical, derivative form of this energy,

$$4\pi E_{\mathbf{dd}} = \mathbf{d}_1 \cdot \boldsymbol{\nabla}_1\, \mathbf{d}_2 \cdot \boldsymbol{\nabla}_2 \frac{1}{|\mathbf{r}_1 - \mathbf{r}_2|}\,, \tag{2.54}$$

and the version of the explicit result in which the direction of $\mathbf{r}$ is chosen as the z-axis:

$$4\pi E_{\mathbf{dd}} = -\frac{2d_{1z}d_{2z} - d_{1x}d_{2x} - d_{1y}d_{2y}}{r^3}\,. \tag{2.55}$$

2.3 Spherical Harmonics

Although the procedure we have been following could be continued to higher multipoles, it rapidly become unweildy. It is preferable to use spherical harmonics, in which the restrictions associated with Laplace's equation are already incorporated. Accordingly, the expansion (2.35) is now presented as

$$r > r' \ : \ \frac{1}{|\mathbf{r} - \mathbf{r}'|} = \sum_{lm} \frac{r'^l}{r^{l+1}} \left(\frac{4\pi}{2l+1} \right)^{1/2} Y_{lm}(\theta,\phi) \left(\frac{4\pi}{2l+1} \right)^{1/2} Y_{lm}(\theta',\phi')^*\,, \tag{2.56}$$

leading the expansion of the potential (2.37)

$$4\pi\phi(\mathbf{r}) = \sum_{lm} \frac{1}{r^{l+1}} \left(\frac{4\pi}{2l+1} \right)^{1/2} Y_{lm}(\theta,\phi)\rho_{lm}\,, \tag{2.57}$$

where (omitting primes) the lmth electric multipole moment is

$$\rho_{lm} = \int (\mathbf{dr})\, r^l \left(\frac{4\pi}{2l+1} \right)^{1/2} Y_{lm}(\theta,\phi)^* \rho(\mathbf{r})\,. \tag{2.58}$$

The connection between this systematic definition of multipole moments and the few already discussed can be read off from the explicit construction of the first few spherical harmonics:

$$l = 0 : \quad \rho_{00} = e\,, \tag{2.59a}$$

$$l = 1 : \quad \begin{cases} \rho_{11} = -2^{-1/2}(d_x - id_y)\,, \\ \rho_{10} = d_z\,, \\ \rho_{1,-1} = 2^{-1/2}(d_x + id_y)\,, \end{cases} \tag{2.59b}$$

$$l = 2 : \quad \begin{cases} \rho_{22} = (24)^{-1/2}(Q_{xx} - Q_{yy} - 2iQ_{xy})\,, \\ \rho_{21} = -6^{-1/2}(Q_{xz} - iQ_{yz})\,, \\ \rho_{20} = \frac{1}{2}Q_{zz} = -\frac{1}{2}(Q_{xx} + Q_{yy})\,, \\ \rho_{2,-1} = 6^{-1/2}(Q_{xz} + iQ_{yz})\,, \\ \rho_{2,-2} = (24)^{-1/2}(Q_{xx} - Q_{yy} + 2iQ_{xy})\,. \end{cases} \tag{2.59c}$$

For a complete expression of the general connection we use the spherical harmonic definition (2.20),

$$\frac{(\boldsymbol{\nu} \cdot \mathbf{r})^l}{l!} = \sum_{m=-l}^{l} \psi_{lm} r^l \left(\frac{4\pi}{2l+1}\right)^{1/2} Y_{lm}(\theta, \phi)^* , \tag{2.60}$$

which enables us to present (2.58) as

$$\sum_m \psi_{lm} \rho_{lm} = \int (d\mathbf{r}) \frac{(\boldsymbol{\nu} \cdot \mathbf{r})^l}{l!} \rho(\mathbf{r}) = \frac{1}{l!} \{\boldsymbol{\nu}\}^l \odot \int (d\mathbf{r}) \{\mathbf{r}\}^l \rho(\mathbf{r}) . \tag{2.61}$$

In the simple example of $l = 1$, this reads, according to (2.21) and (2.22),

$$\frac{\psi_+^2}{2^{1/2}} \rho_{11} + \psi_+\psi_- \rho_{10} + \frac{\psi_-^2}{2^{1/2}} \rho_{1,-1}$$
$$= -\frac{1}{2}\psi_+^2 (d_x - id_y) + \psi_+\psi_- d_z + \frac{1}{2}\psi_-^2 (d_x + id_y) , \tag{2.62}$$

from which the $l = 1$ entries in (2.59b) are recovered.

A necessary preliminary to a general derivation of the fictitious charge density '$\rho(\mathbf{r})$' is the construction of an alternative spherical harmonic representation. It is useful to introduce a notation for the homogeneous solid harmonics of degrees l and $-l - 1$,

$$Y_{lm}(\mathbf{r}) = r^l Y_{lm}\left(\frac{\mathbf{r}}{r}\right) , \tag{2.63a}$$

$$Y_{-l-1,m}(\mathbf{r}) = r^{-l-1} Y_{-l-1,m}\left(\frac{\mathbf{r}}{r}\right) , \tag{2.63b}$$

respectively, where

$$Y_{lm}\left(\frac{\mathbf{r}}{r}\right) = Y_{-l-1,m}\left(\frac{\mathbf{r}}{r}\right) = Y_{lm}(\theta, \phi) . \tag{2.64}$$

This notation is used in writing (2.20) as

$$\frac{(\boldsymbol{\nu}^* \cdot \mathbf{r})^l}{l!} = \sum_{m=-l}^{l} \psi_{lm}^* \left(\frac{4\pi}{2l+1}\right)^{1/2} Y_{lm}(\mathbf{r}) , \tag{2.65a}$$

and

$$\frac{(\boldsymbol{\nu}^* \cdot \mathbf{r})^l}{l!} \frac{1}{r^{2l+1}} = \sum_{m=-l}^{l} \psi_{lm}^* \left(\frac{4\pi}{2l+1}\right)^{1/2} Y_{-l-1,m}(\mathbf{r}) . \tag{2.65b}$$

Now consider [recall $(\boldsymbol{\nu}^*)^2 = 0$]

$$\frac{(\boldsymbol{\nu}^* \cdot \boldsymbol{\nabla})^l}{l!} \frac{1}{r} = -\frac{(\boldsymbol{\nu}^* \cdot \boldsymbol{\nabla})^{l-1}}{l!} \boldsymbol{\nu}^* \cdot \mathbf{r} \frac{1}{r^3}$$

$$= (-1)^l 1 \cdot 3 \cdot 5 \cdots (2l-1) \frac{(\boldsymbol{\nu}^* \cdot \mathbf{r})^l}{l!} \frac{1}{r^{2l+1}}$$

$$= (-1)^l \frac{(2l)!}{2^l l!} \sum_{m=-l}^{l} \psi_{lm}^* \left(\frac{4\pi}{2l+1}\right)^{1/2} Y_{-l-1,m}(\mathbf{r}) . \tag{2.66}$$

But the left-hand side of (2.66) can also be written as

$$\sum_{m=-l}^{l} \psi_{lm}^* \left(\frac{4\pi}{2l+1}\right)^{1/2} Y_{lm}(\boldsymbol{\nabla})\frac{1}{r}, \tag{2.67}$$

which gives us the desired representation,

$$Y_{-l-1,m}(\mathbf{r}) = (-1)^l \frac{2^l l!}{(2l)!} Y_{lm}(\boldsymbol{\nabla})\frac{1}{r}. \tag{2.68}$$

A simple illustration, for $l = m = 1$, is [a common factor of $(3/8\pi)^{1/2}$ is omitted]

$$-\frac{x+iy}{r^3} = \left(\frac{\partial}{\partial x} + i\frac{\partial}{\partial y}\right)\frac{1}{r}. \tag{2.69}$$

Another useful result involves the structure

$$Y_{l'm'}(\boldsymbol{\nabla})Y_{lm}(\mathbf{r})^* \bigg|_{r=0} = \delta_{ll'} Y_{lm'}(\boldsymbol{\nabla})Y_{lm}(\mathbf{r})^* \bigg|_{r=0}, \tag{2.70}$$

which already conveys the fact that it vanishes for $l' = l$. Indeed, for $l' > l$, there are more derivatives than coordinates, while for $l' < l$, some coordinates remain after differentiation, which are then set equal to zero. To study the $l' = l$ situation, we evaluate

$$\frac{(\boldsymbol{\nu}^* \cdot \boldsymbol{\nabla})^l}{l!} \frac{(\boldsymbol{\nu} \cdot \mathbf{r})^l}{l!} = \frac{(\boldsymbol{\nu}^* \cdot \boldsymbol{\nabla})^{l-1}}{l!} \frac{(\boldsymbol{\nu} \cdot \mathbf{r})^{l-1}}{(l-1)!} \boldsymbol{\nu}^* \cdot \boldsymbol{\nu}$$

$$= \frac{(\boldsymbol{\nu}^* \cdot \boldsymbol{\nu})^l}{l!} = \frac{(2l)!}{2^l l!} \sum_{m=-l}^{l} \psi_{lm}^* \psi_{lm}, \tag{2.71}$$

where the last step follows from the definition of the ψ_{lm}, (2.21) and (2.22). Accordingly, we have

$$\sum_{mm'} \psi_{lm'}^* \left(\frac{4\pi}{2l+1}\right)^{1/2} Y_{lm'}(\boldsymbol{\nabla})Y_{lm}(\mathbf{r})^* \left(\frac{4\pi}{2l+1}\right)^{1/2} \psi_{lm} = \frac{(2l)!}{2^l l!} \sum_{m} \psi_{lm}^* \psi_{lm}, \tag{2.72}$$

which, together with (2.70), tells us that

$$Y_{l'm'}(\boldsymbol{\nabla})Y_{lm}(\mathbf{r})^* \bigg|_{r=0} = \delta_{ll'}\delta_{mm'} \frac{1}{4\pi} \frac{(2l+1)!}{2^l l!}. \tag{2.73}$$

In the example of $l = m = l' = m' = 1$, this states that

$$\left(\frac{3}{8\pi}\right)^{1/2} \left(\frac{\partial}{\partial x} + i\frac{\partial}{\partial y}\right) \left(\frac{3}{8\pi}\right)^{1/2} (x - iy) = \frac{3}{4\pi}. \tag{2.74}$$

Now we turn to the expression (2.56), (2.57), this time using (2.68) to write it as

$$4\pi\phi(\mathbf{r}) = \sum_{lm} \rho_{lm} \left(\frac{4\pi}{2l+1}\right)^{1/2} \frac{2^l l!}{(2l)!}(-1)^l Y_{lm}(\boldsymbol{\nabla})\frac{1}{r} , \qquad (2.75)$$

and then proceed to compute the fictitious charge density,

$$`\rho(\mathbf{r})' = -\nabla^2\phi(\mathbf{r}) = \sum_{lm} \rho_{lm} \left(\frac{4\pi}{2l+1}\right)^{1/2} \frac{2^l l!}{(2l)!}(-1)^l Y_{lm}(\boldsymbol{\nabla})\delta(\mathbf{r}) , \qquad (2.76)$$

which is the generalization of (2.42). The moments of this charge density are

$$`\rho_{lm}' = \int (\mathrm{d}\mathbf{r}) \left(\frac{4\pi}{2l+1}\right)^{1/2} Y_{lm}(\mathbf{r})^* `\rho(\mathbf{r})'$$

$$= \int (\mathrm{d}\mathbf{r}) \left(\frac{4\pi}{2l+1}\right)^{1/2} Y_{lm}(\mathbf{r})^* \sum_{l'm'} Y_{l'm'}(-\boldsymbol{\nabla})\delta(\mathbf{r}) \frac{2^{l'} l'!}{(2l')!} \left(\frac{4\pi}{2l'+1}\right)^{1/2} \rho_{l'm'} .$$

$$(2.77)$$

Partial integration then transfers $Y_{l'm'}(-\boldsymbol{\nabla})$, acting on $\delta(\mathbf{r})$, to $Y_{l'm'}(\boldsymbol{\nabla})$, acting on $Y_{lm}(\mathbf{r})^*$, which invokes (2.73) in view of the restriction to $\mathbf{r} = 0$ that is enforced by $\delta(\mathbf{r})$. The combination of factors in (2.73) and (2.77) is such that

$$`\rho_{lm}' = \rho_{lm} , \qquad (2.78)$$

as previously stated.

Before proceeding to the next stage, the generalization of energy expressions, let us again pause to devise yet another spherical harmonic representation. We know from (2.56) that

$$\frac{(-\mathbf{r}' \cdot \boldsymbol{\nabla})^l}{l!} \frac{1}{r} = \sum_{m=-l}^{l} \left(\frac{4\pi}{2l+1}\right)^{1/2} Y_{-l-1,m}(\mathbf{r}) \left(\frac{4\pi}{2l+1}\right)^{1/2} Y_{lm}(\mathbf{r}')^* . \qquad (2.79)$$

On the other hand, inasmuch as $\boldsymbol{\nabla}$ acts as a null vector in this context, we can also apply (2.60), with $\mathbf{r} \rightarrow \mathbf{r}'$, to produce

$$\frac{(-\mathbf{r}' \cdot \boldsymbol{\nabla})^l}{l!} \frac{1}{r} = (-1)^l \sum_{m} \left(\frac{4\pi}{2l+1}\right)^{1/2} Y_{lm}(\mathbf{r}')^* \psi_{lm}(\boldsymbol{\nabla})\frac{1}{r} , \qquad (2.80)$$

where $\psi_{lm}(\boldsymbol{\nabla})$ is constructed from the components of the vector $\boldsymbol{\nabla}$ in the same manner that ψ_{lm} is constructed from the components of $\boldsymbol{\nu}$. Comparison with the right-hand side of (2.79) then yields

$$\left(\frac{4\pi}{2l+1}\right)^{1/2} Y_{-l-1,m}(\mathbf{r}) = (-1)^l \psi_{lm}(\boldsymbol{\nabla})\frac{1}{r} , \qquad (2.81)$$

as an alternative version of (2.68). Which is to say that (with the operand $1/r$ understood)

$$\left(\frac{4\pi}{2l+1}\right)^{1/2} Y_{lm}(\boldsymbol{\nabla}) = \frac{(2l)!}{2^l l!}\psi_{lm}(\boldsymbol{\nabla}) \,. \tag{2.82}$$

Notice that (2.75) now has the simpler appearance,

$$4\pi\phi(\mathbf{r}) = \sum_{lm} \rho_{lm}(-1)^l \psi_{lm}(\boldsymbol{\nabla})\frac{1}{r} \,, \tag{2.83}$$

as follows directly from (2.35) and (2.80), and (2.73) becomes

$$\psi_{l'm'}(\boldsymbol{\nabla})\left(\frac{4\pi}{2l+1}\right)^{1/2} Y_{lm}(\mathbf{r})^* \bigg|_{\mathbf{r}=0} = \delta_{ll'}\delta_{mm'} \,. \tag{2.84}$$

Perhaps, then, we should have begun with $\psi_{lm}(\boldsymbol{\nabla})$, rather than $Y_{lm}(\boldsymbol{\nabla})$, but the latter does seem to be a more immediate concept. Incidentally, a direct path to (2.81) could start from (2.66) by regarding the left-hand side as constructed from the product of two null vectors (Problem 2.5).

The $m = 0$ specialization of (2.82) reads

$$\begin{aligned} P_l(\boldsymbol{\nabla}) &= \frac{(2l)!}{2^l l!}\psi_{l0}(\boldsymbol{\nabla}) \\ &= \frac{(2l)!}{2^l l!}\frac{1}{l!}(\psi_+\psi_- \to \nabla_z)^l \,, \end{aligned} \tag{2.85}$$

and indeed, from the highest powers of the Legendre polynomial as produced by

$$\begin{aligned} P_l(\mu) &= \left(\frac{d}{d\mu}\right)^l \frac{\mu^{2l} - l\mu^{2l-2} + \cdots}{2^l l!} \\ &= \frac{(2l)!}{2^l (l!)^2}\mu^l - \frac{(2l-2)!}{2^l (l-1)!(l-2)!}\mu^{l-2} + \cdots \,, \end{aligned} \tag{2.86}$$

we get

$$P_l(\boldsymbol{\nabla}) = \frac{(2l)!}{2^l (l!)^2}(\nabla_z)^l - \frac{(2l-2)!}{2^l (l-1)!(l-2)!}(\nabla_z)^{l-2}\nabla^2 + \cdots \,, \tag{2.87}$$

which is (2.85) in view of the effective null value of ∇^2. The explicit form that (2.81) yields for $m = 0$ is

$$\frac{1}{r^{l+1}}P_l(\cos\theta) = (-1)^l \frac{1}{l!}\frac{\partial^l}{\partial z^l}\frac{1}{r} \,, \tag{2.88}$$

which can also be seen directly: Both sides of the equations are solutions of Laplace's equation ($r > 0$) that are homogeneous of degree $-l - 1$ and do not involve the azimuthal angle ϕ; then, with $\theta = 0$, $z = r$, the evaluation of the right-hand side completes the identification.

We shall present an application of (2.88), one that is facilitated by using the integral representation, for $z > 0$,

$$\frac{1}{(z^2 + \rho^2)^{1/2}} = \int_0^\infty dk\, J_0(k\rho)e^{-kz}$$

$$= \int_0^{2\pi} \frac{d\phi}{2\pi} \int_0^\infty dk\, e^{-k(z+i\rho\cos\phi)}$$

$$= \int_0^{2\pi} \frac{d\phi}{2\pi} \frac{1}{z + i\rho\cos\phi}. \tag{2.89}$$

(For an alternative derivation, see Problem 2.6.) Now the differentiations in (2.88), in which ρ is held fixed, can be performed, with the result

$$\frac{1}{r^{l+1}} P_l(\cos\theta) = \int_0^{2\pi} \frac{d\phi}{2\pi} \frac{1}{(z + i\rho\cos\phi)^{l+1}}. \tag{2.90}$$

Then the introduction of spherical coordinates for z and ρ gives

$$P_l(\cos\theta) = \int_0^{2\pi} \frac{d\phi}{2\pi} \frac{1}{(\cos\theta + i\sin\theta\cos\phi)^{l+1}}, \tag{2.91}$$

which, as mentioned in the context of (2.26), is Laplace's second integral, produced by the substitution of $l \to -l - 1$ in the latter equation. The need for the restriction $z > 0$, which is $\cos\theta > 0$, must be emphasized; changing the sign of $\cos\theta$ in the integral does not reproduce the known behavior of $P_l(\cos\theta)$ (see Problem 2.2).

If we consider $m > 0$ in (2.81) and use for $m > 0$

$$\psi_{lm}(\nabla) = \frac{1}{[(l+m)!(l-m)!]^{1/2}} \psi_+^{2m}(\psi_+\psi_-)^{l-m}$$

$$= \frac{1}{[(l+m)!(l-m)!]^{1/2}}[-(\nabla_x + i\nabla_y)]^m \nabla_z^{l-m}, \tag{2.92}$$

we get

$$\frac{1}{r^{l+1}} \frac{\Theta_{lm}(\theta)e^{im\phi}}{(l+1/2)^{1/2}} = \frac{(-1)^{l-m}}{[(l+m)!(l-m)!]^{1/2}} \left(\frac{\partial}{\partial x} + i\frac{\partial}{\partial y}\right)^m \left(\frac{\partial}{\partial z}\right)^{l-m} \frac{1}{r}. \tag{2.93}$$

The choice $m = l$ is particularly simple,

$$\frac{1}{r^{l+1}} \frac{1}{(l+1/2)^{1/2}} \Theta_{ll}(\theta)e^{il\phi} = \frac{1}{[(2l)!]^{1/2}} \left(\frac{\partial}{\partial x} + i\frac{\partial}{\partial y}\right)^l \frac{1}{r}, \tag{2.94}$$

where [this is an example of (2.66)]

$$\left(\frac{\partial}{\partial x} + i\frac{\partial}{\partial y}\right)^l \frac{1}{r} = -\left(\frac{\partial}{\partial x} + i\frac{\partial}{\partial y}\right)^{l-1} \frac{x + iy}{r^3}$$

$$= (-1)^l \frac{(2l)!}{2^l l!} \frac{(x + iy)^l}{r^{2l+1}} , \tag{2.95}$$

yields the familiar result

$$\Theta_{ll}(\theta) = (-1)^l \left[\frac{1}{2}(2l + 1)!\right]^{1/2} \frac{\sin^l \theta}{2^l l!} . \tag{2.96}$$

One can extend (2.93) to negative m by complex conjugation, the positive m value appearing in this equation then being interpreted as $|m|$. In effect, (2.93) continues to apply, with the understanding that

$$\left(\frac{\partial}{\partial x} + i\frac{\partial}{\partial y}\right)^{-1} \frac{\partial^2}{\partial z^2} = -\left(\frac{\partial}{\partial x} - i\frac{\partial}{\partial y}\right) . \tag{2.97}$$

For the version of Laplace's second integral that is applicable to $m \neq 0$ see Problem 2.3.

2.4 Multipole Interactions

The generalization of (2.52), the expansion of the energy of a charge distribution in a given externally produced potential $\phi(\mathbf{r})$, begins with the energy expression

$$E = \int (d\mathbf{r})\rho(\mathbf{r})\phi(\mathbf{r}) . \tag{2.98}$$

Now, within the charge distribution ρ, ϕ obeys Laplace's equation inasmuch as its sources are outside of ρ. Accordingly, $\phi(\mathbf{r})$ can be presented as a series of solid harmonics relative to the reference point $\mathbf{R}$ in the interior of ρ:

$$\phi(\mathbf{r}) = \sum_{lm} \left(\frac{4\pi}{2l + 1}\right)^{1/2} Y_{lm}(\mathbf{r} - \mathbf{R})\phi_{lm} . \tag{2.99}$$

The coefficients ϕ_{lm} are produced by an application of the complex conjugate form of (2.84), with $\mathbf{r} \to \mathbf{r} - \mathbf{R}$,

$$\phi_{lm} = \psi_{lm}(\boldsymbol{\nabla})^*\phi(\mathbf{r})\Big|_{\mathbf{r}=\mathbf{R}} = \psi_{lm}(\boldsymbol{\nabla_R})^*\phi(\mathbf{R}) , \tag{2.100}$$

where, indeed, $\phi_{00} = \phi(\mathbf{R})$, the three ϕ_{1m} are combinations of the components of $\mathbf{E}(\mathbf{R})$, and so on. The desired expansion is now realized as

$$E = \sum_{lm} \rho_{lm}^*\phi_{lm} = \sum_{lm} \rho_{lm}\phi_{lm}^* . \tag{2.101}$$

As for the generalization of the force on the charge distribution, (2.45), let us begin with

$$\mathbf{F} = \int (d\mathbf{r})\rho(\mathbf{r})\mathbf{E}(\mathbf{r}) \,, \tag{2.102}$$

and use the expansion

$$\mathbf{E}(\mathbf{r}) = \sum_{l=0}^{\infty} \frac{[(\mathbf{r} - \mathbf{R}) \cdot \boldsymbol{\nabla_R}]^l}{l!} \mathbf{E}(\mathbf{R})$$

$$= \sum_{lm} \left(\frac{4\pi}{2l+1}\right)^{1/2} Y_{lm}(\mathbf{r} - \mathbf{R})^* \psi_{lm}(\boldsymbol{\nabla_R})\mathbf{E}(\mathbf{R}) \,, \tag{2.103}$$

to arrive at

$$\mathbf{F} = \sum_{lm} \rho_{lm}\psi_{lm}(\boldsymbol{\nabla_R})\mathbf{E}(\mathbf{R}) \,. \tag{2.104}$$

Had we employed a similar procedure for E, we would have come to the second version of (2.101). Inasmuch as only the moments of ρ appear in this structure, the use of the fictitious density 'ρ' would yield the same result.

The latter remark also applies to the generalization of the torque,

$$\boldsymbol{\tau} = \int (d\mathbf{r})(\mathbf{r} - \mathbf{R}) \times \rho(\mathbf{r})\mathbf{E}(\mathbf{r}) \,, \tag{2.105}$$

where we encounter the expansion

$$\mathbf{r} \times \mathbf{E}(\mathbf{r}) - \mathbf{R} \times \mathbf{E}(\mathbf{r})$$

$$= \sum_{m} \left(\frac{4\pi}{2l+1}\right)^{1/2} Y_{lm}(\mathbf{r} - \mathbf{R})^* \left[\psi_{lm}(\boldsymbol{\nabla_R})\mathbf{R} \times \mathbf{E}(\mathbf{R})\right.$$

$$\left. - \mathbf{R} \times \psi_{lm}(\boldsymbol{\nabla_R})\mathbf{E}(\mathbf{R})\right] \,. \tag{2.106}$$

This confronts us with the symbolic combination

$$\psi_{lm}(\boldsymbol{\nabla_R})\mathbf{R} - \mathbf{R}\psi_{lm}(\boldsymbol{\nabla_R}) = \frac{\partial}{\partial \boldsymbol{\nabla_R}}\psi_{lm}(\boldsymbol{\nabla_R}) \,, \tag{2.107}$$

in which the derivative with respect to $\boldsymbol{\nabla_R}$ serves to pick out in $\psi_{lm}(\boldsymbol{\nabla_R})$ the gradient component that differentiates $\mathbf{R}$ with unit result. The ensuing expansion is

$$\boldsymbol{\tau} = \sum_{lm} \rho_{lm} \left[\frac{\partial}{\partial \boldsymbol{\nabla_R}}\psi_{lm}(\boldsymbol{\nabla_R})\right] \times \mathbf{E}(\mathbf{R}) \,, \tag{2.108}$$

and we record, for an arbitrary null vector $\boldsymbol{\nu}$, the needed derivatives (see Problem 2.7)

$$\left(\frac{\partial}{\partial \nu_x} + i\frac{\partial}{\partial \nu_y}\right)\psi_{lm} = \frac{|m| - m}{[(l - m)(l - m - 1)]^{1/2}}\psi_{l-1,m+1} , \qquad (2.109a)$$

$$\left(\frac{\partial}{\partial \nu_x} - i\frac{\partial}{\partial \nu_y}\right)\psi_{lm} = -\frac{|m| + m}{[(l + m)(l + m - 1)]^{1/2}}\psi_{l-1,m-1} , \qquad (2.109b)$$

$$\frac{\partial}{\partial \nu_z}\psi_{lm} = \left(\frac{l - |m|}{l + |m|}\right)^{1/2}\psi_{l-1,m} . \qquad (2.109c)$$

While we are at it, let us also note that

$$(\nu_x + i\nu_y)\psi_{lm} = -[(l + m + 2)(l + m + 1)]^{1/2}\psi_{l+1,m+1} , \qquad (2.110a)$$

$$(\nu_x - i\nu_y)\psi_{lm} = [(l - m + 2)(l - m + 1)]^{1/2}\psi_{l+1,m-1} , \qquad (2.110b)$$

$$\nu_z\psi_{lm} = [(l + m + 1)(l - m + 1)]^{1/2}\psi_{l+1,m} ; \qquad (2.110c)$$

both sets enter in checking that

$$\boldsymbol{\nu} \cdot \frac{\partial}{\partial \boldsymbol{\nu}}\psi_{lm} = l\psi_{lm} , \qquad (2.111)$$

the statement of homogeneity for the monomials $\psi_{lm}(\boldsymbol{\nu})$.

Finally, we want to relate the ϕ_{lm} in the energy expression (2.101) to the moments of the charge distribution that produces the potential, thereby leading to the generalization of the dipole–dipole interaction energy (2.53), (2.54) and (2.55). The symmetry of the latter in the properties of the two charge distributions invites us to produce a formulation that exhibits such symmetry. To this end, let $\mathbf{r}_1$ and $\mathbf{r}_2$ be position vectors directed from reference points within the charge distributions ρ_1 and ρ_2, respectively, while $\mathbf{r}$ locates one such center with respect to the other, as shown in Fig. 2.1. Then the

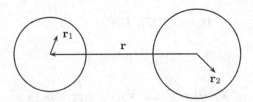

Fig. 2.1. Position vectors locating the charge density within two charge distributions, ρ_1 and ρ_2, and the relative positions of the two distributions

interaction energy of the two nonoverlapping charge distributions is

$$E = \int (\mathrm{d}\mathbf{r}_1)(\mathrm{d}\mathbf{r}_2)\frac{\rho_1(\mathbf{r}_1)\rho_2(\mathbf{r}_2)}{4\pi|\mathbf{r} + \mathbf{r}_1 - \mathbf{r}_2|} , \qquad (2.112)$$

which has the required symmetry under the interchange $1 \to 2$, $\mathbf{r} \to -\mathbf{r}$.

We now proceed to introduce a double expansion

$$\frac{1}{|\mathbf{r} + \mathbf{r}_1 - \mathbf{r}_2|} = \sum_{l_1,l_2=0}^{\infty} \frac{(\mathbf{r}_1 \cdot \nabla)^{l_1}}{l_1!} \frac{(-\mathbf{r}_2 \cdot \nabla)^{l_2}}{l_2!} \frac{1}{r}$$

$$= \sum_{l_1,m_1} \sum_{l_2,m_2} \left(\frac{4\pi}{2l_1+1}\right)^{1/2} Y_{l_1 m_1}(\mathbf{r}_1)^* (-1)^{l_2} \left(\frac{4\pi}{2l_2+1}\right)^{1/2}$$

$$\times Y_{l_2 m_2}(\mathbf{r}_2)^* \psi_{l_1 m_1}(\nabla) \psi_{l_2 m_2}(\nabla) \frac{1}{r} , \tag{2.113}$$

with a twofold application of (2.80). To simplify the combining of the two monomials $\psi_{l_1 m_1}$ and $\psi_{l_2 m_2}$, we employ the notation

$$[lm] = [(l+m)!(l-m)!]^{1/2} = [l,-m] , \tag{2.114}$$

and also

$$L = l_1 + l_2 , \qquad M = m_1 + m_2 , \tag{2.115}$$

so that

$$\psi_{l_1 m_1} \psi_{l_2 m_2} = \frac{[LM]}{[l_1 m_1][l_2 m_2]} \psi_{LM} . \tag{2.116}$$

Then the action of $\psi_{LM}(\nabla)$ on $1/r$, as given by (2.81), and the spatial integrations that produce the individual multipole moments $\rho_{1 l_1 m_1}, \rho_{2 l_2 m_2}$, combine to yield the energy expression

$$E = \sum_{l_1,m_1;l_2,m_2} \frac{1}{r^{L+1}} \left(\frac{4\pi}{2L+1}\right)^{1/2} Y_{LM}(\theta, \phi) [LM] (-1)^{l_1} \frac{\rho_{1 l_1 m_1} \rho_{2 l_2 m_2}}{[l_1 m_1] [l_2 m_2]} . \tag{2.117}$$

Notice that L, which specifies the distance and angle dependence of the contributions, can generally be realized in a number of ways. Of course, for $L = 0$ there is only $l_1 = l_2 = 0$, which leads to the Coulomb charge–charge interaction. With $L = 1$, we have $l_1 = 1$, $l_2 = 0$ and $l_1 = 0$, $l_2 = 1$; the two possibilities of charge–dipole interaction. Then, for $L = 2$, there is $l_1 = 2$, $l_2 = 0$ and $l_1 = 0$, $l_2 = 2$ – the two variants of charge–quadrupole interaction; and $l_1 = l_2 = 1$ – dipole–dipole interaction. In general, the total number is $L + 1$. If, however, we give but a single count to the two variants of the same physical situation, the number of distinct multipole pairs is given by $[L/2] + 1$, where the bracket indicates the largest integer contained in the given number, which is just $\frac{L}{2}$ for L even, and $\frac{1}{2}(L-1)$ for L odd.

We must also comment on the particular form taken by the interaction energy when, as in (2.55), the z-axis is chosen to coincide with the line between the two centers. With $\theta = 0$, only $M = 0$ contributes in (2.117), thus requiring that $m_1 = -m_2 = m$, and the energy expression reduces to

$$E = \sum_{l_1 l_2 m} \frac{L!}{r^{L+1}} (-1)^{l_1} \frac{1}{[l_1 m]} \rho_{1 l_1 m} \frac{1}{[l_2 m]} \rho_{2 l_2 -m} . \tag{2.118}$$

One easily checks that the $l_1 = l_2 = 1$ term reproduces (2.55).

2.5 Problems for Chap. 2

1. Prove Laplace's second integral representation, the result of substitution $l \to -l - 1$ in (2.26), namely (2.91).
2. What happens in the latter representation if the sign of $\cos\theta$ is reversed? Does one still obtain a solution of Laplace's equation?
3. Provide a generalization of Laplace's second integral to $m \neq 0$.
4. Spherical harmonics are defined by (2.20), so show that

$$\frac{\boldsymbol{\nu}^* \cdot \mathbf{r}}{l+1} \sum_{m=-l}^{l} \psi_{lm}^* \sqrt{\frac{4\pi}{2l+1}} r^l Y_{lm}(\theta,\phi)$$

$$= \sum_{m=-l-1}^{l+1} \psi_{l+1,m}^* \sqrt{\frac{4\pi}{2(l+1)+1}} r^{l+1} Y_{l+1,m}(\theta,\phi), \qquad (2.119)$$

where

$$\boldsymbol{\nu} \cdot \mathbf{r} = \frac{1}{2} r(\sin\theta\, e^{i\phi} \psi_-^2 - \sin\theta\, e^{-i\phi} \psi_+^2 + 2\cos\theta\, \psi_+\psi_-) . \qquad (2.120)$$

Write out the explicit constructions this gives for the spherical harmonics of degree $l+1$ in terms of those of degree l. [As a hint, one of the *three* terms in the answer is

$$\sqrt{\frac{2l+3}{2l+1}} \sqrt{(l+1+m)(l+1-m)} \cos\theta\, Y_{lm}(\theta,\phi) .] \qquad (2.121)$$

Begin with $Y_{00} = 1/\sqrt{4\pi}$ and construct the three Y_{1m}, and then find one or more of the five Y_{2m}.
5. Prove (2.81) by regarding the left-hand side of (2.66) as constructed from the product of two null vectors.
6. Provide an alternative derivation of (2.89) by converting the integral over ϕ into a closed contour integral.
7. Verify (2.109a)–(2.110c). Hint: write

$$\psi_+^{l+m} \psi_-^{l-m} = \nu_z^{l-|m|} (-\nu_x - i\nu_y)^{(m+|m|)/2} (\nu_x - i\nu_y)^{(|m|-m)/2} . \qquad (2.122)$$

3

Relativistic Transformations

Most of this book is written in three-dimensional notation, which is entirely appropriate for practical applications. However, electrodynamics was the birthplace of Einstein's special relativity so it is fitting to give here a brief account of the relativistically covariant formulation of the theory. Although special relativity implied a modification of Newton's laws of motion, no change, of course, is made in Maxwell's theory in such a reformulation.

3.1 Four-Dimensional Notation

A space–time coordinate can be represented by a contravariant vector,

$$x^\mu : \quad x^0 = ct , \qquad x^1 = x , \qquad x^2 = y , \qquad x^3 = z , \qquad (3.1)$$

where μ is an index which takes on the values 0, 1, 2, 3. The corresponding covariant vector is

$$x_\mu : \quad x_0 = -ct , \qquad x_1 = x , \qquad x_2 = y , \qquad x_3 = z . \qquad (3.2)$$

The contravariant and covariant vector components are related by the metric tensor $g_{\mu\nu}$,

$$x_\mu = g_{\mu\nu} x^\nu , \qquad (3.3)$$

which uses the Einstein summation convention of summing over repeated covariant and contravariant indices,

$$g_{\mu\nu} x^\nu = \sum_{\nu=0}^{3} g_{\mu\nu} x^\nu . \qquad (3.4)$$

From the above explicit forms for x_μ and x^ν we read off, in matrix form (here the first index labels the rows, the second the columns, both enumerated from

0 to 3)[1]

$$g_{\mu\nu} = \begin{pmatrix} -1 & 0 & 0 & 0 \\ 0 & 1 & 0 & 0 \\ 0 & 0 & 1 & 0 \\ 0 & 0 & 0 & 1 \end{pmatrix}, \qquad (3.5)$$

where evidently $g_{\mu\nu}$ is symmetric,

$$g_{\mu\nu} = g_{\nu\mu} . \qquad (3.6)$$

Similarly,

$$x^{\mu} = g^{\mu\nu} x_{\nu} , \qquad (3.7)$$

where

$$g^{\mu\nu} = \begin{pmatrix} -1 & 0 & 0 & 0 \\ 0 & 1 & 0 & 0 \\ 0 & 0 & 1 & 0 \\ 0 & 0 & 0 & 1 \end{pmatrix}, \qquad g^{\mu\nu} = g^{\nu\mu} . \qquad (3.8)$$

The four-dimensional analog of a rotationally invariant length is the proper length s or the proper time τ:

$$s^2 = -c^2\tau^2 = x^{\mu}x_{\mu} = x^{\mu}g_{\mu\nu}x^{\nu} = x_{\mu}g^{\mu\nu}x_{\nu} = \mathbf{r} \cdot \mathbf{r} - (ct)^2 . \qquad (3.9)$$

Recall the transformation of a scalar field under a coordinate displacement,[2] as in (1.76),

$$\delta\phi(\mathbf{r}) = -\delta\mathbf{r} \cdot \nabla\phi(\mathbf{r}), \qquad \nabla = \frac{\partial}{\partial\mathbf{r}} . \qquad (3.11)$$

The corresponding four-dimensional statement is

$$\delta\phi(x) = -\delta x^{\mu}\partial_{\mu}\phi(x) \qquad \partial_{\mu} = \frac{\partial}{\partial x^{\mu}} , \qquad (3.12)$$

which shows the definition of the covariant gradient operator, so defined in order that $\partial_{\mu}x^{\mu}$ be invariant. The corresponding contravariant gradient is

$$\partial^{\mu} = g^{\mu\nu}\partial_{\nu} = \frac{\partial}{\partial x_{\mu}} . \qquad (3.13)$$

---

[1] In this book, we use what could be referred to as the democratic metric (formerly the East-coast metric), in which the signature is dictated by the larger number of entries.

[2] There is a sign change relative to what appears in Chap. 1. That is because we are now considering passive transformations. Thus, under an infinitesimal coordinate displacement, a scalar field transforms according to $\overline{\phi}(x + \delta x) = \phi(x)$, while $\delta\phi$ is defined at the same coordinate:

$$\delta\phi(x) = \overline{\phi}(x) - \phi(x) = \phi(x - \delta x) - \phi(x) . \qquad (3.10)$$

Using these operators we can write the equation of electric current conservation, (1.14),

$$\mathbf{\nabla} \cdot \mathbf{j} + \frac{\partial}{\partial t} \rho = 0 , \tag{3.14}$$

in the four-dimensional form

$$\partial_\mu j^\mu = 0 , \tag{3.15}$$

where we define the components of the electric current four-vector as

$$j^\mu : \quad j^0 = c\rho , \quad \{j^i\} = \mathbf{j} , \tag{3.16}$$

where we have adopted the convention that Latin indices run over the values 1, 2, 3, corresponding to the three spatial directions. Note that (3.16) is quite analogous to the construction of the position four-vector (3.1).

The invariant interaction term (1.61)

$$\mathcal{L}_{\text{int}} = -\rho\phi + \frac{1}{c}\mathbf{j} \cdot \mathbf{A} \tag{3.17}$$

has the four-dimensional form

$$\frac{1}{c} j^\mu A_\mu , \tag{3.18}$$

where

$$A_\mu = g_{\mu\nu} A^\nu , \quad A^0 = \phi , \quad \{A^i\} = \mathbf{A} . \tag{3.19}$$

The four-dimensional generalization of

$$\mathbf{B} = \mathbf{\nabla} \times \mathbf{A} \tag{3.20}$$

is the tensor construction

$$F_{\mu\nu} = \partial_\mu A_\nu - \partial_\nu A_\mu , \tag{3.21}$$

where the antisymmetric field strength tensor

$$F_{\mu\nu} = -F_{\nu\mu} \tag{3.22}$$

contains the magnetic field components as

$$F_{23} = B_1 , \quad F_{31} = B_2 , \quad F_{12} = B_3 , \tag{3.23}$$

which may be presented more succinctly as

$$F_{ij} = \epsilon_{ijk} B^k , \tag{3.24}$$

which uses the totally antisymmetric Levi-Cività symbol:

$$\epsilon_{123} = \epsilon_{231} = \epsilon_{312} = -\epsilon_{213} = -\epsilon_{132} = -\epsilon_{321} = 1 , \tag{3.25}$$

all other components being zero. The construction (3.21) includes the other potential statement (1.48),

$$\mathbf{E} = -\boldsymbol{\nabla}\phi - \frac{1}{c}\frac{\partial}{\partial t}\mathbf{A} , \tag{3.26}$$

provided

$$F_{0i} = -E_i . \tag{3.27}$$

Alternatively, with

$$F^\mu{}_\nu = g^{\mu\lambda}F_{\lambda\nu} , \tag{3.28}$$

we have

$$F^0{}_i = E_i . \tag{3.29}$$

Maxwell's equations with only electric currents present are summarized by

$$\partial_\nu F^{\mu\nu} = \frac{1}{c}j^\mu , \tag{3.30a}$$

$$\partial_\lambda F_{\mu\nu} + \partial_\mu F_{\nu\lambda} + \partial_\nu F_{\lambda\mu} = 0 , \tag{3.30b}$$

where

$$F^{\mu\nu} = F^\mu{}_\lambda g^{\lambda\nu} = g^{\mu\kappa}F_{\kappa\lambda}g^{\lambda\nu} . \tag{3.31}$$

It is convenient to define a dual field strength tensor by

$$^*F^{\mu\nu} = \frac{1}{2}\epsilon^{\mu\nu\kappa\lambda}F_{\kappa\lambda} = -^*F^{\nu\mu} , \tag{3.32}$$

where $\epsilon^{\mu\nu\kappa\lambda}$ is the four-dimensional totally antisymmetric Levi-Cività symbol, which therefore vanishes if any two of the indices are equal, normalized by

$$\epsilon^{0123} = +1 . \tag{3.33}$$

We now have

$$^*F^{01} = F_{23} = B_1 , \quad ^*F^{02} = F_{31} = B_2 , \quad ^*F^{03} = F_{12} = B_3 , \tag{3.34}$$

and

$$^*F^{23} = F_{01} = -E_1 , \quad ^*F^{31} = F_{02} = -E_2 , \quad ^*F^{12} = F_{03} = -E_3 , \tag{3.35}$$

so indeed the dual transformation corresponds to the replacement

$$\mathbf{E} \to \mathbf{B} , \qquad \mathbf{B} \to -\mathbf{E} . \tag{3.36}$$

[This is a special case of the duality rotation (1.219).] Note that two dual operations brings you back to the beginning:

$$^*(^*F^{\mu\nu}) = -F^{\mu\nu} . \tag{3.37}$$

Using the dual, Maxwell's equations including both the electric (j^μ) and the magnetic $(^*j^\mu)$ currents (called j_e and j_m in the problems in Chap. 1) are given by

$$\partial_\nu F^{\mu\nu} = \frac{1}{c}j^\mu , \qquad \partial_\nu {}^*F^{\mu\nu} = \frac{1}{c} {}^*j^\mu , \tag{3.38}$$

where both currents must be conserved,

$$\partial_\mu j^\mu = c\partial_\mu \partial_\nu F^{\mu\nu} = 0 , \tag{3.39a}$$

$$\partial_\mu {}^*j^\mu = c\partial_\mu \partial_\nu {}^*F^{\mu\nu} = 0 , \tag{3.39b}$$

because of the symmetry in μ and ν of $\partial_\mu \partial_\nu$ and the antisymmetry of $F^{\mu\nu}$ and $^*F^{\mu\nu}$.

We had earlier in (3.9) introduced the proper time. The corresponding differential statement is

$$d\tau = \frac{1}{c}\sqrt{-dx^\mu dx_\mu} = dt\sqrt{1 - \frac{v^2}{c^2}} , \tag{3.40}$$

which is an invariant time interval. The particle equations of motion using τ as the time parameter read (see Problem 3.1)

$$m_0 \frac{d^2 x^\mu}{d\tau^2} = \frac{e}{c}F^\mu{}_\nu \frac{dx^\nu}{d\tau} , \tag{3.41}$$

to which is to be added $\frac{g}{c} {}^*F^\mu{}_\nu dx^\nu/d\tau$ if the particle possesses magnetic charge g. We can write down three alternative forms for the action of the particle:

$$W_{12} = \int_2^1 \left(-m_0 c^2 d\tau + \frac{e}{c}A_\mu dx^\mu \right) \tag{3.42a}$$

$$= \int_2^1 d\tau \left[\frac{1}{2}m_0 \left(\frac{dx^\mu}{d\tau}\frac{dx_\mu}{d\tau} - c^2 \right) + \frac{e}{c}A_\mu \frac{dx^\mu}{d\tau} \right] \tag{3.42b}$$

$$= \int_2^1 d\tau \left[p_\mu \left(\frac{dx^\mu}{d\tau} - v^\mu \right) + \frac{1}{2}m_0 \left(v^\mu v_\mu - c^2 \right) + \frac{e}{c}A_\mu v^\mu \right] . \tag{3.42c}$$

In the last two forms, τ is an independent parameter, with the added requirement that each generator G [recall the action principle states $\delta W_{12} = G_1 - G_2$] is independent of $\delta\tau$. In the third version, where x^μ, v^μ, and p^μ are independent dynamical variables, it is a consequence of the action principle that

$$v^\mu = \frac{dx^\mu}{d\tau} , \qquad p^\mu = m_0 v^\mu + \frac{e}{c}A^\mu , \qquad v^\mu v_\mu = -c^2 , \tag{3.43a}$$

$$\frac{dp^\mu}{d\tau} = \frac{e}{c}\partial^\mu A_\lambda v^\lambda . \tag{3.43b}$$

The invariant Lagrange function for the electromagnetic field (1.60) is

$$\mathcal{L}_f = -\frac{1}{4}F^{\mu\nu}F_{\mu\nu} = \frac{E^2 - B^2}{2} . \tag{3.44}$$

The energy–momentum, or stress tensor, subsumes the energy density, the momentum density (or energy flux vector) and the three-dimensional stress tensor:

$$T^{\mu\nu} = T^{\nu\mu} = F^{\mu\lambda} F^{\nu}{}_{\lambda} - g^{\mu\nu} \frac{1}{4} F^{\kappa\lambda} F_{\kappa\lambda} \; ; \tag{3.45}$$

It has the property of being traceless:

$$T^{\mu}{}_{\mu} = g_{\mu\nu} T^{\mu\nu} = 0 \; , \tag{3.46}$$

and has the following explicit components:

$$T^{00} = U \; , \qquad T^{0}{}_{k} = \frac{1}{c} S_k = c G_k \; , \qquad T_{ij} = \mathsf{T}_{ij} \; , \tag{3.47}$$

in terms of the energy density, (1.20a), the energy flux vector (1.20b) or momentum density (1.20c), and the stress tensor (1.20d). It satisfies the equation

$$\partial_{\nu} T^{\mu\nu} = -F^{\mu\nu} \frac{1}{c} j_{\nu} \; , \tag{3.48}$$

which restates the energy and momentum conservation laws (1.44a) and (1.44b).

3.2 Field Transformations

A Lorentz transformation, or more properly a *boost,* is a transformation that mixes the time and space coordinates without changing the invariant distance s^2. An infinitesimal transformation of this class is

$$\delta\mathbf{r} = \delta\mathbf{v} t \; , \qquad \delta t = \frac{1}{c^2} \delta\mathbf{v} \cdot \mathbf{r} \; , \tag{3.49}$$

where $-\delta\mathbf{v}$ is the velocity with which the new coordinate frame moves relative to the old one. (It is assumed that the two coordinate frames coincide at $t = 0$.) In terms of the four-vector position, $x^{\mu} = (ct, \mathbf{r})$, we can write this result compactly as

$$\delta x^{\mu} = \delta\omega^{\mu\nu} x_{\nu} \; , \tag{3.50}$$

where the only nonzero components of the transformation parameter $\delta\omega^{\mu\nu}$ are

$$\delta\omega^{0i} = -\delta\omega^{i0} = \frac{\delta v^i}{c} \; . \tag{3.51}$$

Ordinary rotations of course also preserve s^2, so they must be included in the transformations (3.50), and they are, corresponding to $\delta\omega^{\mu\nu}$ having no time components, and spatial components

$$\delta\omega^{ij} = -\epsilon^{ijk} \delta\omega_k \; , \tag{3.52}$$

so, as in (1.81),

$$\delta\mathbf{r} = \delta\boldsymbol{\omega} \times \mathbf{r} \,. \tag{3.53}$$

In fact, the only property $\delta\omega^{\mu\nu}$ must have in order to preserve the invariant length s^2 is antisymmetry:

$$\delta\omega^{\mu\nu} \doteq -\delta\omega^{\nu\mu} \,, \tag{3.54}$$

for

$$\delta(x^\mu x_\mu) = 2\delta\omega^{\mu\nu} x_\mu x_\nu = 0 \,, \tag{3.55}$$

and a scalar product, such as that in $j^\mu A_\mu$ is similarly invariant. Any infinitesimal transformation with this property we will dub a Lorentz transformation.

Now consider the transformation of a four-vector field, such as the vector potential, $A^\mu = (\phi, \mathbf{A})$. This field undergoes the same transformation as given by the coordinate four-vector, but one must also transform to the new coordinate representing the same physical point. That is, under a Lorentz transformation,

$$A^\mu(x) \to \overline{A}^\mu(\overline{x}) = A^\mu(x) + \delta\omega^{\mu\nu} A_\nu(x) \,, \tag{3.56}$$

where

$$\overline{x}^\mu = x^\mu + \delta x^\mu = x^\mu + \delta\omega^{\mu\nu} x_\nu \,. \tag{3.57}$$

So that the transformation may be considered a field variation only, we define the change in the field at the same coordinate value (which refers to different physical points in the two frames):

$$\begin{aligned}
\delta A^\mu(x) &= \overline{A}^\mu(x) - A^\mu(x) \\
&= A^\mu(x - \delta x) + \delta\omega^{\mu\nu} A_\nu(x) - A^\mu(x) \\
&= -\delta x^\nu \partial_\nu A^\mu(x) + \delta\omega^{\mu\nu} A_\nu(x) \,.
\end{aligned} \tag{3.58}$$

The four-vector current $j^\mu = (c\rho, \mathbf{j})$ must transform in the same way:

$$\delta j^\mu = -\delta x^\nu \partial_\nu j^\mu(x) + \delta\omega^{\mu\nu} j_\nu(x) \,. \tag{3.59}$$

A scalar field, $\lambda(x)$, on the other hand only undergoes the coordinate transformation:

$$\lambda(x) \to \overline{\lambda}(\overline{x}) = \lambda(x) \,, \tag{3.60}$$

so

$$\delta\lambda(x) = -\delta x^\nu \partial_\nu \lambda(x) \,. \tag{3.61}$$

Because a vector potential can be changed by a gauge transformation,

$$A^\mu \to A^\mu + \partial^\mu \lambda \,, \tag{3.62}$$

without altering any physical quantity, in particular the field strength tensor $F^{\mu\nu}$, the transformation law for the vector potential must follow by differentiating that of λ, and indeed it does.

What about the transformation property of the field strength tensor? Again, it follows by direct differentiation:

$$\delta F_{\mu\nu} = \delta(\partial_\mu A_\nu - \partial_\nu A_\mu)$$
$$= -\delta x^\lambda \partial_\lambda F_{\mu\nu} - (\partial_\mu \delta x^\lambda)\partial_\lambda A_\nu + (\partial_\nu \delta x^\lambda)\partial_\lambda A_\mu + \delta\omega_{\nu\lambda}\partial_\mu A^\lambda - \delta\omega_{\mu\lambda}\partial_\nu A^\lambda$$
$$= -\delta x^\lambda \partial_\lambda F_{\mu\nu} + \delta\omega_\mu{}^\lambda F_{\lambda\nu} + \delta\omega_\nu{}^\lambda F_{\mu\lambda} \,. \tag{3.63}$$

So we see that each index of a tensor transforms like that of a vector. From this it is easy to work out how the components of the electric and magnetic fields transform under a boost (3.51). Apart from the coordinate change – which just says we are evaluating fields at the same physical point – we see [cf. (1.230)]

$$`\delta'\mathbf{E} = -\frac{\delta\mathbf{v}}{c} \times \mathbf{B} \,, \tag{3.64a}$$

$$`\delta'\mathbf{B} = \frac{\delta\mathbf{v}}{c} \times \mathbf{E} \,. \tag{3.64b}$$

The proof of the Lorentz invariance for the relativistic Lagrangian is now immediate. That is,

$$\delta\mathcal{L} = -\delta x^\lambda \partial_\lambda \mathcal{L} \,, \tag{3.65}$$

which just says that $\overline{\mathcal{L}}(\overline{x}) = \mathcal{L}(x)$, implying that $\delta W = \delta \int (dx)\mathcal{L}(x) = 0$. We have already remarked that $`\delta'\mathcal{L}_{\text{int}} = 0$. The invariance of the field Lagrangian (3.44) is simply the statement

$$`\delta'\mathcal{L}_f = \delta\omega_{\mu\nu} F^{\mu\lambda} F^\nu{}_\lambda = 0 \,, \tag{3.66}$$

and the particle action in (3.42a)–(3.42c) is manifestly invariant.

3.3 Problems for Chap. 3

1. Show that the time and space components of (3.41) are equivalent to the equations of motion (1.45a) and (1.45b) provided the relativistic form of the particle kinetic energy and momentum, (1.18a) and (1.18b), are employed.
2. Derive the first form of the particle action (3.42a) from the relativistic particle Lagrangian $-m_0 c^2 \sqrt{1 - v^2/c^2}$ and the interaction (3.17).
3. Obtain the equations resulting from variations of the second form of the particle action (3.42b) with respect to both x^μ and τ variations, and verify that these are as expected.
4. A covariant form for the current vector of a moving point charge e is the proper-time integral

$$\frac{1}{c}j^\mu(x) = \int_{-\infty}^{\infty} d\tau \, e \, \frac{dx^\mu(\tau)}{d\tau} \delta(x - x(\tau)) \,. \tag{3.67}$$

Verify that
$$\partial_\mu j^\mu = 0 \,, \tag{3.68}$$
under the assumption that the charge is infinitely remote at $\tau = \pm\infty$. Show that
$$\int (\mathrm{d}\mathbf{r}) \frac{1}{c} j^0(x) = e \,, \tag{3.69}$$
provided $\mathrm{d}x^0(\tau)/\mathrm{d}\tau$ is always positive. The stress tensor $T^{\mu\nu}$ for a mass point is given analogously by
$$T^{\mu\nu}(x) = \int_{-\infty}^{\infty} \mathrm{d}\tau \, m_0 c \frac{\mathrm{d}x^\mu(\tau)}{\mathrm{d}\tau} \frac{\mathrm{d}x^\nu(\tau)}{\mathrm{d}\tau} \delta(x - x(\tau)) \,. \tag{3.70}$$

Verify that
$$\partial_\nu T^{\mu\nu}(x) = 0 \,, \tag{3.71}$$
provided the particle is unaccelerated $(\mathrm{d}^2 x^\mu(\tau)/\mathrm{d}\tau^2 = 0)$. Then show that
$$\int (\mathrm{d}\mathbf{r}) \, T^{0\nu} = m_0 c \frac{\mathrm{d}x^\nu(\tau)}{\mathrm{d}\tau} \,. \tag{3.72}$$

Does this comprise the expected values for the energy and momentum (multiplied by c) of a uniformly moving particle?

5. Suppose the particle of the previous problem is accelerated – it carries charge e and moves in an electromagnetic field. Use the covariant equations of motion (3.41) to show that
$$\partial_\nu T^{\mu\nu}_{\mathrm{part}} = \frac{1}{c} F^\mu{}_\nu j^\nu \,. \tag{3.73}$$

What do you conclude by comparison with the corresponding divergence of the electromagnetic stress tensor (3.48)?

6. The next several problems refer to a purely electromagnetic model of the electron described first in [15]. A spherically symmetrical distribution of charge e at rest has the potentials $\phi = e f(r^2)$, $\mathbf{A} = \mathbf{0}$, where, at distances large compared with its size, $f(r^2) \sim 1/\sqrt{r^2}$. As observed in a frame in uniform relative motion, the potentials are
$$A^\mu(x) = \frac{e}{c} v^\mu f(\xi^2) \,, \qquad \xi^\mu = x^\mu + \frac{v^\mu}{c}\left(\frac{v^\lambda}{c} x_\lambda\right) \,, \tag{3.74}$$
where
$$v^\lambda \xi_\lambda = 0 \,, \qquad \xi^2 = x^2 + \left(\frac{v^\lambda}{c} x_\lambda\right)^2 \,. \tag{3.75}$$

Check that for motion along the z-axis with velocity v,
$$\xi^2 = x^2 + y^2 + \frac{(z - vt)^2}{1 - v^2/c^2} \,, \tag{3.76}$$
as could be inferred from Problem 31.1 of [9]. Compute the field strengths $F^{\mu\nu}$ and evaluate the electromagnetic field stress tensor (3.45).

7. From the previous problem, use the field equation (3.30a) to produce $j^\mu(x)$. Check that $\partial_\mu j^\mu = 0$. Construct $F^{\mu\nu}\frac{1}{c}j_\nu$ and note that its vector nature lets one write

$$F^{\mu\nu}\frac{1}{c}j_\nu = -\partial^\mu t(\xi^2) \, . \tag{3.77}$$

Exhibit $t(\xi^2)$ for the example $f(\xi^2) = (\xi^2 + a^2)^{-1/2}$. Inasmuch as the field tensor obeys (3.48)

$$\partial_\nu T_f^{\mu\nu} = -F^{\mu\nu}\frac{1}{c}j_\nu = \partial^\mu t \, , \tag{3.78}$$

one has realized a divergenceless electromagnetic tensor:

$$T^{\mu\nu} = T_f^{\mu\nu} - g^{\mu\nu}t \, , \qquad \partial_\nu T^{\mu\nu} = 0 \, . \tag{3.79}$$

It is the basis of a purely electromagnetic relativistic model of mass. There is, however, an ambiguity, because from (3.75)

$$\partial_\nu \left(\frac{v^\mu}{c}\frac{v^\nu}{c}t(\xi^2) \right) = 0 \, . \tag{3.80}$$

Therefore,

$$T^{\mu\nu} = T_f^{\mu\nu} - \left(g^{\mu\nu} + \frac{v^\mu}{c}\frac{v^\nu}{c} \right) t \, , \tag{3.81}$$

for example, is also a possible electromagnetic tensor. Choice (3.79) has the property that the momentum density of the moving system (multiplied by c) is just that of the field,

$$T^{0k} = T_f^{0k} = (\mathbf{E} \times \mathbf{B})_k \, . \tag{3.82}$$

Choice (3.81) is such that the energy density of the system at rest is just that of the field,

$$\mathbf{v} = \mathbf{0} : \quad T^{00} = T_f^{00} = \frac{E^2}{2} \, . \tag{3.83}$$

One cannot have both. That requires $t = 0$; that is, no charge. The system then is an electromagnetic pulse – it moves at the speed c.

8. Without specializing $f(\xi^2)$, integrate over all space (by introducing the variable $z' = (z - vt)/\sqrt{1 - v^2/c^2}$ to show that, whether one uses tensor (3.79) or (3.81),

$$E = \int (\mathrm{dr})\, T^{00} = \frac{mc^2}{\sqrt{1 - v^2/c^2}} \, , \qquad p_k = \frac{1}{c}\int (\mathrm{dr})\, T^0{}_k = \frac{mv_k}{\sqrt{1 - v^2/c^2}} \, . \tag{3.84}$$

What numerical factor relates m in scheme (3.79) to that in scheme (3.81)?

9. Repeat the action discussion following from (3.42c) with $m_0 = 0$ and unspecified $f(\xi^2)$. What mass emerges?

10. Verify that the Maxwell equations involving magnetic currents, the second set in (3.38), can also be given by

$$\partial_\lambda F_{\mu\nu} + \partial_\mu F_{\nu\lambda} + \partial_\nu F_{\lambda\mu} = \epsilon_{\mu\nu\lambda\kappa} \frac{1}{c} {}^*j^\kappa .$$ (3.85)

11. A particle has velocity components $v_x = \frac{dx}{dt}$ and $v_z = \frac{dz}{dt}$ in one coordinate frame. There is a second frame with relative velocity $\mathbf{v}$ along the z-axis. What are the velocity components $v'_x = \frac{dx'}{dt'}$ and $v'_z = \frac{dz'}{dt'}$ in this frame? Give a simple interpretation of the v'_x result for $v_z = 0$.

12. Let the motion referred to in the previous problem be that of light, moving at angle θ with respect to the z-axis. Find $\cos\theta'$ and $\sin\theta'$ in terms of $\cos\theta$ and $\sin\theta$. Check that $\cos^2\theta' + \sin^2\theta' = 1$. Exhibit θ' explicitly when $\beta = v/c \ll 1$.

13. The infinitesimal transformation contained in (3.51)

$$\delta\mathbf{p} = \frac{\delta\mathbf{v}}{c}\frac{E}{c} , \qquad \frac{\delta E}{c} = \frac{\delta\mathbf{v}}{c}\cdot\mathbf{p} ,$$ (3.86)

identify the four-vector of momentum $p^\mu = (E/c, \mathbf{p})$ What is the value of the invariant $p^\mu p_\mu$ for a particle of rest mass m_0? Apply the analog of the space–time transformation equations

$$t' = \frac{t + \mathbf{v}\cdot\mathbf{r}/c^2}{\sqrt{1 - v^2/c^2}} , \qquad \mathbf{v}\cdot\mathbf{r}' = \mathbf{v}\cdot\frac{\mathbf{r} + \mathbf{v}t}{\sqrt{1 - v^2/c^2}}$$ (3.87)

to find the energy and momentum of a moving particle from their values when the particle is at rest.

14. A body of mass M is at rest relative to one observer. Two photons, each of energy ϵ, moving in opposite directions along the x-axis, fall on the body, and are absorbed. Since the photons carry equal and opposite momenta, no net momentum is transferred to the body, and it remains at rest. Another observer is moving slowly along the y-axis. Relative to him, the two photons and the body, both before and after the absorption act, have a common velocity $\mathbf{v}$ ($|\mathbf{v}| \ll c$) along the y-axis, Reconcile conservation of the y-component of momentum with the fact that the velocity of the body does *not* change when the photons are absorbed.

15. Show, very simply, that $\mathbf{B}$, the magnetic field of a uniformly moving charge is $\frac{1}{c}\mathbf{v} \times \mathbf{E}$. Then consider two charges, moving with a common velocity $\mathbf{v}$ along parallel tracks, and show that the magnetic force between them is *opposite* to the electric force, and smaller by a factor of v^2/c^2. (This is an example of the rule that like *charges* repel, like *currents* attract.) Can you derive the same result by Lorentz transforming the equation of motion in the common rest frame of the two charges? (Hint: Coordinates perpendicular to the line of relative motion are unaffected by the transformation.)

16. This continues Problems 1.13–1.17. The relativistic formula

$$dN = \frac{(dk)}{k^0} \frac{\alpha}{4\pi^2} \left(\frac{v_1}{kv_1} - \frac{v_2}{kv_2} \right)^2 \qquad (3.88)$$

describes the number of photons emitted with momentum k^μ under the deflection of a particle: $v_1^\mu \rightarrow v_2^\mu$. (Here, the scalar product is denoted by $kv_1 = k^\mu v_{1\mu}$, etc.) More realistic is a collision of two particles, with masses m_a, m_b and charges e_a and e_b. When, as a result of a collision in which the particle velocities change from v_{a2}, v_{b2} to v_{a1}, v_{b1}, what is dN? Suppose this collision satisfies the conservation of energy–momentum, $(p_a^\mu + p_b^\mu)_1 = (p_a^\mu + p_b^\mu)_2$. Rewrite your expression for dN in terms of the p^μ rather than the v^μ. What follows if it should happen that e/kp has the same value for both particles, before and after the collision? Connect the nonrelativistic limit of this circumstance with Problem 1.13. Verify that this special circumstance does hold relativistically in the head-on, center-of-mass collision, where all momenta are of equal magnitude, for radiation *perpendicular* to the line of motion of the particles, provided $e_a/E_a = e_b/E_b$. What restriction does this impose on the energy if the particles are identical (same charge and rest mass)?

17. Use the fact that

$$k_\mu \left(\frac{v_1^\mu}{kv_1} - \frac{v_2^\mu}{kv_2} \right) = 0 \qquad (3.89)$$

for example to show that

$$\left(\frac{v_1}{kv_1} - \frac{v_2}{kv_2} \right)^2 = \left(\mathbf{n} \times \left(\frac{\mathbf{v_1}}{kv_1} - \frac{\mathbf{v_2}}{kv_2} \right) \right)^2 . \qquad (3.90)$$

Repeat the calculation of Problem 1.16, using this form and show the identity of the two results.

18. From the response of a particle momentum to an infinitesimal Lorentz transformation (3.86), find the infinitesimal change of the particle velocity $\mathbf{V}$ when $\mathbf{V}$ and $\delta\mathbf{v}$ are in the same direction. Compare your result with the implication of the formula for the relativistic addition of velocities.

19. Light travels at the speed c/n in a stationary, nondispersive medium. What is the speed of light when this medium is moving at speed v parallel or antiparallel to the direction of the light? To what does this simplify when $v/c \ll 1$?

20. An infinitesimal Lorentz transformation (boost) is characterized by a parameter $\delta\theta = \delta v/c$. Assuming that $\delta\mathbf{v}$ lies along the z-direction, construct and solve the first-order differential equations obeyed by $ct(\theta) \pm z(\theta)$. What do the solutions tell you about the relation between θ and v/c? How does the addition of velocity formula read in terms of the corresponding θs? (The angle θ is often referred to as the "rapidity.")

21. The frequency ω and the propagation vector $\mathbf{k}$ of a plane wave form a four-vector: $k^\mu = (\omega/c, \mathbf{k})$. Check that $k^\mu k_\mu = 0$ and that $\exp(ik_\mu x^\mu) =$

$\exp[i(\mathbf{k} \cdot \mathbf{r} - \omega t)]$. Use Lorentz transformations to show that radiation, of frequency ω, propagating at an angle θ with respect to the z-axis, will, to an observer moving with relative velocity $v = \beta c$ along the z-axis, have the frequency

$$\omega' = \frac{1}{\sqrt{1 - \beta^2}} \omega (1 - \beta \cos \theta) \tag{3.91}$$

(this is the Doppler effect) and an angle relative to the z-axis given by

$$\cos \theta' = \frac{\cos \theta - \beta}{1 - \beta \cos \theta} \tag{3.92}$$

(this is aberration). Find θ' explicitly for $|\beta| \ll 1$.

22. By writing the angle relation (3.92) as

$$\cos \theta - \cos \theta' = \beta(1 - \cos \theta \cos \theta') , \tag{3.93}$$

show that

$$\tan \frac{1}{2} \theta' = \sqrt{\frac{1 + \beta}{1 - \beta}} \tan \frac{1}{2} \theta , \tag{3.94}$$

or, replacing the angle θ for the direction of travel by the angle $\alpha = \pi - \theta$ for the direction of arrival,

$$\tan \frac{1}{2} \alpha' = \sqrt{\frac{1 - \beta}{1 + \beta}} \tan \frac{1}{2} \alpha . \tag{3.95}$$

23. An ellipse of eccentricity β is inscribed in a circle. The major axis of the ellipse lies along the x-axis, the origin of which is the center of the circle. A line drawn from the origin to a point on the circle makes an angle α with the x-axis. Now one finds a related point on the ellipse by moving down, perpendicularly to the x-axis, from the point on the circle. A line drawn from the left-hand focus of the ellipse to this point on the ellipse makes an angle α' with the x-axis. Show that the relation between α and α' is that of (3.95).

24. Show that the four-potential produced by a charged particle with four-velocity v^μ is

$$A^\mu(x) = \frac{1}{4\pi} \int ds' \eta(x^0 - x^{0\prime}(s')) 2\delta[(x - x'(s'))^2] e v^\mu(s')$$

$$= -\frac{e}{4\pi} \frac{v^\mu(s')}{(x - x'(s'))v(s')} . \tag{3.96}$$

Here η is the unit step function (the corresponding capital letter looks like the initial letter of Heaviside)

$$\eta(x) = \begin{cases} 1 , & x > 0 , \\ 0 , & x < 0 . \end{cases} \tag{3.97}$$

Make explicit what is left implicit in the result (3.96). Write the result in $3 + 1$ dimensional notation and compare with the Liénard–Wiechert potentials (1.125) and (1.126).

25. A charge at rest scatters radiation with unchanged frequency. Give a relativistically invariant form to this statement. Then deduce that radiation of frequency ω_0, moving in the direction of unit vector $\mathbf{n}_0$, which is scattered by a charge with velocity $\mathbf{v}$ into the direction of unit vector $\mathbf{n}$, has the frequency

$$\omega = \omega_0 \frac{1 - \mathbf{n}_0 \cdot \frac{\mathbf{v}}{c}}{1 - \mathbf{n} \cdot \frac{\mathbf{v}}{c}} . \tag{3.98}$$

26. Find the total scattering cross section for the scattering of radiation by a charge that is moving with velocity $\mathbf{v}$ in the direction of the incident radiation. Assume that $\beta = v/c$ is small so that is an essentially nonrelativistic calculation.

27. Repeat the above using a relativistic calculational method. Check the consistency of the result with that of the $\beta \ll 1$ calculation.

28. The classical statement that light is scattered with unchanged frequency by a charge at rest appears generally as, in four-vector notation,

$$v^2 = -c^2 : \qquad kv = k_0 v , \qquad \text{or} \qquad v(k - k_0) = 0 . \tag{3.99}$$

Now think of a *photon* scattered by a charged particle. If initial momenta are denoted by a subscript 0, the statement of energy–momentum conservation reads

$$(\hbar k + p)^\mu = (\hbar k_0 + p_0)^\mu , \tag{3.100}$$

where

$$k^2 = k_0^2 = 0 , \qquad p^2 = p_0^2 = -m_0^2 , \tag{3.101}$$

where m_0 is the rest mass of the particle. Show that

$$(p + p_0)(k - k_0) = 0 , \tag{3.102}$$

or in terms of four-velocities, given by $p^\mu = m_0 v^\mu$, $p_0^\mu = m_0 v_0^\mu$, that

$$\frac{1}{2}(v + v_0)(k - k_0) = 0 . \tag{3.103}$$

Thus the classical result (3.99) appears when the difference between v^μ and v_0^μ can be neglected in

$$\left(\frac{1}{2}(v + v_0) \right)^2 + \left(\frac{1}{2}(v - v_0) \right)^2 = -c^2 . \tag{3.104}$$

The nearest quantum equivalent to the classical rest frame occurs when $\mathbf{p}_0 = -\mathbf{p}$. In that frame let $\mathbf{k}_0$ and $\mathbf{k}$ each make an angle $\frac{1}{2}\theta$ with respect to the plane perpendicular to $\mathbf{p}$. Check that (3.102) implies $\omega = \omega_0$. What is the value of $|\mathbf{p}| = |\mathbf{p}_0|$ in terms of ω_0 and the photon scattering angle θ? Under what circumstances can the equal and opposite velocities $\mathbf{v}$ and $\mathbf{v}_0$ be regarded as negligible? (This is the underlying principle of the Free Electron Laser [16].)

29. Integrate the invariant $(dk)2\delta(k^2)$, where $(dk) = dk^0(d\mathbf{k})$, over all $k^0 > 0$ to arrive at the invariant

$$\frac{(d\mathbf{k})}{|\mathbf{k}|} = \frac{\omega\, d\omega}{c^2} d\Omega\,, \qquad \omega = kc\,, \tag{3.105}$$

in which $d\Omega$ is an element of solid angle. Use the Doppler effect formula (3.91) to deduce the solid angle transformation law,

$$d\Omega' = \frac{1 - \beta^2}{(1 - \beta\cos\theta)^2} d\Omega\,. \tag{3.106}$$

Then get it directly from the aberration formula (3.92). What did you assume about the azimuthal angle ϕ, and why? Check that the above relation is consistent with the requirement that $\int d\Omega' = 4\pi$.

30. Let v^μ be the four-vector velocity $\gamma(c, \mathbf{v})$, $\gamma = (1 - v^2/c^2)^{-1/2}$ of a physical system. Use the invariance of $k^\mu v_\mu$, in relating ω', a frequency observed when the system is at rest, to quantities measured when the system is in motion along the z-axis with velocity v. Compare with a result found in Problem 3.21. Show that the invariant

$$I = \frac{dp^\mu}{d\tau}\frac{dp_\mu}{d\tau} - \left(mc\frac{k_\mu dp^\mu/d\tau}{k_\nu p^\nu}\right)^2\,, \tag{3.107a}$$

is written as

$$I = \left(\frac{E}{mc^2}\right)^2\left[\dot{\mathbf{p}}^2 - \left(\frac{1}{c}\dot{E}\right)^2 - \left(\frac{mc^2}{E}\right)^2\frac{(\mathbf{n}\cdot\dot{\mathbf{p}} - \dot{E}/c)^2}{(1 - \mathbf{n}\cdot\mathbf{p}c/E)^2}\right]\,, \tag{3.107b}$$

where m is the rest mass, and $\mathbf{n}$ is the direction of $\mathbf{k}$, which appears in the angular distribution given in (3.111).

31. Verify that the energy radiated per unit time into a unit solid angle, by a system that is momentarily at rest, is given, in any coordinate frame, by the invariant expression

$$-\frac{d^2p^\mu}{d\tau\, d\Omega'}v_\mu\,, \tag{3.108}$$

where v^μ is the velocity four-vector of the system; $d\Omega'$ refers to the rest frame. Then use the relation between the momentum and the energy of the radiation moving in a given direction (unit vector $\mathbf{n}$) to write the above radiation quantity, for a system moving with velocity $\mathbf{v}$, as

$$\frac{d^2E}{dt\, d\Omega'}\frac{1 - \mathbf{n}\cdot\mathbf{v}/c}{1 - v^2/c^2}\,. \tag{3.109}$$

32. The power radiated in the direction $\mathbf{n}$, per unit solid angle $d\Omega'$, by an accelerated charge that is momentarily at rest, is given by (1.167) or

$$\frac{e^2}{(4\pi)^2 c^3}(\mathbf{n} \times \dot{\mathbf{v}})^2 = \frac{e^2}{(4\pi)^2 c^3}[\dot{\mathbf{v}}^2 - (\mathbf{n} \cdot \dot{\mathbf{v}})^2] . \tag{3.110}$$

Now combine this with the results of (3.109), (3.106), and (3.107b) to produce the power radiated into a solid angle $d\Omega$,

$$\frac{dP}{d\Omega} = \frac{e^2}{(4\pi)^2 m^2 c^3}\left(\frac{mc^2}{E}\right)^2 \left[\frac{\dot{\mathbf{p}}^2 - (\dot{E}/c)^2}{(1 - \mathbf{n} \cdot \mathbf{p}c/E)^3} - \left(\frac{mc^2}{E}\right)^2 \frac{(\mathbf{n} \cdot \dot{\mathbf{p}} - \dot{E}/c)^2}{(1 - \mathbf{n} \cdot \mathbf{p}c/E)^5}\right] .$$
$$\tag{3.111}$$

This result is the same as that given in (1.210) upon substituting there $\mathbf{v} = \mathbf{p}c^2/E$.

4

Variational Principles for Harmonic Time Dependence

Throughout the remainder of the book we shall be primarily concerned with fields possessing a simple harmonic time dependence. Since the time dependence of all fields and currents must be the same, it is useful to suppress the common time factor with the aid of complex notation. We shall write the real quantity $A(t)$ as

$$A(t) = \frac{1}{2} \left(A e^{-i\omega t} + A^* e^{i\omega t} \right) = \operatorname{Re} A e^{-i\omega t} , \tag{4.1}$$

where A^* is the complex conjugate of the amplitude A, that is, the sign of i is reversed, and Re symbolizes the operation of constructing the real part of the following complex number. Linear relations between quantities, such as

$$A(t) + B(t) = C(t) , \tag{4.2a}$$

$$\frac{d}{dt} F(t) = G(t) , \tag{4.2b}$$

can then be written as

$$A + B = C , \tag{4.3a}$$

$$-i\omega F = G , \tag{4.3b}$$

with the understanding that the exponential time factor must be supplied and the real part obtained to give the equation meaning. Quadratic quantities, such as $A(t)B(t)$, will vary with time, consisting partly of a constant term, and partly of terms oscillating harmonically with angular frequency 2ω,

$$A(t)B(t) = \frac{1}{4} \left(AB^* + A^*B \right) + \frac{1}{4} \left(AB e^{-2i\omega t} + A^*B^* e^{2i\omega t} \right) . \tag{4.4}$$

We shall invariably be concerned only with the constant part, or time average of these products, for the rapidly oscillating component of physical quantities like electric flux is not readily susceptible of observation. Hence

$$\overline{A(t)B(t)} = \frac{1}{4}\left(AB^* + A^*B\right) = \frac{1}{2}\operatorname{Re} AB^* = \frac{1}{2}\operatorname{Re} A^*B \,. \qquad (4.5)$$

In particular

$$\overline{(A(t))^2} = \frac{1}{2}|A|^2. \qquad (4.6)$$

With these conventions, the harmonic form of the Maxwell equations (1.1a) and (1.1b) is, in SI units,[1]

$$\boldsymbol{\nabla} \times \mathbf{H} = -\mathrm{i}\omega\varepsilon\mathbf{E} + \mathbf{J} \,, \qquad (4.7a)$$

$$\boldsymbol{\nabla} \times \mathbf{E} = \mathrm{i}\omega\mu\mathbf{H} \,, \qquad (4.7b)$$

and the equation of charge conservation (1.14) reads

$$\boldsymbol{\nabla} \cdot \mathbf{J} = \mathrm{i}\omega\rho \,. \qquad (4.8)$$

The remaining set of Maxwell equations, the divergence equations,

$$\boldsymbol{\nabla} \cdot \varepsilon\mathbf{E} = \rho \,, \qquad \boldsymbol{\nabla} \cdot \mu\mathbf{H} = 0 \,, \qquad (4.9)$$

are immediate consequences of the curl equations for nonstatic fields. For a conducting medium it is useful to indicate explicitly the conduction current density $\sigma\mathbf{E}$,

$$\boldsymbol{\nabla} \times \mathbf{H} = (\sigma - \mathrm{i}\omega\varepsilon)\mathbf{E} + \mathbf{J} \,, \qquad (4.10)$$

where $\mathbf{J}$ now denotes the impressed current density, arising from all other causes. The average intensity of energy flow (1.203b) is

$$\overline{\mathbf{S}} = \frac{1}{2}\operatorname{Re} \mathbf{E} \times \mathbf{H}^* \,, \qquad (4.11)$$

while the average density of electromagnetic energy is (proved in Problem 6.2)

$$\overline{U} = \frac{1}{4}\left(\frac{\partial\omega\varepsilon}{\partial\omega}|\mathbf{E}|^2 + \frac{\partial\omega\mu}{\partial\omega}|\mathbf{H}|^2\right) \,. \qquad (4.12)$$

4.1 Variational Principles

The Maxwell equations in complex, or harmonic, form can be derived from a variational principle without the intervention of auxiliary potentials. Indeed, almost no further use of potentials will be made in the sequel. The variational principle can be cast into a variety of forms of which a rather general example is

$$\delta L \equiv \delta \int_V (\mathrm{d}\mathbf{r}) \left(\frac{1}{2}\mathrm{i}\omega\mu H^2 - \frac{1}{2}\mathrm{i}\omega\varepsilon E^2 - \mathbf{E}\cdot\boldsymbol{\nabla}\times\mathbf{H} + \frac{1}{2}\sigma E^2 + \mathbf{J}\cdot\mathbf{E}\right) \,, \qquad (4.13)$$

---

[1] Note that because the sign of ω is without significance, we may, without loss of generality, take $\omega \geq 0$.

in which $\mathbf{E}$ and $\mathbf{H}$ are subject to independent variations within the volume V, and $\mathbf{J}$, the impressed current density, is a prescribed function of position. In performing the variation the components of $\mathbf{H}$ tangential to the surface S bounding the region V are to be regarded as prescribed functions,

$$\mathbf{n} \times \delta \mathbf{H} = \mathbf{0} \quad \text{on} \quad S\,, \tag{4.14}$$

where $\mathbf{n}$ is the normal to the surface. The direct evaluation of the variation gives

$$\int_V (\mathrm{d}\mathbf{r}) \left[\delta \mathbf{H} \cdot (\mathrm{i}\omega\mu\mathbf{H} - \boldsymbol{\nabla} \times \mathbf{E}) + \delta \mathbf{E} \cdot (-\mathrm{i}\omega\varepsilon\mathbf{E} + \sigma\mathbf{E} + \mathbf{J} - \boldsymbol{\nabla} \times \mathbf{H}) \right]$$

$$- \oint_S \mathrm{d}S \, (\mathbf{n} \times \delta\mathbf{H}) \cdot \mathbf{E} = 0\,, \tag{4.15}$$

on employing the identity

$$\mathbf{A} \cdot \boldsymbol{\nabla} \times \mathbf{B} - \mathbf{B} \cdot \boldsymbol{\nabla} \times \mathbf{A} = \boldsymbol{\nabla} \cdot (\mathbf{B} \times \mathbf{A})\,. \tag{4.16}$$

The surface integral vanishes in virtue of the restricted nature of the variation (4.14) and the condition that the volume integral equals zero for arbitrary $\delta\mathbf{E}$ and $\delta\mathbf{H}$ yields the two curl equations (4.10) and (4.7b), which are the fundamental set of the Maxwell equations. In the surface condition imposed on the variation, we have an indication of the uniqueness theorem, which states that the fields within a region are completely determined by the value of the magnetic field tangential to the surface enclosing the region.

It often occurs that the fields within a region are required subject to the boundary condition that the electric field components tangential to the bounding surface be zero,

$$\mathbf{n} \times \mathbf{E} = \mathbf{0} \quad \text{on} \quad S\,. \tag{4.17}$$

On writing the surface integral contribution to the variation (4.15) as

$$\oint_S \mathrm{d}S \, \delta\mathbf{H} \cdot \mathbf{n} \times \mathbf{E}\,, \tag{4.18}$$

it is apparent that the requirement that the Lagrangian be stationary for completely arbitrary variations, in the interior of the region and on the surface, will yield both the Maxwell equations and the boundary condition (4.17). More general boundary conditions of this type are of the form

$$\mathbf{n} \times \mathbf{E} = \mathsf{Z} \cdot \mathbf{H} \quad \text{on} \quad S\,, \tag{4.19}$$

where Z is a symmetrical dyadic possessing no components normal to the surface S, that is, the boundary condition is a relation between tangential components of $\mathbf{E}$ and $\mathbf{H}$. To derive this boundary condition from a variational principle, the Lagrangian must be extended to

$$L = \int_V (\mathrm{d}\mathbf{r}) \left(\frac{1}{2} \mathrm{i}\omega\mu H^2 - \frac{1}{2} \mathrm{i}\omega\varepsilon E^2 - \mathbf{E} \cdot \nabla \times \mathbf{H} + \frac{1}{2}\sigma E^2 + \mathbf{J} \cdot \mathbf{E} \right)$$
$$- \oint_S \mathrm{d}S \frac{1}{2} \mathbf{H} \cdot \mathbf{Z} \cdot \mathbf{H} . \tag{4.20}$$

The Maxwell equations for the interior of the region are obtained as before, but the surface integral contribution to the variation is now

$$\oint_S \mathrm{d}S \, \delta\mathbf{H} \cdot (\mathbf{n} \times \mathbf{E} - \mathbf{Z} \cdot \mathbf{H}) , \tag{4.21}$$

and the requirement that this vanish for arbitrary surface variations yields the desired boundary condition (4.19).

The roles of electric and magnetic fields relative to the surface conditions can be interchanged by replacing the term $\mathbf{E} \cdot \nabla \times \mathbf{H}$ in the Lagrangian by $\mathbf{H} \cdot \nabla \times \mathbf{E}$. The sole effect of this alteration is that the surface integral contribution to the variation now reads

$$- \oint_S \mathrm{d}S \, \mathbf{n} \times \delta\mathbf{E} \cdot \mathbf{H} = \oint_S \mathrm{d}S \, \delta\mathbf{E} \cdot (\mathbf{n} \times \mathbf{H}) , \tag{4.22}$$

which vanishes if $\mathbf{n} \times \mathbf{E}$ is prescribed on the surface, indicating another aspect of the uniqueness theorem. If boundary conditions of the type

$$\mathbf{n} \times \mathbf{H} = -\mathbf{Y} \cdot \mathbf{E} \quad \text{on} \quad S \tag{4.23}$$

are given, with $\mathbf{Y}$ a symmetrical dyadic relating tangential components of $\mathbf{E}$ and $\mathbf{H}$, the appropriate Lagrangian is

$$L = \int_V (\mathrm{d}\mathbf{r}) \left(\frac{1}{2} \mathrm{i}\omega\mu H^2 - \frac{1}{2} \mathrm{i}\omega\varepsilon E^2 - \mathbf{H} \cdot \nabla \times \mathbf{E} + \frac{1}{2}\sigma E^2 + \mathbf{J} \cdot \mathbf{E} \right)$$
$$+ \oint_S \mathrm{d}S \frac{1}{2} \mathbf{E} \cdot \mathbf{Y} \cdot \mathbf{E} . \tag{4.24}$$

The Lagrangian can be simplified by adopting one of the Maxwell equations as a defining equation and employing the variational principle to derive the other equation. Thus, if $\mathbf{H}$ is defined in terms of $\mathbf{E}$ by (4.7b) or

$$\nabla \times \mathbf{E} = \mathrm{i}\omega\mu\mathbf{H} , \tag{4.25}$$

the variational principle (4.24) with $\mathbf{Y} = 0$ becomes

$$\delta \int_V (\mathrm{d}\mathbf{r}) \left(-\frac{1}{2} \mathrm{i}\omega\mu H^2 - \frac{1}{2} \mathrm{i}\omega\varepsilon E^2 + \frac{1}{2}\sigma E^2 + \mathbf{J} \cdot \mathbf{E} \right)$$
$$= \int_V (\mathrm{d}\mathbf{r}) \, \delta\mathbf{E} \cdot (-\mathrm{i}\omega\varepsilon\mathbf{E} + \sigma\mathbf{E} + \mathbf{J} - \nabla \times \mathbf{H})$$
$$- \oint_S \mathrm{d}S \, \mathbf{n} \times \delta\mathbf{E} \cdot \mathbf{H} = 0 , \tag{4.26}$$

in which $\mathbf{H}$ is varied according to its defining equation (4.25), yields the second Maxwell equation (4.10) if $\mathbf{n} \times \mathbf{E}$ is prescribed on the boundary. The discussion of homogeneous boundary conditions is unaltered. Note that the correct Lagrangian for $\mathbf{E}$ as the only independent variable is identical with the general Lagrangian save that $\mathbf{H}$ is now considered a function of $\mathbf{E}$. In a similar way, if $\mathbf{E}$ is defined in terms of $\mathbf{H}$ by (4.10)

$$\nabla \times \mathbf{H} = (\sigma - i\omega\varepsilon)\mathbf{E} + \mathbf{J} , \tag{4.27}$$

the appropriate variational principle is from (4.13)

$$\delta \int_V (\mathrm{d}r) \left(\frac{1}{2} i\omega\mu H^2 + \frac{1}{2} i\omega\varepsilon E^2 - \frac{1}{2}\sigma E^2 \right)$$
$$= \int_V (\mathrm{d}r)\, \delta\mathbf{H} \cdot (i\omega\mu\mathbf{H} - \nabla \times \mathbf{E}) - \oint_S \mathrm{d}S\, \mathbf{n} \times \delta\mathbf{H} \cdot \mathbf{E} = 0 . \tag{4.28}$$

A somewhat different type of variational principle can be employed when the medium is dissipationless, $\sigma = 0$, $\varepsilon = \varepsilon^*$, $\mu = \mu^*$. Consider the pure imaginary Lagrangian

$$L = \int_V (\mathrm{d}r) \Big(-i\omega\mu\mathbf{H}^* \cdot \mathbf{H} - i\omega\varepsilon\mathbf{E}^* \cdot \mathbf{E} + \mathbf{H}^* \cdot \nabla \times \mathbf{E} - \mathbf{H} \cdot \nabla \times \mathbf{E}^*$$
$$+ \mathbf{J} \cdot \mathbf{E}^* - \mathbf{J}^* \cdot \mathbf{E} \Big) , \tag{4.29}$$

which is to be regarded as a function of the four independent variables $\mathbf{E}$, $\mathbf{E}^*$, $\mathbf{H}$, $\mathbf{H}^*$. Performing the required variations,

$$\delta L = \int_V \Big[\delta\mathbf{H}^* \cdot (-i\omega\mu\mathbf{H} + \nabla \times \mathbf{E}) - \delta\mathbf{H} \cdot (i\omega\mu\mathbf{H}^* + \nabla \times \mathbf{E}^*)$$
$$+ \delta\mathbf{E}^* \cdot (-i\omega\varepsilon\mathbf{E} + \mathbf{J} - \nabla \times \mathbf{H}) - \delta\mathbf{E} \cdot (i\omega\varepsilon\mathbf{E}^* + \mathbf{J}^* - \nabla \times \mathbf{H}^*) \Big]$$
$$+ \oint_S \mathrm{d}S\, (\mathbf{n} \times \delta\mathbf{E} \cdot \mathbf{H}^* - \mathbf{n} \times \delta\mathbf{E}^* \cdot \mathbf{H}) , \tag{4.30}$$

we obtain, as the stationary conditions for prescribed $\mathbf{n} \times \mathbf{E}$ and $\mathbf{n} \times \mathbf{E}^*$ on the boundary surface, both curl equations (4.25) and (4.27), and their complex conjugates. Tangential boundary conditions of the type

$$\mathbf{n} \times \mathbf{H} = i\mathsf{B} \cdot \mathbf{E} \quad \text{on} \quad S \tag{4.31}$$

with B a real symmetrical dyadic, can be derived from a variational principle by the addition of

$$- \oint_S \mathrm{d}S\, i\mathbf{E}^* \cdot \mathsf{B} \cdot \mathbf{E} \tag{4.32}$$

to the Lagrangian. If $\mathbf{H}$ and $\mathbf{H}^*$ are defined in terms of $\mathbf{E}$ and $\mathbf{E}^*$ by

$$\nabla \times \mathbf{E} = \mathrm{i}\omega\mu\mathbf{H}, \quad \nabla \times \mathbf{E}^* = -\mathrm{i}\omega\mu\mathbf{H}^*, \qquad (4.33)$$

the simplified form of the Lagrangian (4.29) is

$$L = \int_V (\mathrm{d}\mathbf{r}) \left(\mathrm{i}\omega\mu\mathbf{H}^* \cdot \mathbf{H} - \mathrm{i}\omega\varepsilon\mathbf{E}^* \cdot \mathbf{E} + \mathbf{J} \cdot \mathbf{E}^* - \mathbf{J}^* \cdot \mathbf{E} \right). \qquad (4.34)$$

The appropriate Lagrangian if $\mathbf{n} \times \mathbf{H}$ and $\mathbf{n} \times \mathbf{H}^*$ are prescribed on S is

$$L = \int_V (\mathrm{d}\mathbf{r}) \left(-\mathrm{i}\omega\mu\mathbf{H}^* \cdot \mathbf{H} - \mathrm{i}\omega\varepsilon\mathbf{E}^* \cdot \mathbf{E} + \mathbf{E} \cdot \nabla \times \mathbf{H}^* - \mathbf{E}^* \cdot \nabla \times \mathbf{H} \right. $$
$$\left. + \mathbf{J} \cdot \mathbf{E}^* - \mathbf{J}^* \cdot \mathbf{E} \right). \qquad (4.35)$$

To derive boundary conditions of the form

$$\mathbf{n} \times \mathbf{E} = -\mathrm{i}\mathsf{X} \cdot \mathbf{H} \quad \text{on} \quad S, \qquad (4.36)$$

with X a real symmetrical dyadic relating tangential components, it is necessary to augment the Lagrangian by the term

$$-\oint_S \mathrm{d}S\, \mathrm{i}\mathbf{H}^* \cdot \mathsf{X} \cdot \mathbf{H}. \qquad (4.37)$$

If $\mathbf{E}$ and $\mathbf{E}^*$ are defined in terms of $\mathbf{H}$ and $\mathbf{H}^*$ by

$$\nabla \times \mathbf{H} = -\mathrm{i}\omega\varepsilon\mathbf{E} + \mathbf{J}, \quad \nabla \times \mathbf{H}^* = \mathrm{i}\omega\varepsilon\mathbf{E}^* + \mathbf{J}^*, \qquad (4.38)$$

the Lagrangian (4.35) reduces to

$$L = \int_V (\mathrm{d}\mathbf{r}) \left(-\mathrm{i}\omega\mu\mathbf{H}^* \cdot \mathbf{H} + \mathrm{i}\omega\varepsilon\mathbf{E}^* \cdot \mathbf{E} \right), \qquad (4.39)$$

as a function of $\mathbf{H}$ and $\mathbf{H}^*$, which has the expected familiar form.

4.2 Boundary Conditions

Thus far, no restriction has been imposed upon the macroscopic parameters ε, μ, and σ, which can be considered arbitrary continuous functions of position. However, spatial variations of these quantities usually arise through the juxtaposition of two different homogeneous media, the entire region of spatial dependence being confined to the immediate vicinity of the contact surface. It is usually possible to neglect the thickness of the transition layer, on a macroscopic scale, and thus treat the problem in terms of two completely homogeneous media, the effect of the region of transition being expressed by conditions relating the fields on opposite sides of the mutual boundary. The appropriate boundary conditions can easily be derived from the variational principle. The variation of the Lagrangian will contain volume contributions from the interiors of each of the media, which we shall distinguish by subscripts 1 and 2, and from the region of the transition layer. The stationary requirement on the Lagrangian will yield the Maxwell equations, with appropriate values of the material constants, for variations within each of the substances; the boundary conditions are an expression of the stationary character of the Lagrangian for variations within the transition region. The parameters ε, μ, and σ will be supposed to vary rapidly but continuously in the transition layer, and accordingly the fields, although subject to sharp changes within the region, remain finite. The variations of the fields are finite, continuous functions of position and therefore such contributions to the variation in the transition layer as $\int (\mathrm{d}\mathbf{r})\,\delta\mathbf{H} \cdot i\omega\mu\mathbf{H}$ are essentially proportional to the thickness of the layer and may be discarded. However, when derivatives of the field occur the volume integral can be independent of thickness, and therefore the essential terms to be considered are, from (4.15)

$$-\int (\mathrm{d}\mathbf{r})(\delta\mathbf{H} \cdot \boldsymbol{\nabla} \times \mathbf{E} + \delta\mathbf{E} \cdot \boldsymbol{\nabla} \times \mathbf{H}) = -\int (\mathrm{d}\mathbf{r})\boldsymbol{\nabla} \cdot (\mathbf{E} \times \delta\mathbf{H} + \mathbf{H} \times \delta\mathbf{E})$$

$$-\int (\mathrm{d}\mathbf{r})(\mathbf{E} \cdot \boldsymbol{\nabla} \times \delta\mathbf{H} + \mathbf{H} \cdot \boldsymbol{\nabla} \times \delta\mathbf{E})\,, \tag{4.40}$$

provided the impressed current density $\mathbf{J}$ is finite in the boundary region. In the rearranged form indicated, the second integral contains the finite derivatives of the field variations and will therefore be neglected. The final result, then, is

$$-\oint \mathrm{d}S\,(\mathbf{n} \times \mathbf{E} \cdot \delta\mathbf{H} + \mathbf{n} \times \mathbf{H} \cdot \delta\mathbf{E}) = \int_{S_{12}} \mathrm{d}S\left[(\mathbf{n}_1 \times \mathbf{E}_1 + \mathbf{n}_2 \times \mathbf{E}_2) \cdot \delta\mathbf{H}\right.$$

$$\left. + (\mathbf{n}_1 \times \mathbf{H}_1 + \mathbf{n}_2 \times \mathbf{H}_2) \cdot \delta\mathbf{E}\right], \tag{4.41}$$

an integral extended over the closed surface surrounding the discontinuity region, which consists of the two surfaces on opposite sides of the contact surface S_{12}, as expressed by the second integral. In the latter, $\mathbf{n}_1$ and $\mathbf{n}_2$ are unit normals at a common point on S_{12}, drawn outward from the respective regions. The stationary condition now requires that

$$\mathbf{n}_1 \times \mathbf{E}_1 + \mathbf{n}_2 \times \mathbf{E}_2 = \mathbf{n}_1 \times (\mathbf{E}_1 - \mathbf{E}_2) = 0 \quad \text{on} \quad S\,, \tag{4.42a}$$

$$\mathbf{n}_1 \times \mathbf{H}_1 + \mathbf{n}_2 \times \mathbf{H}_2 = \mathbf{n}_1 \times (\mathbf{H}_1 - \mathbf{H}_2) = 0 \quad \text{on} \quad S\,, \tag{4.42b}$$

or, the components of the electric and magnetic field tangential to the boundary surface are continuous on traversing the surface.

Other boundary conditions can be derived from these fundamental ones. A small area of the surface S_{12} can be considered a plane surface with a fixed normal. Hence, on the side of S_{12} adjacent to region 1,

$$i\omega\mu_1\mathbf{n}_1 \cdot \mathbf{H}_1 = \mathbf{n}_1 \cdot \boldsymbol{\nabla} \times \mathbf{E}_1 = -\boldsymbol{\nabla} \cdot \mathbf{n}_1 \times \mathbf{E}_1\,, \tag{4.43}$$

in which the divergence involves only derivatives tangential to the surface. On adding a similar equation for the other side of S_{12} amd employing the continuity of the tangential electric field across the interface between the media, we obtain

$$\mu_1 \mathbf{n}_1 \cdot \mathbf{H}_1 + \mu_2 \mathbf{n}_2 \cdot \mathbf{H}_2 = \mathbf{n}_1 \cdot (\mu_1 \mathbf{H}_1 - \mu_2 \mathbf{H}_2) = 0 , \qquad (4.44)$$

or, the component of the magnetic induction normal to the surface is continuous. In a similar way, from

$$(\sigma_1 - i\omega\varepsilon_1)\mathbf{n}_1 \cdot \mathbf{E}_1 = -\boldsymbol{\nabla} \cdot \mathbf{n}_1 \times \mathbf{H}_1 \qquad (4.45)$$

valid in the absence of impressed currents at the interface, one finds

$$\left(\varepsilon_1 + i\frac{\sigma_1}{\omega}\right) \mathbf{n}_1 \cdot \mathbf{E}_1 + \left(\varepsilon_2 + i\frac{\sigma_2}{\omega}\right) \mathbf{n}_2 \cdot \mathbf{E}_2$$
$$= \mathbf{n}_1 \cdot \left[\left(\varepsilon_1 + i\frac{\sigma_1}{\omega}\right) \mathbf{E}_1 - \left(\varepsilon_2 + i\frac{\sigma_2}{\omega}\right) \mathbf{E}_2\right] = 0 . \,(4.46)$$

If both substances are nonconductors ($\sigma_1 = \sigma_2 = 0$), this relation states the continuity of the normal component of the electric displacement. For conducting media, the normal component of $\mathbf{D}$ is discontinuous save in the exceptional circumstance

$$\frac{\sigma_1}{\varepsilon_1} = \frac{\sigma_2}{\varepsilon_2} , \qquad (4.47)$$

that is, when the charge relaxation times, ε/σ, of the two substances are identical (see Problem 4.1). On integrating the divergence equation

$$\boldsymbol{\nabla} \cdot \mathbf{D} = \rho \qquad (4.48)$$

over the infinitesimal volume formed by two small surfaces drawn on either side of S_{12}, it becomes apparent that a discontinuity in the normal component of $\mathbf{D}$ implies that a finite amount of charge per unit area is distributed in the transition layer. Indeed,

$$\mathbf{n}_1 \cdot \mathbf{D}_1 + \mathbf{n}_2 \cdot \mathbf{D}_2 = \mathbf{n}_1 \cdot (\mathbf{D}_1 - \mathbf{D}_2) = -\tau , \qquad (4.49)$$

where τ is the surface density of charge. Hence, from (4.46),

$$\tau = \frac{i}{\omega}(\sigma_1 \mathbf{n}_1 \cdot \mathbf{E}_1 + \sigma_2 \mathbf{n}_2 \cdot \mathbf{E}_2) = -\mathbf{n}_1 \cdot \mathbf{E}_1 \frac{\varepsilon_1\sigma_2 - \varepsilon_2\sigma_1}{\sigma_2 - i\omega\varepsilon_2}$$
$$= -\mathbf{n}_2 \cdot \mathbf{E}_2 \frac{\varepsilon_2\sigma_1 - \varepsilon_1\sigma_2}{\sigma_1 - i\omega\varepsilon_1} . \qquad (4.50)$$

The first of these relations when written in terms of the conduction current density $\mathbf{J} = \sigma\mathbf{E}$, reads

$$\mathbf{n}_1 \cdot \mathbf{J}_1 + \mathbf{n}_2 \cdot \mathbf{J}_2 = \mathbf{n}_1 \cdot (\mathbf{J}_1 - \mathbf{J}_2) = -i\omega\tau , \qquad (4.51)$$

which is an immediate consequence of the charge conservation theorem (4.8)

$$\mathbf{\nabla} \cdot \mathbf{J} = i\omega\rho. \tag{4.52}$$

The idealization of perfect conductors ($\sigma = \infty$) often affords an excellent first approximation in the description of good conductors. Within a perfect conductor the electric field intensity, and in consequence the magnetic field intensity, must be zero, else the conduction current density $\mathbf{J} = \sigma\mathbf{E}$ would be infinite and infinite power would be dissipated in heat. At the interface between an ordinary medium (region 1) and a perfect conductor (region 2), boundary conditions of particular simplicity prevail. Continuity of the tangential electric field and the normal magnetic induction requires that

$$\mathbf{n}_1 \times \mathbf{E}_1 = 0\,, \quad \mathbf{n}_1 \cdot \mathbf{H}_1 = 0 \quad \text{on} \quad S_{12}\,. \tag{4.53}$$

With regard to the tangential magnetic field, it must be realized that a finite surface current will flow in the infinitely thin region in which the conductivity changes from the infinite value it possesses in the perfect conductor to the finite value ascribed to region 2. The conduction current density is therefore infinite in the transition layer and the continuity proof for the tangential magnetic field must be modified. Consider a small effectively plane area on the interface and let $\mathbf{t}$ be an arbitrary constant vector tangential to the surface. On integrating the equation

$$\mathbf{t} \cdot \mathbf{\nabla} \times \mathbf{H} = \mathbf{\nabla} \cdot (\mathbf{H} \times \mathbf{t}) = -i\omega\varepsilon\mathbf{E} \cdot \mathbf{t} + \mathbf{J} \cdot \mathbf{t} \tag{4.54}$$

over the infinitesimal volume bounded by the two sides of the small surface, we obtain

$$-\mathbf{n}_1 \cdot \mathbf{H}_1 \times \mathbf{t} = \mathbf{K} \cdot \mathbf{t}\,, \tag{4.55}$$

or, since $\mathbf{t}$ is arbitrary,

$$\mathbf{n}_1 \times \mathbf{H}_1 = -\mathbf{K}\,, \tag{4.56}$$

where $\mathbf{K}$ is the surface current density tangential to S_{12}. The volume integral of the tangential electric displacement has been discarded, for the latter is finite within the transition layer. The normal component of $\mathbf{D}$ in region 1 is related to the surface charge density by

$$\mathbf{n}_1 \cdot \mathbf{D}_1 = -\tau\,. \tag{4.57}$$

The relation between the normal component of the conduction current and the charge density must be altered by the presence of surface currents. The appropriate result can be obtained by considering

$$\mathbf{\nabla} \cdot \mathbf{K} = \mathbf{n}_1 \cdot \mathbf{\nabla} \times \mathbf{H}_1 = -i\omega\mathbf{n}_1 \cdot \mathbf{D}_1 + \mathbf{n}_1 \cdot \mathbf{J}_1\,, \tag{4.58}$$

in which $\mathbf{\nabla} \cdot \mathbf{K}$ involves only derivatives tangential to the interface. Hence

$$\mathbf{\nabla} \cdot \mathbf{K} - i\omega\tau = \mathbf{n}_1 \cdot \mathbf{J}_1\,. \tag{4.59}$$

If medium 1 is a nonconductor, $\mathbf{n}_1 \cdot \mathbf{J}_1 = 0$, and this result is a conservation theorem for the surface charge. When the medium adjacent to the perfect conductor has a finite conductivity, surface charge is lost at the rate

$$-\mathbf{n}_1 \cdot \mathbf{J}_1 = -\sigma_1 \mathbf{n}_1 \cdot \mathbf{E}_1 = \frac{\sigma_1}{\varepsilon_1}\tau \,, \tag{4.60}$$

per unit area, where again we see the appearance of the charge relaxation time ε/σ.

A useful concept is that of a perfect conductor in the form of an infinitely thin sheet. If a surface of this character, which we shall call an electric wall, is inserted in a pre-existing field in such a fashion that at every point on the surface the normal is in the direction of the established electric field, the field structure will in no way be affected. Such an operation is advantageous when, by symmetry considerations, the proper position for an electric wall is immediately apparent, for the space occupied by the field can then be divided in two parts with a consequent simplification in treatment.

An impressed current sheet is another concept of some utility. It is understood as an infinitely thin surface bearing a finite surface current. The boundary conditions relating fields on opposite sides of the surface (regions 1 and 2) can be obtained from the variational principle. In our previous considerations of the transition layer between two media, the current density was assumed finite and therefore made a negligible contribution to the variation of the Lagrangian arising from field variations in the transition layer. If, however, a surface current density $\mathbf{K}$ flows on the interface, the variation of the Lagrangian must be supplemented by the term

$$\int_{S_{12}} \mathrm{d}S \, \mathbf{K} \cdot \delta \mathbf{E} \,, \tag{4.61}$$

and therefore the stationary requirement on the Lagrangian yields, instead of (4.42b)

$$\mathbf{n}_1 \times (\mathbf{E}_1 - \mathbf{E}_2) = \mathbf{0} \,, \tag{4.62a}$$
$$\mathbf{n}_1 \times (\mathbf{H}_1 - \mathbf{H}_2) = -\mathbf{K} \,, \tag{4.62b}$$

that is, on crossing the current sheet the tangential magnetic field is discontinuous by an amount equal to the surface current density at that point; the tangential electric field is continuous. Of course, these considerations are equally applicable to the conduction current sheet flowing on the surface of a perfect conductor, in which case

$$\mathbf{n}_2 \times \mathbf{E}_2 = \mathbf{n}_2 \times \mathbf{H}_2 = \mathbf{0} \,, \tag{4.63}$$

and so

$$\mathbf{n}_1 \times \mathbf{E}_1 = \mathbf{0} \,, \tag{4.64a}$$
$$\mathbf{n}_1 \times \mathbf{H}_1 = -\mathbf{K} \,, \tag{4.64b}$$

coinciding with (4.53) and (4.56).

4.3 Babinet's Principle

The Maxwell equations for a current-free region are essentially symmetrical in the electric and magnetic field intensities.[2] More precisely, if $\mathbf{E}$ and $\mathbf{H}$ form a solution of the equations in a homogeneous medium characterized by the constants ε and μ, then the fields

$$\mathbf{E}' = \sqrt{\frac{\mu}{\varepsilon}}\mathbf{H} , \qquad \mathbf{H}' = -\sqrt{\frac{\varepsilon}{\mu}}\mathbf{E} \tag{4.65}$$

are also a solution of the equations. This symmetry aspect of the field equations we shall call Babinet's principle. It is useful to extend Babinet's principle to the general situation by introducing hypothetical magnetic currents $\mathbf{J}_m$ as the analog of electric currents $\mathbf{J}_e$. The generalized Maxwell equations are

$$\nabla \times \mathbf{H} = -i\omega\varepsilon\mathbf{E} + \mathbf{J}_e , \tag{4.66a}$$

$$\nabla \times \mathbf{E} = i\omega\mu\mathbf{H} - \mathbf{J}_m . \tag{4.66b}$$

The sign of $\mathbf{J}_m$ is so chosen that the magnetic charge density, related to the magnetic current density by the conservation equation,

$$\nabla \cdot \mathbf{J}_m = i\omega\rho_m , \tag{4.67}$$

obeys

$$\nabla \cdot \mathbf{B} = \rho_m , \tag{4.68a}$$

the analog of

$$\nabla \cdot \mathbf{D} = \rho_e . \tag{4.68b}$$

The extended field equations are invariant under the substitutions (ε and μ are constant)

$$\mathbf{E}' = \sqrt{\frac{\mu}{\varepsilon}}\mathbf{H} , \quad \mathbf{J}'_e = \sqrt{\frac{\varepsilon}{\mu}}\mathbf{J}_m , \quad \rho'_e = \sqrt{\frac{\varepsilon}{\mu}}\rho_m , \tag{4.69a}$$

$$\mathbf{H}' = -\sqrt{\frac{\varepsilon}{\mu}}\mathbf{E} , \quad \mathbf{J}'_m = -\sqrt{\frac{\mu}{\varepsilon}}\mathbf{J}_e , \quad \rho'_m = -\sqrt{\frac{\mu}{\varepsilon}}\rho_e . \tag{4.69b}$$

If the material constants ε and μ are real, the field equations (4.66a), (4.66b), (4.68a), and (4.68b) are also invariant under the substitutions

$$\mathbf{E}' = \sqrt{\frac{\mu}{\varepsilon}}\mathbf{H}^* , \quad \mathbf{J}'_e = -\sqrt{\frac{\varepsilon}{\mu}}\mathbf{J}_m^* , \quad \rho'_e = \sqrt{\frac{\varepsilon}{\mu}}\rho_m^* , \tag{4.70a}$$

$$\mathbf{H}' = \sqrt{\frac{\varepsilon}{\mu}}\mathbf{E}^* , \quad \mathbf{J}'_m = -\sqrt{\frac{\mu}{\varepsilon}}\mathbf{J}_e^* , \quad \rho'_m = \sqrt{\frac{\mu}{\varepsilon}}\rho_e^* , \tag{4.70b}$$

---

[2] This is what we referred to as electromagnetic duality in the Problems in Chap. 1 [see (1.219)].

which can be considered the resultant of the first transformation and

$$\mathbf{E}' = -\mathbf{E}^* , \quad \mathbf{J}'_e = \mathbf{J}^*_e , \quad \rho'_e = -\rho^*_e , \tag{4.71a}$$

$$\mathbf{H}' = \mathbf{H}^* , \quad \mathbf{J}'_m = -\mathbf{J}^*_m , \quad \rho'_m = \rho^*_m \tag{4.71b}$$

applied in succession. The latter transformation, which leaves the field equations unchanged in form if ε and μ are real, can be considered a reversal in the sense of time. It is to be expected that only in nondissipative media is a possible state of the system thus obtained. That the transformation, (4.71a) and (4.71b), is a time reversal operation can be seen on noting that the substitution reverses the sense of energy flow (4.11),

$$\overline{\mathbf{S}}' = \operatorname{Re} \frac{1}{2}\mathbf{E}' \times \mathbf{H}'^* = \operatorname{Re} \frac{1}{2}(-\mathbf{E}^*) \times \mathbf{H} = -\overline{\mathbf{S}} , \tag{4.72}$$

while the average energy flux is unaltered under the Babinet substitution, (4.69a) and (4.69b),

$$\overline{\mathbf{S}}' = \operatorname{Re}\mathbf{H} \times (-\mathbf{E}^*) = \overline{\mathbf{S}} . \tag{4.73}$$

The general field equations can be derived from a variational principle by replacing $\mathbf{J} \cdot \mathbf{E}$ in (4.13) by

$$\mathbf{J}_e \cdot \mathbf{E} - \mathbf{J}_m \cdot \mathbf{H} , \tag{4.74}$$

or, for the variational principle appropriate to nondissipative media (4.29), on replacing $\mathbf{J} \cdot \mathbf{E}^* - \mathbf{J}^* \cdot \mathbf{E}$ with

$$(\mathbf{J}_e \cdot \mathbf{E}^* - \mathbf{J}^*_e \cdot \mathbf{E}) + (\mathbf{J}_m \cdot \mathbf{H}^* - \mathbf{J}^*_m \cdot \mathbf{H}) . \tag{4.75}$$

It will be noted that the extended Lagrangians are invariant [save for an alteration in sign in (4.13) modified by (4.74)] under the Babinet transformations (4.69a) and (4.69b). The Lagrangian (4.29) including (4.75) is also invariant under a time-reversal substitution (4.71a) and (4.71b).

The principal application of magnetic currents is in the form of current sheets. The conditions relating the fields on opposite sides of a surface bearing a magnetic surface current density $\mathbf{K}_m$ are

$$\mathbf{n}_1 \times (\mathbf{E}_1 - \mathbf{E}_2) = \mathbf{K}_m , \tag{4.76a}$$

$$\mathbf{n}_1 \times (\mathbf{H}_1 - \mathbf{H}_2) = \mathbf{0} . \tag{4.76b}$$

More generally, if a surface supports electric and magnetic surface currents, the discontinuity equations are

$$\mathbf{n}_1 \times (\mathbf{E}_1 - \mathbf{E}_2) = \mathbf{K}_m , \tag{4.77a}$$

$$\mathbf{n}_1 \times (\mathbf{H}_1 - \mathbf{H}_2) = -\mathbf{K}_e . \tag{4.77b}$$

A magnetic wall, as the analog of an electric wall, is an infinitely thin sheet to which the magnetic field must be normal and the electric field tangential,

$$\mathbf{n} \times \mathbf{H} = 0 , \qquad \mathbf{n} \cdot \mathbf{E} = 0 . \tag{4.78}$$

The structure of space containing electromagnetic fields can also be simplified by the introduction of magnetic walls if the proper position to insert the barrier without disturbing the field can be ascertained by considerations of symmetry. It should be noted that if Babinet's principle is applied to the fields in a region bounded by electric and magnetic walls, the interchange of the electric and magnetic fields also applies to the character of the bounding surfaces. It is sometimes possible to regard the magnetic walls obtained from electric barriers by the Babinet principle as the expression of symmetry conditions in a larger space, and thus obtain a solution to a new (approximately) physically realizable problems. We will see some applications of Babinet's principle in practical problems in Chap. 13, as well as in Chap. 16.

4.4 Reciprocity Theorems

A number of fundamental theorems can be obtained from the variational principle by considering special types of variations. As a first example, let $\mathbf{E}_a$ and $\mathbf{H}_a$ be the fields in a region occupied by impressed electric and magnetic current densities $\mathbf{J}_e^a$ and $\mathbf{J}_m^a$. Consider a small variation

$$\delta \mathbf{E} = \lambda \mathbf{E}_b , \qquad \delta \mathbf{H} = \lambda \mathbf{H}_b , \tag{4.79}$$

where $\mathbf{E}_b$ and $\mathbf{H}_b$ are the proper fields corresponding to another distribution of currents $\mathbf{J}_e^b$, $\mathbf{J}_m^b$, and λ is a small, arbitrary parameter. The variation of the Lagrangian (4.13), including the magnetic current term (4.74), is

$$\delta L = \lambda \int_V (\mathrm{d}\mathbf{r})(i\omega\mu\mathbf{H}_a \cdot \mathbf{H}_b - i\omega\varepsilon\mathbf{E}_a \cdot \mathbf{E}_b - \mathbf{E}_a \cdot \boldsymbol{\nabla} \times \mathbf{H}_b - \mathbf{E}_b \cdot \boldsymbol{\nabla} \times \mathbf{H}_a$$
$$+ \sigma\mathbf{E}_a \cdot \mathbf{E}_b + \mathbf{J}_e^a \cdot \mathbf{E}_b - \mathbf{J}_m^a \cdot \mathbf{H}_b) . \tag{4.80}$$

On the other hand, this must equal, from (4.15),

$$\lambda \oint_S \mathrm{d}S\, \mathbf{n} \cdot \mathbf{E}_a \times \mathbf{H}_b . \tag{4.81}$$

Interchanging a and b and subtracting the resulting equation, we obtain

$$\int_V (\mathrm{d}\mathbf{r}) \left[(\mathbf{J}_e^a \cdot \mathbf{E}_b - \mathbf{J}_e^b \cdot \mathbf{E}_a) - (\mathbf{J}_m^a \cdot \mathbf{H}_b - \mathbf{J}_m^b \cdot \mathbf{H}_a) \right]$$
$$= \oint_S \mathrm{d}S\, \mathbf{n} \cdot (\mathbf{E}_a \times \mathbf{H}_b - \mathbf{E}_b \times \mathbf{H}_a) , \tag{4.82}$$

or, in differential form,

$$\boldsymbol{\nabla} \cdot (\mathbf{E}_a \times \mathbf{H}_b - \mathbf{E}_b \times \mathbf{H}_a) = (\mathbf{J}_e^a \cdot \mathbf{E}_b - \mathbf{J}_e^b \cdot \mathbf{E}_a) - (\mathbf{J}_m^a \cdot \mathbf{H}_b - \mathbf{J}_m^b \cdot \mathbf{H}_a) . \tag{4.83}$$

This result we shall call the reciprocity theorem. A related theorem can be derived with the fields obtained from $\mathbf{E}_b$ and $\mathbf{H}_b$ by the operation of time reversal (4.71a) and (4.71b), for σ, ε, μ real,

$$\delta\mathbf{E} = -\lambda\mathbf{E}_b^* , \qquad \delta\mathbf{H} = \lambda\mathbf{H}_b^* . \qquad (4.84)$$

This variation implies that

$$
\begin{aligned}
&\int_V (\mathrm{d}\mathbf{r}) \Big(i\omega\mu\mathbf{H}_a \cdot \mathbf{H}_b^* + i\omega\varepsilon\mathbf{E}_a \cdot \mathbf{E}_b^* + \mathbf{E}_b^* \cdot \nabla \times \mathbf{H}_a - \mathbf{E}_a \cdot \nabla \times \mathbf{H}_b^* \\
&\qquad - \sigma\mathbf{E}_a \cdot \mathbf{E}_b^* - \mathbf{J}_e^a \cdot \mathbf{E}_b^* - \mathbf{J}_m^a \cdot \mathbf{H}_b^* \Big) \\
&= \int_V (\mathrm{d}\mathbf{r}) \Big(i\omega\mu\mathbf{H}_a \cdot \mathbf{H}_b^* - i\omega\varepsilon\mathbf{E}_a \cdot \mathbf{E}_b^* - \sigma\mathbf{E}_a \cdot \mathbf{E}_b^* - \mathbf{J}_e^{b*} \cdot \mathbf{E}_a - \mathbf{J}_m^a \cdot \mathbf{H}_b^* \Big) \\
&= \oint_S \mathrm{d}S\, \mathbf{n} \cdot \mathbf{E}_a \times \mathbf{H}_b^* .
\end{aligned}
\qquad (4.85)
$$

Here, in the second line we have used the Maxwell equation (4.66a). Adding a similar equation with a and b interchanged and i replaced by $-i$, we get, in differential form

$$
\begin{aligned}
\nabla \cdot (\mathbf{E}_a \times \mathbf{H}_b^* + \mathbf{E}_b^* \times \mathbf{H}_a) &= -2\sigma\mathbf{E}_a \cdot \mathbf{E}_b^* - (\mathbf{J}_e^a \cdot \mathbf{E}_b^* + \mathbf{J}_e^{b*} \cdot \mathbf{E}_a) \\
&\quad - (\mathbf{J}_m^a \cdot \mathbf{H}_b^* + \mathbf{J}_m^{b*} \cdot \mathbf{H}_a) .
\end{aligned}
\qquad (4.86)
$$

In a nondissipative medium ($\sigma = 0$), the two reciprocity theorems (4.83) and (4.86) are essentially identical – see (4.71a) and (4.71b). Further aspects of reciprocity will be explored later – in particular, in Chap. 13, reciprocity between in and out currents and fields will be used to derive the symmetry of the S-matrix.

4.5 Problems for Chap. 4

In these problems, Heaviside–Lorentz units are used.

1. Charge density $\rho(\mathbf{r},t)$ is placed in a medium with conductivity σ and permeability ε. Show that the charge migrates to the surface of the region at a characteristic rate $\sigma/\varepsilon = \gamma$, where we may call $1/\gamma$ the charge relaxation time.

2. Derive the formula for the time average energy density in a dispersive medium by direct calculation, employing as a simple model a gas with N atoms per unit volume, each atom containing an electron of charge e and mass m oscillating harmonically with angular frequency ω_0. The permittivity for such a medium is given by the plasma formula,

$$\varepsilon(\omega) = 1 + \frac{Ne^2}{m} \frac{1}{\omega_0^2 - \omega^2} . \qquad (4.87)$$

3. The Lorenz gauge potentials for a particle of charge e moving uniformly in a dispersionless dielectric medium are determined by

$$\mathbf{A} = \varepsilon \frac{\mathbf{v}}{c} \phi , \quad -\left(\nabla^2 - \frac{\varepsilon}{c^2} \frac{\partial^2}{\partial t^2} \right) \phi = \frac{e}{\varepsilon} \delta(\mathbf{r} - \mathbf{v}t) . \tag{4.88}$$

Verify this. (Warning: The Lorenz gauge condition differs somewhat from its vacuum form.) Let $\mathbf{v}$ point along the z-axis. Write the frequency transform of the differential equation, solve it, and perform the inversion that gives $\phi(\mathbf{r}, t)$. Consider the circumstances of Čerenkov radiation and show that ϕ, $\mathbf{A}$ and therefore the fields are zero if

$$vt - z < \rho\sqrt{\beta^2 \varepsilon - 1} , \quad \rho = |\mathbf{r}_\perp| . \tag{4.89}$$

Recognize in this the conical surface of the trailing "shock wave," and demonstrate that the radiation moving perpendicularly to that surface has the expected direction.

4. A charged particle moves at speed $v = \beta c$ along the axis of a dielectric cylinder. Suppose that Čerenkov radiation of some frequency is emitted. What fraction of this radiation passes into the surrounding vacuum through the cylindrical surface of radius $R \gg \lambda$?

5. Start from the macroscopic Maxwell equations, Fourier transformed in time,

$$\nabla \times \mathbf{H}(\mathbf{r}, \omega) = -\frac{i\omega}{c} \mathbf{D}(\mathbf{r}, \omega) + \frac{1}{c} \mathbf{J}(\mathbf{r}, \omega) , \quad \nabla \cdot \mathbf{J}(\mathbf{r}, \omega) - i\omega \rho(\mathbf{r}, \omega) = 0 ,$$
$$\tag{4.90a}$$

$$-\nabla \times \mathbf{E}(\mathbf{r}, \omega) = -\frac{i\omega}{c} \mathbf{B}(\mathbf{r}, \omega) , \quad \nabla \cdot \mathbf{B}(\mathbf{r}, \omega) = 0 . \tag{4.90b}$$

Assume that $\mu = 1$, so $\mathbf{H} = \mathbf{B}$, and that $\varepsilon(\omega)$ is independent of $\mathbf{r}$. Show that $\mathbf{E}(\mathbf{r}, \omega)$ obeys

$$-\left(\nabla^2 + \frac{\omega^2 \varepsilon}{c^2} \right) \mathbf{E} = \frac{i\omega}{c^2} \left(\mathbf{J} + \frac{c^2}{\omega^2 \varepsilon} \nabla \nabla \cdot \mathbf{J} \right) . \tag{4.91}$$

Write the solution of this equation that represents the retarded time boundary condition. How is the latter related to the outgoing wave boundary condition?

6. The total energy transferred from the current to the electromagnetic field is (why?)

$$\mathcal{E} = -\int_{-\infty}^{\infty} dt \int (d\mathbf{r}) \, \mathbf{E}(\mathbf{r}, t) \cdot \mathbf{J}(\mathbf{r}, t) ; \tag{4.92}$$

write this in terms of time Fourier transforms. Then substitute the solution for $\mathbf{E}$ found in the previous problem and arrive at an expression for $\mathcal{E}$ as a frequency integral, and a double spectral integral.

7. Consider the current of a point electric charge e, moving at a uniform velocity $\mathbf{v}$. (Of course, since the current transfers energy to the electromagnetic field, the charge gradually slows down, but that need not be made explicit here.) What is $\mathbf{J}(\mathbf{r}, \omega)$? (Assume that the point charge is moving along the z-axis.) Arrive at an expression for $\mathcal{E}$ that is a frequency integral and a double integral over z and z'. By introducing the average of z and z', and the difference $\zeta = z - z'$, identify the rate $\mathcal{R}$ at which the charge loses energy, per unit distance, as

$$\mathcal{R} = -\int_{-\infty}^{\infty} \frac{d\omega}{2\pi} i\omega \frac{2e^2}{c^2} \left(1 - \frac{1}{\beta^2 \varepsilon(\omega)}\right) \int_0^\infty d\zeta \frac{\cos(\omega\zeta/v)\, e^{i\omega\sqrt{\varepsilon(\omega)}\zeta/c}}{\zeta},$$

$$(4.93)$$

which breaks up into the following imaginary and real parts, $\mathcal{R} = \mathcal{R}_1 + \mathcal{R}_2$,

$$\mathcal{R}_1 = -\int_{-\infty}^{\infty} \frac{d\omega}{2\pi} i\omega \frac{2e^2}{c^2} \left(1 - \frac{1}{\beta^2 \varepsilon}\right) \int_0^\infty d\zeta \frac{\cos(\omega\zeta/v)\cos(\omega\sqrt{\varepsilon}\zeta/c)}{\zeta},$$

$$(4.94a)$$

$$\mathcal{R}_2 = \int_{-\infty}^{\infty} \frac{d\omega}{2\pi} \omega \frac{2e^2}{c^2} \left(1 - \frac{1}{\beta^2 \varepsilon}\right) \int_0^\infty d\zeta \frac{\cos(\omega\zeta/v)\sin(\omega\sqrt{\varepsilon}\zeta/c)}{\zeta}.$$

$$(4.94b)$$

The divergence of the ζ integral in $\mathcal{R}_1$ is a failure of the classical theory; quantum mechanical uncertainty makes it finite. Notice that if $\varepsilon(\omega)$ were an even, real function of ω, $\mathcal{R}_1$ would be zero. It is not zero because ε has an odd, imaginary part, produced by dissipation. Thus $\mathcal{R}_1$ represents energy transferred to the medium. For $\mathcal{R}_2$, it is sufficient to let $\varepsilon(\omega)$ be an even real function. Show that this gives the Čerenkov effect.

8. Repeat the above analysis for a uniformly moving magnetic charge g.

5

Transmission Lines

5.1 Dissipationless Line

In this brief chapter we shall consider some simple elements of low-frequency transmission lines, and a prolegomenon to the considerations, in the following chapter, of waveguides. Following the preceding chapter, we will here, and in most of the following chapters, assume harmonic time dependence through the implicit factor $e^{-i\omega t}$.

A transmission line is a two-conductor system, with translational symmetry along an axis, the z-axis, as shown in Fig. 5.1. We assume that the

$I(z)$ $-I(z)$

Fig. 5.1. Cross section of two-conductor transmission line. The z-direction is out of the page; $I(z)$ flows out of the page, $-I(z)$ flows into the page

surrounding medium has constant electrical properties, characterized by a permittivity ε and a permeability μ. To begin, we will suppose the conductors have infinite conductivity. This means that the tangential electric field must vanish at the surface of the conductors. Here we will treat only the lowest electromagnetic mode of the system, the so-called T (or TEM) mode. The higher TE and TM modes will be the subject of the next chapter. This means that the electric and magnetic fields lie entirely in the x–y plane; there are no longitudinal fields. It is convenient to decompose the fields into their transverse and longitudinal dependence, which we do by writing

$$\mathbf{E} = -\boldsymbol{\nabla}_\perp \varphi(x,y) V(z) \,, \tag{5.1a}$$

$$\mathbf{H} = -\mathbf{e} \times \boldsymbol{\nabla}_\perp \varphi(x,y) I(z) \,, \tag{5.1b}$$

where we have introduced "current" and "voltage" functions $I(z)$ and $V(z)$. Here, $\mathbf{e}$ is the unit vector in the direction of the line, that is, the z-axis. Of course, this decomposition leaves the normalization of the latter functions undetermined. We will find it convenient to normalize the transverse functions by

$$\int d\sigma (\boldsymbol{\nabla}_\perp \varphi)^2 = 1 \,, \tag{5.2}$$

where the integration extends over the entire plane bounded by the two conductors.

The Maxwell equations determine these transmission line functions. Gauss' law $\boldsymbol{\nabla} \cdot \mathbf{E} = 0$ tells us immediately that φ is a harmonic function:

$$\nabla_\perp^2 \varphi = 0 \,. \tag{5.3}$$

Because the conductors are assumed perfect, the potential φ must be constant on either surface, but of course will assume different values on the two conductors. The two curl equations, (4.7a) and (4.7b), outside the conductors, supply the equations that determine the voltage and current functions,

$$\frac{d}{dz} V(z) = i\omega\mu I(z) \,, \tag{5.4a}$$

$$\frac{d}{dz} I(z) = i\omega\varepsilon V(z) \,. \tag{5.4b}$$

From the form of these equations, we see that our system has an immediate interpretation in terms of circuit elements. That is, there is a shunt capacitance $C_\perp$ per unit length of ε, and a series inductance L_s per unit length of μ. The general dissipationless circuit diagram describing a general line is described as follows. An ideal two-conductor transmission line can be thought of as a series of elements, each of which consists of a series inductance L_s and capacitance C_s, and a shunt inductance $L_\perp$ and capacitance $C_\perp$, as illustrated in Fig. 5:2. Let the length of each element be Δz. Then the voltage drop across the element is, for a given frequency ω,

$$\Delta V = i\omega L_s \Delta z I + \frac{\Delta z}{i\omega C_s} I \,, \tag{5.5a}$$

from which we infer a series impedance per unit length

$$-Z_s = i\omega L_s + \frac{1}{i\omega C_s} \,. \tag{5.5b}$$

Similarly, because the current shorted between the two conductors is

$$\Delta I = i\omega C_\perp \Delta z V + \frac{\Delta z}{i\omega L_\perp} V \,, \tag{5.6a}$$

the shunt admittance per unit length is

$$-Y_\perp = i\omega C_\perp + \frac{1}{i\omega L_\perp} . \tag{5.6b}$$

Here, only two of the parameters are finite:

$$C_\perp = \varepsilon , \qquad L_s = \mu . \tag{5.7}$$

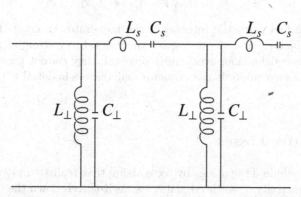

Fig. 5.2. Transmission line represented in terms of equivalent series and shunt inductances and capacitances. Represented here are two elements, each of length Δz, which are repeated indefinitely

Consider now a propagating wave,

$$I(z) = I e^{i\kappa z} , \qquad V(z) = V e^{i\kappa z} . \tag{5.8}$$

The dispersion relation following then from (5.4a) and (5.4b) is

$$\kappa^2 = \omega^2 \mu \varepsilon = \frac{\omega^2}{c^2} = k^2 , \tag{5.9}$$

which exhibits no cutoff; that is, waves of arbitrarily low frequency may be transmitted. Moreover, from the ratio of V to I we infer the characteristic impedance of the line,

$$\frac{V}{I} = \frac{\kappa}{\omega \varepsilon} = \sqrt{\frac{\mu}{\varepsilon}} = \zeta , \tag{5.10}$$

which is simply the characteristic impedance of the medium.

Evidently, there is considerable ambiguity in defining the impedance of the line. It might appear more natural to define Z as the ratio of the voltage between the conductors, $\Delta\varphi\, V$, to the current flowing in one of the conductors, as defined by Ampére's law, $\oint_C ds \cdot \mathbf{H}$. That is,

$$Z' = \frac{\Delta\varphi\, V(z)}{\oint_C ds\, (\partial_n \varphi) I(z)} , \tag{5.11}$$

where the circuit encloses just one of the conductors, and the derivative is with the respect to the normal to the surface of that conductor. We might also define the inductance as the ratio of the magnetic flux crossing a plane defined by a line connecting the two conductors and extended a unit length in the z-direction, to the current, which gives for the inductance per unit length

$$L' = \frac{\mu \int dl\, \partial_l \varphi}{\oint_C ds\,(\partial_n \varphi)} = \frac{\mu \Delta \varphi}{\oint_C ds\,(\partial_n \varphi)} = \frac{Z'}{c}, \tag{5.12}$$

where in the first form the integral in the numerator is along the line joining the conductors, and the derivative there is taken tangential to that line. Although these definitions seem quite physical, they cannot generally be extended to the waveguide situation, as we shall discuss in detail in the following chapter.

5.2 Resistive Losses

Now let us include dissipation, by recognizing that realistic materials possess a finite conductivity σ, so the signal is lost as it travels down the transmission line. We can calculate the resistance per unit length by examining the flux of energy into the conductors,

$$P_{\text{diss}} = \frac{1}{2}\text{Re} \oint_C ds(\mathbf{E} \times \mathbf{H}^*) \cdot \mathbf{n}, \tag{5.13}$$

where the integral extends over the boundary of the conductors, $\mathbf{n}$ being the normal to that boundary. We will treat this problem perturbatively, by the usual relation between the tangential electric field and the magnetic field at the surface (e.g., see (42.19) of [9]),

$$\mathbf{n} \times \mathbf{E} = \zeta \frac{k\delta}{2}(1 - i)\mathbf{H}, \tag{5.14}$$

where the skin depth is $\delta = (2/\mu\omega\sigma)^{1/2}$. This says immediately that

$$P_{\text{diss}} = \frac{1}{2}\mathcal{R}|I|^2, \tag{5.15}$$

where the resistance per unit length is

$$\mathcal{R} = \frac{\zeta k\delta}{2} \oint_C ds\,(\partial_n \varphi)^2. \tag{5.16}$$

Because of this, the series impedance of the line becomes

$$Z_s = \mathcal{R} - i\omega L_s, \tag{5.17}$$

and the shunt admittance is unchanged,

$$Y_\perp = -\mathrm{i}\omega C_\perp \ . \tag{5.18}$$

The modified transmission line equations,

$$\frac{\mathrm{d}V}{\mathrm{d}z} = -Z_s I \ , \qquad \frac{\mathrm{d}I}{\mathrm{d}z} = -Y_\perp V \ , \tag{5.19}$$

then imply that the propagation constant is given by

$$\kappa^2 = -Z_s Y_\perp \ . \tag{5.20}$$

Regarding the resistance losses per unit length as small, we find

$$\kappa = \omega\sqrt{L_s C_\perp} + \mathrm{i}\frac{\mathcal{R}}{2}\sqrt{\frac{C_\perp}{L_s}} \ , \tag{5.21}$$

or inserting the values $C_\perp = \varepsilon$ and $L_s = \mu$,

$$\kappa = k + \mathrm{i}\frac{\alpha}{2} \ , \qquad \alpha = \frac{\mathcal{R}}{\zeta} \ . \tag{5.22}$$

The power transmitted down the line, then, is proportional to $\mathrm{e}^{-\alpha z}$, so α is called the attenuation constant.

5.3 Example: Coaxial Line

The simplest example of a two-conductor transmission line is a coaxial cable, with inner radius a and outer radius b. In that case, the normalized potential function is

$$\varphi = \frac{\ln r}{\sqrt{2\pi \ln b/a}} \ . \tag{5.23}$$

The corresponding electric and magnetic fields are

$$\mathbf{E} = -\frac{1}{\sqrt{2\pi \ln b/a}}\frac{\hat{\mathbf{r}}}{r}V(z) \ , \tag{5.24a}$$

$$\mathbf{H} = -\frac{1}{\sqrt{2\pi \ln b/a}}\frac{\hat{\boldsymbol{\phi}}}{r}I(z) \ . \tag{5.24b}$$

The alternative impedance Z' (5.11) is computed to be

$$Z' = \frac{\zeta}{2\pi}\ln\frac{b}{a} \ , \tag{5.25}$$

which differs from $Z = \zeta$ by a factor

$$N = \frac{1}{2\pi}\ln\frac{b}{a}, \tag{5.26}$$

while the alternative inductance per unit length (5.12) is

$$L'_s = \frac{\mu}{2\pi} \ln \frac{b}{a} \,, \tag{5.27}$$

again differs from $L_s = \mu$ by the same factor N. The attenuation constant, or the resistance per unit length, when a finite conductivity is included, is

$$\alpha = \frac{\mathcal{R}}{\zeta} = \frac{k\delta}{2} \frac{1}{\ln \frac{b}{a}} \left(\frac{1}{a} + \frac{1}{b} \right) \,, \tag{5.28}$$

where $k\delta = 2/\zeta\sigma\delta$, where we have assumed that the two conductors have the same skin depth.

5.4 Cutoff Frequencies

We have treated only the lowest T mode of the line in this chapter. In the following chapter we will consider the higher modes; for a hollow waveguide, in contrast to a coaxial cable, only the latter exist, and there is no T mode. These higher modes cannot propagate if they have wavelengths longer than a scale set by the dimensions of the guide, so there is a cutoff wavelength, or a cutoff wavenumber below which no signal can be propagated. We can see how this comes about by considering the general line equations (5.19) with the general series impedance (5.5b) and shunt admittance (5.6b). The corresponding propagation constant is (5.20), or

$$\kappa^2 = \omega^2 L_s C_\perp - \left(\frac{C_\perp}{C_s} + \frac{L_s}{L_\perp} \right) + \frac{1}{\omega^2 C_s L_\perp} \,. \tag{5.29}$$

For a propagating mode, this must be positive. This will be true only if ω^2 exceeds the larger of $1/L_s C_s$ and $1/L_\perp C_\perp$, or is smaller than either of these quantities. It will turn out that there are precisely two types of modes, E-modes, so called because they possess a longitudinal electric field in addition to transverse fields, and H-modes, due to the presence of a longitudinal component of the magnetic field. The corresponding circuit parameters are

$$\text{E mode:} \quad C_\perp = \varepsilon \,, \quad L_s = \mu \,, \quad C_s = \frac{\varepsilon}{\gamma^2} \,, \quad L_\perp = \infty \,, \tag{5.30a}$$

$$\text{H mode:} \quad C_\perp = \varepsilon \,, \quad L_s = \mu \,, \quad C_s = \infty \,, \quad L_\perp = \frac{\mu}{\gamma^2} \,, \tag{5.30b}$$

where the quantity γ^2 is a characteristic eigenvalue of the transverse Laplacian operator, corresponding to some typical inverse length squared. Therefore, in either case, to have propagation, ω must exceed γc, or the intrinsic wavelength of the radiation must be smaller than a cutoff wavelength

$$\lambda_c = \frac{2\pi}{\gamma} \,. \tag{5.31}$$

5.5 Problems for Chap. 5

1. Consider a transmission line composed of two parallel rectangular conductors, separated by a distance a, and with dimension $b \gg a$ in the transverse direction perpendicular to the separation. Calculate the potential function φ, and compute the alternative impedance, and with dissipation included, the attenuation constant.

2. Calculate the resistance per unit length for a single wire of radius ρ and conductivity σ and compare with the result found for a coaxial cable, (5.28).

6

Waveguides and Equivalent Transmission Lines

A waveguide is a device for transferring electromagnetic energy from one point to another without appreciable loss. In its simplest form it consists of a hollow metallic tube of rectangular or circular cross section, within which electromagnetic waves can propagate. The two-conductor transmission line discussed in Chap. 5 is a particular type of waveguide, with special properties. The simple physical concept implied by these examples may be extended to include any region within which one-dimensional propagation of electromagnetic waves can occur. It is the purpose of this chapter to establish the theory of simple waveguides, expressed in the general transmission line nomenclature sketched in Chap. 5.

This chapter will be devoted to the theory of uniform waveguides – cylindrical metallic tubes which have the same cross section in any plane perpendicular to the axis of the guide. Initially, the simplifying assumption will be made that the metallic walls of the waveguide are perfectly conducting. (We will consider the effects of finite conductivity in Sect. 13.6.) Since the field is then entirely confined to the interior of the waveguide, the guide is completely described by specifying the curve C which defines a cross section σ of the inner waveguide surface S. The curve C may be a simple closed curve, corresponding to a hollow waveguide, or two unconnected curves, as in a coaxial line. Particular simple examples will be the subject of Chaps. 7 and 8.

6.1 Transmission Line Formulation

We first consider the problem of finding the possible fields that can exist within a waveguide, in the absence of any impressed currents. This is equivalent to seeking the solutions of the Maxwell equations (4.7a) and (4.7b)

$$\nabla \times \mathbf{E} = ik\zeta\mathbf{H} , \tag{6.1a}$$

$$\nabla \times \mathbf{H} = -ik\eta\mathbf{E} , \tag{6.1b}$$

where we have defined (SI units)

$$k = \omega\sqrt{\varepsilon\mu} = \frac{\omega}{c} ,\tag{6.2a}$$

c being the speed of light in the medium inside the guide, and introduced the abbreviations for the intrinsic impedance or admittance of the medium,

$$\zeta = \sqrt{\frac{\mu}{\varepsilon}} , \qquad \eta = \sqrt{\frac{\varepsilon}{\mu}} = \zeta^{-1} .\tag{6.2b}$$

These equations are to be solved subject to the boundary condition (4.53)

$$\mathbf{n} \times \mathbf{E} = \mathbf{0} \quad \text{on } S ,\tag{6.3}$$

where $\mathbf{n}$ is the unit normal to the surface S of the guide. Recall that the other two Maxwell equations, in charge-free regions,

$$\boldsymbol{\nabla} \cdot \mathbf{D} = 0 , \qquad \boldsymbol{\nabla} \cdot \mathbf{B} = 0 ,\tag{6.4}$$

are contained within these equations, as is the boundary condition $\mathbf{n} \cdot \mathbf{B} = 0$. The medium filling the waveguide is assumed to be uniform and nondissipative. In view of the cylindrical nature of the boundary surface, it is convenient to separate the field equations into components parallel to the axis of the guide, which we take as the z-axis, and components transverse to the guide axis. This we achieve by scalar and vector multiplication with $\mathbf{e}$, a unit vector in the z-direction, thus obtaining

$$\boldsymbol{\nabla} \cdot \mathbf{e} \times \mathbf{E} = -ik\zeta H_z ,\tag{6.5a}$$

$$\boldsymbol{\nabla} \cdot \mathbf{e} \times \mathbf{H} = ik\eta E_z ,\tag{6.5b}$$

and

$$\boldsymbol{\nabla} E_z - \frac{\partial}{\partial z}\mathbf{E} = ik\zeta \mathbf{e} \times \mathbf{H} ,\tag{6.6a}$$

$$\boldsymbol{\nabla} H_z - \frac{\partial}{\partial z}\mathbf{H} = -ik\eta \mathbf{e} \times \mathbf{E} .\tag{6.6b}$$

On substituting (6.5b) [(6.5a)] into (6.6a) [(6.6b)], one recasts the latter into the form

$$\frac{\partial}{\partial z}\mathbf{E} = ik\zeta \left(1 + \frac{1}{k^2}\boldsymbol{\nabla}\boldsymbol{\nabla}\right) \cdot \mathbf{H} \times \mathbf{e} ,\tag{6.7a}$$

$$\frac{\partial}{\partial z}\mathbf{H} = ik\eta \left(1 + \frac{1}{k^2}\boldsymbol{\nabla}\boldsymbol{\nabla}\right) \cdot \mathbf{e} \times \mathbf{E} ,\tag{6.7b}$$

in which 1 denotes the unit dyadic. This set of equations is fully equivalent to the original field equations, for it still contains (6.5a) and (6.5b) as its z-component. The transverse components of (6.7a) and (6.7b) constitute a system of differential equations to determine the transverse components of the electric and magnetic fields. These equations are in transmission line form, but with the series impedance and the shunt admittance per unit length appearing as dyadic differential operators. The subsequent analysis has for its aim the replacement of the operator transmission line equations by an infinite set of ordinary differential equations. This is performed by successively suppressing the vectorial aspect of the equations and the explicit dependence on x and y, the coordinates in a transverse plane.

Any two-component vector field, such as the transverse part of the electric field $\mathbf{E}_\perp$, can be represented as a linear combination of two vectors derived from a potential function and a stream function, respectively. Thus

$$\mathbf{E}_\perp = -\boldsymbol{\nabla}_\perp V' + \mathbf{e} \times \boldsymbol{\nabla} V'' , \tag{6.8}$$

where $V'(\mathbf{r})$ and $V''(\mathbf{r})$ are two arbitrary scalar functions and $\boldsymbol{\nabla}_\perp$ indicates the transverse part of the gradient operator. In a similar way, we write

$$\mathbf{H}_\perp = -\mathbf{e} \times \boldsymbol{\nabla} I' - \boldsymbol{\nabla}_\perp I'' , \tag{6.9a}$$

or

$$\mathbf{H} \times \mathbf{e} = -\boldsymbol{\nabla}_\perp I' + \mathbf{e} \times \boldsymbol{\nabla} I'' , \tag{6.9b}$$

with $I'(\mathbf{r})$ and $I''(\mathbf{r})$ two new arbitrary scalar functions. This general representation can be obtained by constructing the two-component characteristic vectors (eigenvectors) of the operator $1 + \frac{1}{k^2}\boldsymbol{\nabla}\boldsymbol{\nabla}$. Such vectors must satisfy the eigenvector equation in the form

$$\boldsymbol{\nabla}_\perp \boldsymbol{\nabla} \cdot \mathbf{A}_\perp = \gamma \mathbf{A}_\perp . \tag{6.10}$$

Hence, either $\boldsymbol{\nabla} \cdot \mathbf{A}_\perp = 0$ and $\gamma = 0$, implying that $\mathbf{A}_\perp$ is the curl of a vector directed along the z-axis; or $\boldsymbol{\nabla} \cdot \mathbf{A}_\perp \neq 0$, and $\mathbf{A}_\perp$ is the gradient of a scalar function. The most general two-component vector $\mathbf{A}_\perp$ is a linear combination of these two types, and $\mathbf{e} \times \mathbf{A}$ is still of the same form, as it must be. In consequence of these observations, the substitution of the representation (6.8) and (6.9a) into the differential equations (6.7a) and (6.7b) will produce a set of equations in which every term has one or the other of these forms. This yields a system of four scalar differential equations, which are grouped into two pairs,

$$\frac{\partial}{\partial z}I' = \mathrm{i}k\eta V' , \quad \frac{\partial}{\partial z}V' = \mathrm{i}k\zeta \left(1 + \frac{1}{k^2}\nabla_\perp^2\right) I' , \tag{6.11a}$$

$$\frac{\partial}{\partial z}I'' = \mathrm{i}k\eta \left(1 + \frac{1}{k^2}\nabla_\perp^2\right) V'' , \quad \frac{\partial}{\partial z}V'' = \mathrm{i}k\zeta I'' , \tag{6.11b}$$

where $\nabla_\perp^2$ is the Laplacian for the transverse coordinates x and y. (Any constant annihilated by $\boldsymbol{\nabla}_\perp$ is excluded because it would not contribute to the electric and magnetic fields.) The longitudinal field components may now be written as

$$ik\eta E_z = \nabla_\perp^2 I' \,, \tag{6.12a}$$
$$ik\zeta H_z = \nabla_\perp^2 V'' \,. \tag{6.12b}$$

The net effect of these operations is the decomposition of the field into two independent parts derived, respectively, from the scalar functions V', I', and V'', I''. Note that the first type of field in general possesses a longitudinal component of electric field, but no longitudinal magnetic field, while the situation is reversed with the second type of field. For this reason, the various field configurations derived from V' and I' are designated as E modes, while those obtained from V'' and I'' are called H modes; the nomenclature in each case specifies the nonvanishing z-component of the field.[1]

The scalar quantities involved in (6.11a) and (6.11b) are functions of x, y, and z. The final step in the reduction to one-dimensional equations consists in representing the x, y dependence of these functions by an expansion in the complete set of functions forming the eigenfunctions of $\nabla_\perp^2$. For the E mode, let these functions be $\varphi_a(x, y)$, satisfying

$$(\nabla_\perp^2 + \gamma_a'^2)\varphi_a(x, y) = 0 \,, \tag{6.13}$$

and subject to boundary conditions, which we shall shortly determine. On substituting the expansion

$$V'(x, y, z) = \sum_a \varphi_a(x, y)V_a'(z) \,, \tag{6.14a}$$
$$I'(x, y, z) = \sum_a \varphi_a(x, y)I_a'(z) \,, \tag{6.14b}$$

into (6.11a), we immediately obtain the transmission line equations

$$\frac{\mathrm{d}}{\mathrm{d}z}I_a'(z) = ik\eta V_a'(z) \,, \tag{6.15a}$$
$$\frac{\mathrm{d}}{\mathrm{d}z}V_a'(z) = ik\zeta \left(1 - \frac{\gamma_a'^2}{k^2}\right) I_a'(z) \,. \tag{6.15b}$$

In a similar way, we introduce another set of eigenfunctions for $\nabla_\perp^2$:

$$(\nabla_\perp^2 + \gamma_a''^2)\psi_a(x, y) = 0 \,. \tag{6.16}$$

and expand the H-mode quantities in terms of them:

---

[1] A more common terminology for E modes are TM modes, meaning "transverse magnetic"; and for H modes, TE modes, for "transverse electric." Still another notation is $\perp$ for "perpendicular," referring to H modes, and $\parallel$ for "parallel," referring to E modes.

$$V''(x, y, z) = \sum_a \psi_a(x, y) V_a''(z) , \tag{6.17a}$$

$$I''(x, y, z) = \sum_a \psi_a(x, y) I_a''(z) . \tag{6.17b}$$

The corresponding differential equations are

$$\frac{d}{dz} I_a''(z) = ik\eta \left(1 - \frac{\gamma_a''^2}{k^2} \right) V_a''(z) , \tag{6.18a}$$

$$\frac{d}{dz} V_a''(z) = ik\zeta I_a''(z) . \tag{6.18b}$$

The boundary conditions on the electric field require that

$$E_z = 0, \qquad E_s = 0 \quad \text{on} \quad S , \tag{6.19}$$

where E_s is the component of the electric field tangential to the boundary curve C. These conditions imply from (6.12a) and (6.8) that

$$\nabla_\perp^2 I' = 0, \quad \frac{\partial}{\partial s} V' = \frac{\partial}{\partial n} V'' = 0 \quad \text{on} \quad S , \tag{6.20}$$

where $\frac{\partial}{\partial n}$ is the derivative normal to the surface of the waveguide S, and $\frac{\partial}{\partial s}$ is the circumferential derivative, tangential to the curve C. Since these equations must be satisfied for all z, they impose the following requirements on the functions φ_a and ψ_a:

$$\gamma_a'^2 \varphi_a = 0 , \quad \frac{\partial}{\partial s} \varphi_a = 0 , \quad \frac{\partial}{\partial n} \psi_a = 0 \quad \text{on} \quad C . \tag{6.21}$$

If we temporarily exclude the possibility $\gamma_a' = 0$, the second E-mode boundary condition is automatically included in the first statement, that $\varphi_a = 0$ on the boundary curve C. Hence, E modes are derived from scalar functions defined by

$$\left(\nabla_\perp^2 + \gamma_a'^2 \right) \varphi_a(x, y) = 0 , \tag{6.22a}$$
$$\varphi_a(x, y) = 0 \quad \text{on} \quad C , \tag{6.22b}$$

while H modes are derived from functions satisfying

$$\left(\nabla_\perp^2 + \gamma_a''^2 \right) \psi_a(x, y) = 0 , \tag{6.23a}$$
$$\frac{\partial}{\partial n} \psi_a(x, y) = 0 \quad \text{on} \quad C , \tag{6.23b}$$

These equations are often encountered in physics. For example, they describe the vibrations of a membrane bounded by the curve C, which is either rigidly clamped at the boundary [(6.22b)], or completely free [(6.23b)]. Mathematically, these are referred to a Dirichlet and Neumann boundary conditions, respectively. Each equation defines an infinite set of eigenfunctions and eigen-values φ_a, γ'_a and ψ_a, γ''_a. Hence, a waveguide possesses a twofold infinity of possible modes of electromagnetic oscillation, each completely characterized by one of these scalar functions and its attendant eigenvalue.

We shall now show that the discarded possibility, $\gamma'_a = 0$, cannot occur for hollow waveguides, but does correspond to an actual field configuration in two-conductor lines, being in fact the T mode discussed in Chap. 5. The scalar function φ associated with $\gamma'_a = 0$ satisfies Laplace's equation

$$\nabla^2_\perp \varphi(x, y) = 0 , \tag{6.24}$$

and is restricted by the second boundary condition, $\frac{\partial}{\partial s}\varphi(x, y) = 0$ on C, or

$$\varphi(x, y) = \text{ constant on } C . \tag{6.25}$$

Since φ satisfies Laplace's equation, we deduce that

$$\oint_C ds\, \varphi \frac{\partial}{\partial n}\varphi = \int_\sigma d\sigma \, (\boldsymbol{\nabla}_\perp \varphi)^2 , \tag{6.26}$$

in which the line integral is taken around the curve C and the surface integral is extended over the guide cross section σ. For a hollow waveguide with a cross section bounded by a single closed curve on which

$$\varphi = \text{constant} = \varphi_0 , \tag{6.27}$$

we conclude

$$\oint_C ds\, \varphi \frac{\partial}{\partial n}\varphi = \varphi_0 \oint_C ds \frac{\partial}{\partial n}\varphi = \varphi_0 \int_\sigma d\sigma \, \nabla^2_\perp \varphi = 0 , \tag{6.28}$$

and therefore from (6.26) $\boldsymbol{\nabla}_\perp \varphi = 0$ everywhere within the guide, which im-plies that all field components vanish, effectively denying the existence of such a mode. If, however, the contour C consists of two unconnected curves C_1 and C_2, as in a coaxial line, the boundary condition, $\frac{\partial}{\partial s}\varphi = 0$ on C, requires that φ be constant on each contour

$$\varphi = \varphi_1 \text{ on } C_1, \qquad \varphi = \varphi_2 \text{ on } C_2 , \tag{6.29}$$

but does not demand that $\varphi_1 = \varphi_2$. Hence

$$\oint_C ds\, \varphi \frac{\partial}{\partial n}\varphi = \varphi_1 \oint_{C_1} ds \frac{\partial}{\partial n}\varphi + \varphi_2 \oint_{C_2} ds \frac{\partial}{\partial n}\varphi = (\varphi_1 - \varphi_2) \oint_{C_1} ds \frac{\partial}{\partial n}\varphi , \tag{6.30}$$

since

$$0 = \oint_C ds \frac{\partial}{\partial n}\varphi = \oint_{C_1} ds \frac{\partial}{\partial n}\varphi + \oint_{C_2} ds \frac{\partial}{\partial n}\varphi , \tag{6.31}$$

and the preceding proof fails if $\varphi_1 \neq \varphi_2$. The identification with the T mode is completed by noting [(6.12a)] that $E_z = H_z = 0$.

The preceding discussion has shown that the electromagnetic field within a waveguide consists of a linear superposition of an infinite number of completely independent field configurations, or modes. Each mode has a characteristic field pattern across any section of the guide, and the amplitude variations of the fields along the guide are specified by "currents" and "voltages" which satisfy transmission line equations. We shall summarize our results by collecting together the fundamental equations describing a typical E mode and H mode (omitting distinguishing indices for simplicity).

- E mode:

$$\mathbf{E}_\perp = -\boldsymbol{\nabla}_\perp \varphi(x,y) V(z) \,, \tag{6.32a}$$

$$\mathbf{H}_\perp = -\mathbf{e} \times \boldsymbol{\nabla} \varphi(x,y) I(z) \,, \tag{6.32b}$$

$$E_z = \mathrm{i}\zeta \frac{\gamma^2}{k} \varphi(x,y) I(z) \,, \tag{6.32c}$$

$$H_z = 0 \,, \tag{6.32d}$$

$$(\nabla_\perp^2 + \gamma^2)\varphi(x,y) = 0 \,, \quad \varphi(x,y) = 0 \text{ on } C \,, \tag{6.32e}$$

$$\frac{\mathrm{d}}{\mathrm{d}z} I(z) = \mathrm{i}k\eta V(z) \,, \tag{6.32f}$$

$$\frac{\mathrm{d}}{\mathrm{d}z} V(z) = \mathrm{i}k\zeta \left(1 - \frac{\gamma^2}{k^2} \right) I(z) \,. \tag{6.32g}$$

- H mode:

$$\mathbf{E}_\perp = \mathbf{e} \times \boldsymbol{\nabla} \psi(x,y) V(z) \,, \tag{6.33a}$$

$$\mathbf{H}_\perp = -\boldsymbol{\nabla}_\perp \psi(x,y) I(z) \,, \tag{6.33b}$$

$$E_z = 0 \,, \tag{6.33c}$$

$$H_z = \mathrm{i}\eta \frac{\gamma^2}{k} \psi(x,y) V(z) \,, \tag{6.33d}$$

$$(\nabla_\perp^2 + \gamma^2)\psi(x,y) = 0 \,, \quad \frac{\partial}{\partial n}\psi(x,y) = 0 \text{ on } C \,, \tag{6.33e}$$

$$\frac{\mathrm{d}}{\mathrm{d}z} I(z) = \mathrm{i}k\eta \left(1 - \frac{\gamma^2}{k^2} \right) V(z) \,, \tag{6.33f}$$

$$\frac{\mathrm{d}}{\mathrm{d}z} V(z) = \mathrm{i}k\zeta I(z) \,. \tag{6.33g}$$

The T mode in a two-conductor line is to be regarded as an E mode with $\gamma = 0$, and the boundary condition replaced by $\frac{\partial}{\partial s}\varphi = 0$. It may also be considered an H mode with $\gamma = 0$.

The transmission line equations for the two mode types, written as

- E mode:

$$\frac{\mathrm{d}}{\mathrm{d}z} I(z) = \mathrm{i}\omega\varepsilon V(z) \,, \tag{6.34a}$$

$$\frac{\mathrm{d}}{\mathrm{d}z} V(z) = \left(\mathrm{i}\omega\mu + \frac{\gamma^2}{\mathrm{i}\omega\varepsilon} \right) I(z) \,, \tag{6.34b}$$

- H mode:

$$\frac{\mathrm{d}}{\mathrm{d}z}I(z) = \left(\mathrm{i}\omega\varepsilon + \frac{\gamma^2}{\mathrm{i}\omega\mu}\right)V(z)\,, \tag{6.35a}$$

$$\frac{\mathrm{d}}{\mathrm{d}z}V(z) = \mathrm{i}\omega\mu I(z)\,, \tag{6.35b}$$

are immediately recognized as the equations of the E and H type for the distributed parameter circuits discussed in Chap. 5. The E-mode equivalent transmission line has distributed parameters per unit length specified by a shunt capacitance $C = \varepsilon$, and series inductance $L = \mu$, and a series capacitance $C' = \varepsilon/\gamma^2$. The H-mode line distributed parameters are a series inductance $L = \mu$, a shunt capacitance $C = \varepsilon$, and a shunt inductance $L'' = \mu/\gamma^2$, all per unit length.[2] Thus, if we consider a progressive wave, $I \propto \mathrm{e}^{\mathrm{i}\kappa z}$, with $V = ZI$, the propagation constant κ and characteristic impedance $Z = 1/Y$ associated with the two types of lines are

- E mode:

$$\kappa = \sqrt{k^2 - \gamma^2}\,, \tag{6.36a}$$

$$Z = \zeta\frac{\kappa}{k}\,, \tag{6.36b}$$

- H mode:

$$\kappa = \sqrt{k^2 - \gamma^2}\,, \tag{6.36c}$$

$$Y = \eta\frac{\kappa}{k}\,. \tag{6.36d}$$

We may again remark on the filter property of these transmission lines, which is discussed in Chap. 5. Actual transport of energy along a waveguide in a particular mode can only occur if the wavenumber k exceeds the quantity γ associated with the mode. The eigenvalue γ is therefore referred to as the cutoff or critical wavenumber for the mode. Other quantities related to the cutoff wavenumber are the cutoff wavelength,

$$\lambda_c = \frac{2\pi}{\gamma}\,, \tag{6.37}$$

and the cutoff (angular) frequency

$$\omega_c = \gamma(\epsilon\mu)^{-1/2}\,. \tag{6.38}$$

When the frequency exceeds the cutoff frequency for a particular mode, the wave motion on the transmission line, indicating the field variation along the guide, is described by an associated wavelength

---

[2] The vacuum value of the universal series inductance and shunt capacitance is $L_0 = \mu_0 = 1.257\,\mu\mathrm{H/m}$ and $C_0 = \varepsilon_0 = 8.854\,\mathrm{pF/m}$, respectively.

$$\lambda_g = \frac{2\pi}{\kappa} \,, \tag{6.39}$$

which is called the guide wavelength. The relation between the guide wavelength, intrinsic wavelength, and cutoff wavelength for a particular mode is, according to (6.36a) or (6.36c),

$$\frac{1}{\lambda_g} = \sqrt{\frac{1}{\lambda^2} - \frac{1}{\lambda_c^2}} \,, \tag{6.40a}$$

or

$$\lambda_g = \frac{\lambda}{\sqrt{1 - \left(\frac{\lambda}{\lambda_c}\right)^2}} \,. \tag{6.40b}$$

Thus, at cutoff ($\lambda = \lambda_c$), the guide wavelength is infinite and becomes imaginary at longer wavelengths, indicating attenuation, while at very short wavelengths ($\lambda \ll \lambda_c$), the guide wavelength is substantially equal to the intrinsic wavelength of the guide medium. Correspondingly, the characteristic impedance for an E (H) mode is zero (infinite) at the cutoff frequency and is imaginary at lower frequencies in the manner typical of a capacitance (inductance). The characteristic impedance approaches the intrinsic impedance of the medium $\zeta = \sqrt{\mu/\varepsilon}$ for very short wavelengths. For $\varepsilon = \varepsilon_0$, $\mu = \mu_0$, the latter reduces to the impedance of free space,

$$\zeta_0 = \frac{1}{\eta_0} = \sqrt{\frac{\mu_0}{\varepsilon_0}} = 376.7\,\Omega \,. \tag{6.41}$$

The existence of a cutoff frequency for each mode involves the implicit statement that γ^2 is real and positive; γ is positive by definition. A proof is easily supplied for both E and H modes with the aid of the identity

$$\oint_C \mathrm{d}s\, f^* \frac{\partial}{\partial n} f = \int_\sigma \mathrm{d}\sigma\, |\boldsymbol{\nabla}_\perp f|^2 - \gamma^2 \int_\sigma \mathrm{d}\sigma\, |f|^2 \,, \tag{6.42}$$

where f stands for either an E-mode function φ or an H-mode function ψ. In either event, the line integral vanishes and

$$\gamma^2 = \frac{\int_\sigma \mathrm{d}\sigma\, |\boldsymbol{\nabla}_\perp f|^2}{\int_\sigma \mathrm{d}\sigma\, |f|^2} \,, \tag{6.43}$$

which establishes the theorem. It may be noted that we have admitted, in all generality, that f may be complex. However, with the knowledge that γ^2 is real, it is evident from the form of the defining wave equation and boundary conditions that real mode functions can always be chosen.

The impedance (admittance) at a given point on the transmission line describing a particular mode,

$$Z(z) = \frac{1}{Y(z)} = \frac{V(z)}{I(z)} , \tag{6.44}$$

determines the ratio of the transverse electric and magnetic field components at that point. According to (6.32a) and (6.32b), an E-mode magnetic field is related to the transverse electric field by

$$\text{E mode:} \quad \mathbf{H} = Y(z)\mathbf{e} \times \mathbf{E} , \tag{6.45a}$$

which is a general vector relation since it correctly predicts that $H_z = 0$. The analogous H-mode relation is

$$\text{H mode:} \quad \mathbf{E} = -Z(z)\mathbf{e} \times \mathbf{H} . \tag{6.45b}$$

For either type of mode, the connections between the rectangular components of the transverse fields are

$$E_x = Z(z)H_y, \quad E_y = -Z(z)H_x , \tag{6.46a}$$
$$H_x = -Y(z)E_y, \quad H_y = Y(z)E_x . \tag{6.46b}$$

In the particular case of a progressive wave propagating (or attenuating) in the positive z-direction, the impedance at every point equals the characteristic impedance of the line, $Z(z) = Z$. The analogous relation $Z(z) = -Z$ describes a wave progressing in the negative direction.

6.2 Hertz Vectors

The reduction of the vector field equations to a set of transmission line equations, as set forth in Sect. 6.1, requires four scalar functions of z for its proper presentation. However, it is often convenient to eliminate two of these functions and exhibit the general electromagnetic field as derived from two scalar functions of position, which appear in the role of single component Hertz vectors. (Recall Sect. 1.5.1.) On eliminating the functions $V'(\mathbf{r})$ and $I''(\mathbf{r})$ with the aid of (6.11a) and (6.11b), the transverse components of $\mathbf{E}$ and $\mathbf{H}$, (6.8) and (6.9a), become

$$\mathbf{E}_\perp = \frac{i}{k}\zeta\boldsymbol{\nabla}_\perp \frac{\partial}{\partial z}I' + \mathbf{e} \times \boldsymbol{\nabla}V'' , \tag{6.47a}$$

$$\mathbf{H}_\perp = -\mathbf{e} \times \boldsymbol{\nabla}I' + \frac{i}{k}\eta\boldsymbol{\nabla}_\perp \frac{\partial}{\partial z}V'' , \tag{6.47b}$$

which can be combined with the expressions for the longitudinal field components, (6.12a) and (6.12b), into general vector equations[3]

---

[3] Equations (6.48a) and (6.48b) agree with (1.144a) and (1.144b) when the wave equation is satisfied, with the identification $\boldsymbol{\Pi}' = \boldsymbol{\Pi}_e$, $\boldsymbol{\Pi}'' = \boldsymbol{\Pi}_m$.

$$\mathbf{E} = \nabla \times (\nabla \times \mathbf{\Pi}') + ik\zeta \nabla \times \mathbf{\Pi}'' , \tag{6.48a}$$

$$\mathbf{H} = -ik\eta \nabla \times \mathbf{\Pi}' + \nabla \times (\nabla \times \mathbf{\Pi}'') . \tag{6.48b}$$

The electric and magnetic Hertz vectors that appear in this formulation only possess z-components, which are given by

$$\Pi_z' = \frac{i}{k}\zeta I' , \qquad \Pi_z'' = \frac{i}{k}\eta V'' . \tag{6.49}$$

The Maxwell equations are completely satisfied if the Hertz vector components satisfy the scalar wave equation:

$$(\nabla^2 + k^2)\Pi_z' = 0 , \qquad (\nabla^2 + k^2)\Pi_z'' = 0 , \tag{6.50}$$

which is verified by eliminating V' and I'' from (6.11a) and (6.11b). For a particular E mode, the scalar function I' is proportional to the longitudinal electric field and

$$\text{E mode:} \quad \Pi_z' = \frac{1}{\gamma^2}E_z . \tag{6.51a}$$

Similarly, the other Hertz vector is determined:

$$\text{H mode:} \quad \Pi_z'' = \frac{1}{\gamma^2}H_z . \tag{6.51b}$$

Hence the field structure of an E or H mode can be completely derived from the corresponding longitudinal field component.

6.3 Orthonormality Relations

We turn to an examination of the fundamental physical quantities associated with the electromagnetic field in a waveguide – energy density and energy flux. In the course of the investigation we shall also derive certain orthogonal properties possessed by the electric and magnetic field components of the various modes. Inasmuch as these relations are based on similar orthogonal properties of the scalar functions φ_a and ψ_a, we preface the discussion by a derivation of the necessary theorems. Let us consider two E-mode functions φ_a and φ_b, and construct the identity

$$\int_C ds\, \varphi_a \frac{\partial}{\partial n}\varphi_b = \int_\sigma d\sigma\, \nabla_\perp \varphi_a \cdot \nabla_\perp \varphi_b - \gamma_b'^2 \int_\sigma d\sigma\, \varphi_a \varphi_b . \tag{6.52}$$

If we temporarily exclude the T mode of a two-conductor guide, the line integral vanishes by virtue of the boundary condition. On interchanging φ_a and φ_b, and subtracting the resulting equation, we obtain

$$(\gamma_a'^2 - \gamma_b'^2) \int_\sigma d\sigma\, \varphi_a \varphi_b = 0 , \tag{6.53}$$

which demonstrates the orthogonality of two mode functions with different eigenvalues. In consequence of the vanishing of the surface integral in (6.52), we may write this orthogonal relation as

$$\int_\sigma d\sigma \, \boldsymbol{\nabla}_\perp \varphi_a \cdot \boldsymbol{\nabla}_\perp \varphi_b = 0 \, , \quad \gamma_a' \neq \gamma_b' \, . \tag{6.54}$$

If more than one linearly independent mode function is associated with a particular eigenvalue – a situation which is referred to as "degeneracy" – no guarantee of orthogonality for these eigenfunctions is supplied by (6.53). However, a linear combination of degenerate eigenfunctions is again an eigenfunction, and such linear combinations can always be arranged to have the orthogonal property. In this sense, the orthogonality theorem (6.54) is valid for all pairs of different eigenfunctions. The theorem is also valid for the T mode of a two-conductor system. To prove this, we return to (6.52) and choose the mode a as an ordinary E mode ($\varphi_a = 0$ on C), and the mode b as the T mode ($\gamma_b' = 0$); the desired relation follows immediately. Note, however, that in this situation orthogonality in the form $\int d\sigma \, \varphi_a \varphi_b = 0$ is not obtained. Finally, then, the orthogonal relation, applicable to all E modes, is

$$\int_\sigma d\sigma \, \boldsymbol{\nabla}_\perp \varphi_a \cdot \boldsymbol{\nabla}_\perp \varphi_b = \delta_{ab} \, , \tag{6.55}$$

which also contains a convention regarding the normalization of the E-mode functions:

$$\int_\sigma d\sigma \, (\boldsymbol{\nabla}_\perp \varphi_a)^2 = 1 \, , \tag{6.56}$$

a convenient choice for the subsequent discussion. With the exception of the T mode, the normalization condition can also be written

$$\gamma_a'^2 \int_\sigma d\sigma \, \varphi_a^2 = 1 \, . \tag{6.57}$$

The corresponding derivation for H modes proceeds on identical lines, with results expressed by

$$\int_\sigma d\sigma \, \boldsymbol{\nabla}_\perp \psi_a \cdot \boldsymbol{\nabla}_\perp \psi_b = \delta_{ab} \, , \tag{6.58}$$

which contains the normalization convention

$$\int_\sigma d\sigma \, (\boldsymbol{\nabla}_\perp \psi_a)^2 = \gamma_a''^2 \int_\sigma d\sigma \, \psi_a^2 = 1 \, . \tag{6.59}$$

As we shall now see, no statement of orthogonality between E and H modes is required.

6.4 Energy Density and Flux

The energy quantities with which we shall be concerned are the linear energy densities (i.e., the energy densities per unit length) obtained by integrating the volume densities across a section of the guide. It is convenient to consider separately the linear densities associated with the various electric and magnetic components of the field. Thus the linear electric energy density connected with the longitudinal electric field is

$$W_{E_z} = \frac{\varepsilon}{2} \int_\sigma d\sigma \left[\mathrm{Re}\left(E_z e^{-i\omega t} \right) \right]^2 = \frac{\varepsilon}{4} \int_\sigma d\sigma \, |E_z|^2 \,, \tag{6.60}$$

where the oscillating terms are omitted due to time-averaging [(4.6)]. On inserting the general superposition of individual E-mode fields [cf. (6.32c)],

$$E_z = \frac{i\zeta}{k} \sum_a \gamma_a'^2 \varphi_a(x,y) I_a'(z) \,, \tag{6.61}$$

we find

$$W_{E_z} = \frac{\varepsilon}{4} \frac{\zeta^2}{k^2} \sum_a \gamma_a'^2 \, |I_a'(z)|^2 \,, \tag{6.62}$$

in which the orthogonality and normalization (6.57) of the E-mode functions has been used. The orthogonality of the longitudinal electric fields possessed by different E modes is thus a trivial consequence of the corresponding property of the scalar functions φ_a. The longitudinal electric field energy density can also be written

$$W_{E_z} = \frac{1}{4} \sum_a \frac{1}{\omega^2 C_a'} \, |I_a'(z)|^2 \,, \tag{6.63}$$

by introducing the distributed series capacitance, $C_a' = \varepsilon/\gamma_a'^2$, associated with the transmission line that describes the ath E mode. In a similar way, the linear energy density

$$W_{H_z} = \frac{\mu}{4} \int_\sigma d\sigma \, |H_z|^2 \tag{6.64}$$

derived from the longitudinal magnetic field [(6.33d)]

$$H_z = \frac{i\eta}{k} \sum_a \gamma_a''^2 \psi_a(x,y) V_z''(z) \tag{6.65}$$

reads

$$W_{H_z} = \frac{\mu}{4} \frac{\eta^2}{k^2} \sum_a \gamma_a''^2 \, |V_a''(z)|^2 \,, \tag{6.66}$$

in consequence of the normalization condition (6.59) for ψ_a and the orthogonality of the longitudinal magnetic fields of different H modes. The insertion of the distributed shunt inductance characteristic of the ath H-mode transmission line, $L_a'' = \mu/\gamma_a''^2$, transforms this energy density expression into

$$W_{H_z} = \frac{1}{4} \sum_a \frac{1}{\omega^2 L_a''} |V_a''(z)|^2 .$$

(6.67)

To evaluate the linear energy density associated with the transverse electric field

$$W_{\mathbf{E}_\perp} = \frac{\varepsilon}{4} \int_\sigma d\sigma \, |\mathbf{E}_\perp|^2 ,$$

(6.68)

it is convenient to first insert the general representation (6.8), thus obtaining

$$W_{\mathbf{E}_\perp} = \frac{\varepsilon}{4} \left[\int_\sigma d\sigma \, |\nabla_\perp V'|^2 + \int_\sigma d\sigma \, |\mathbf{e} \times \nabla V''|^2 \right.$$

$$\left. - 2\mathrm{Re} \int_\sigma d\sigma \, \nabla_\perp V' \cdot \mathbf{e} \times \nabla V''^* \right] .$$

(6.69)

The last term of this expression, representing the mutual energy of the E and H modes, may be proved to vanish by the following sequence of equations:

$$\int_\sigma d\sigma \, \nabla_\perp V' \cdot \mathbf{e} \times \nabla V''^* = - \int_\sigma d\sigma \, \nabla_\perp V''^* \cdot \mathbf{e} \times \nabla V'$$

$$= - \int_\sigma d\sigma \, \nabla_\perp \cdot (V''^* \mathbf{e} \times \nabla V')$$

$$= - \oint_C ds \, V''^* \mathbf{n} \cdot \mathbf{e} \times \nabla V' = \oint_C ds \, V''^* \frac{\partial}{\partial s} V' = 0 ,$$

(6.70)

in which the last step involves the generally valid boundary condition, $\frac{\partial}{\partial s} V' = 0$ on C, see (6.21). (A proof employing the boundary condition $V' = 0$ on C would not apply to the T mode.) It has thus been shown that the transverse electric field of an E mode is orthogonal to the transverse electric field of an H mode. For the transverse electric field energy density of the E mode, we have from (6.14a)

$$\frac{\varepsilon}{4} \int_\sigma d\sigma \, |\nabla_\perp V'|^2 = \frac{\varepsilon}{4} \sum_a |V_a'(z)|^2 ,$$

(6.71)

as an immediate consequence of the orthonormality condition (6.55), which demonstrates the orthogonality of the transverse electric fields of different E modes. Similarly, from (6.17a) the transverse electric field energy density of the H modes:

$$\frac{\varepsilon}{4} \int_\sigma d\sigma \, |\mathbf{e} \times \nabla V''|^2 = \frac{\varepsilon}{4} \int_\sigma d\sigma \, |\nabla_\perp V''|^2 = \frac{\varepsilon}{4} \sum_a |V''(z)|^2 ,$$

(6.72)

is a sum of individual mode contributions, indicating the orthogonality of the transverse electric fields of different H modes. Finally, the transverse electric field energy density is the sum of (6.71) and (6.72), or

$$W_{\mathbf{E}_\perp} = \frac{1}{4}\sum_a C|V_a'(z)|^2 + \frac{1}{4}\sum_a C|V_a''(z)|^2 \,, \tag{6.73}$$

where $C = \varepsilon$ is the distributed shunt capacitance common to all E- and H-mode transmission lines.

The discussion of the transverse magnetic field energy density,

$$W_{\mathbf{H}_\perp} = \frac{\mu}{4}\int_\sigma d\sigma\,|\mathbf{H}_\perp|^2 = \frac{\mu}{4}\int_\sigma d\sigma\,|\mathbf{H}\times\mathbf{e}|^2 \,, \tag{6.74}$$

is precisely analogous and requires no detailed treatment, for in virtue of (6.8) and (6.9b) it is merely necessary to make the substitutions $V' \to I'$, $V'' \to I''$ (and $\varepsilon \to \mu$, of course) to obtain the desired result. The boundary condition upon which the analog of (6.70) depends now reads $\frac{\partial}{\partial s}I' = 0$ on C, which is again an expression of the E-mode boundary condition. Hence

$$W_{\mathbf{H}_\perp} = \frac{1}{4}\sum_a L|I_a'(z)|^2 + \frac{1}{4}\sum_a L|I_a''(z)|^2 \,, \tag{6.75}$$

where $L = \mu$ is the distributed series inductance characteristic of all mode transmission lines. The orthogonality of the transverse magnetic fields associated with two different modes, which is contained in the result, may also be derived from the previously established transverse electric field orthogonality with the aid of the relations between transverse field components that is exhibited in (6.45a) and (6.45b).

The complex power flowing along the waveguide is obtained from the longitudinal component of the complex Poynting vector [see (4.11)] by integration across a guide section:

$$P = \frac{1}{2}\int_\sigma d\sigma\,\mathbf{E}\times\mathbf{H}^*\cdot\mathbf{e} = \frac{1}{2}\int_\sigma d\sigma\,\mathbf{E}\cdot(\mathbf{H}\times\mathbf{e})^* \,, \tag{6.76}$$

whence from (6.8) and (6.9b)

$$\begin{aligned}
P = \frac{1}{2}\Bigg[&\int_\sigma d\sigma\,\boldsymbol{\nabla}_\perp V'\cdot\boldsymbol{\nabla}_\perp I'^* + \int_\sigma d\sigma\,\mathbf{e}\times\boldsymbol{\nabla}V''\cdot\mathbf{e}\times\boldsymbol{\nabla}I''^* \\
&- \int_\sigma d\sigma\,\boldsymbol{\nabla}V'\cdot\mathbf{e}\times\boldsymbol{\nabla}I''^* - \int_\sigma d\sigma\,\mathbf{e}\times\boldsymbol{\nabla}V''\cdot\boldsymbol{\nabla}I'^* \Bigg] \\
= \frac{1}{2}&\sum_a V_a'(z)I_a'(z)^* + \frac{1}{2}\sum_a V_a''(z)I_a''(z)^* \,, \tag{6.77}
\end{aligned}$$

which uses (6.14a), (6.14b) and (6.17a), (6.17b), the orthonormality relations (6.55) and (6.58), and the analog of (6.70). Hence the complex power flow is a sum of individual mode contributions, each having the proper transmission line form; it will now be evident that the normalization conditions (6.56) and (6.59) were adopted in anticipation of this result. The orthogonality that is implied by expression (6.77) is a simple consequence of the orthogonality property of transverse electric fields, since $\mathbf{H} \times \mathbf{e}$ for an individual mode is proportional to the corresponding transverse electric field.

6.5 Impedance Definitions

It will have been noticed that the linear energy densities associated with the different field components are in full agreement with the energies stored per unit length in the various elements of the distributed parameter circuits. Thus the E- and H-type circuits give a complete pictorial description of the electromagnetic properties of E and H modes in the usual sense: Capacitance and inductance represent electric and magnetic energy; series elements are associated with longitudinal electric and transverse magnetic fields (longitudinal displacement and conduction currents); shunt elements describe transverse electric and longitudinal magnetic fields (transverse displacement and conduction currents). [See (6.63) and (6.75) for series and (6.67) and (6.73) for shunt.] The air of precise definition attached to the line parameters, however, is spurious. We are at liberty to multiply a transmission line voltage by a constant and divide the associated current by the same constant without violating the requirement that the complex power has the transmission line form. Thus, let a mode voltage and current be replaced by $N^{-1/2}V(z)$ and $N^{1/2}I(z)$, respectively, implying that the new voltage and current are obtained from the old definitions through multiplication by $N^{1/2}$ and $N^{-1/2}$, respectively. In order to preserve the form of the energy expressions, the inductance parameters must be multiplied by N, and the capacitance parameters divided by N. It follows from these statements that the characteristic impedance must be multiplied by N, in agreement with its significance as a voltage–current ratio. The propagation constant is unaffected by this alteration, of course. We may conclude that one of the basic quantities that specifies the transmission line, characteristic impedance, remains essentially undefined by any considerations thus far introduced. The same situation arose in the field analysis of the two-conductor transmission line and it was shown that a natural definition for the characteristic impedance could be obtained by ascribing the customary physical meaning to either the current or voltage, the same result being obtained in either event. This somewhat artificial procedure was employed in order to emphasize the rather different character of waveguide fields for, as we shall now show, a precise definition of characteristic impedance can be obtained by ascribing a physical significance to either the current or the voltage, depending on the type of mode, but not to both simultaneously.

An E mode is essentially characterized by E_z, from which all other field components can be derived. Associated with the longitudinal electric field is an electric displacement current $\dot{\mathbf{D}}$, the current density being [(6.32c)]

$$-\mathrm{i}\omega\varepsilon E_z = \gamma^2\varphi(x,y)N^{1/2}I(z) . \tag{6.78}$$

In addition, there is a longitudinal electric conduction current on the metal walls, with the surface density (4.64b), according to (6.32b),

$$-(\mathbf{n} \times \mathbf{H})_z = \frac{\partial}{\partial n}\varphi(x,y)N^{1/2}I(z), \quad (x,y) \in S.$$ (6.79)

The total (conduction plus displacement) longitudinal electric current is zero,[4] and it is natural to identify the total current flowing in the positive direction with $I(z)$, which leads to the following equation for N:

$$N^{1/2}\left[\gamma^2 \int_{+} d\sigma\,\varphi + \int_{+} ds\,\frac{\partial}{\partial n}\varphi\right] = 1,$$ (6.80)

where the surface and line integrals, denoted by $\int_{+}$, are to be conducted over those regions where φ and $\frac{\partial}{\partial n}\varphi$ are positive. This equation is particularly simple for the lowest E mode in any hollow waveguide, that is, the mode of minimum cutoff frequency, for this mode has the property, established in Problem 6.13, that the scalar function φ is nowhere negative, and vanishes only on the boundary. It follows that the (outward) normal derivative on the boundary cannot be positive. Hence the displacement current flows entirely in the positive direction, and the conduction current entirely in the negative direction. Consequently,

$$N = \frac{1}{\gamma^4 \left(\int_{\sigma} d\sigma\,\varphi\right)^2} = \frac{1}{\gamma^2}\frac{\int_{\sigma} d\sigma\,\varphi^2}{\left(\int_{\sigma} d\sigma\,\varphi\right)^2},$$ (6.81)

on employing the normalization condition for φ, (6.57), to express N in a form that is independent of the absolute scale of the function φ, Therefore, for the lowest E mode in any guide, a natural choice of characteristic impedance is, from (6.36b),

$$Z = \zeta\frac{\kappa}{k}\frac{1}{\gamma^2}\frac{\int_{\sigma} d\sigma\,\varphi^2}{\left(\int_{\sigma} d\sigma\,\varphi\right)^2}.$$ (6.82)

For the other E modes, the φ normalization condition can be used in an analogous way to obtain

$$Z = \zeta\frac{\kappa}{k}\frac{1}{\gamma^2}\frac{\int_{\sigma} d\sigma\,\varphi^2}{\left(\int_{+} d\sigma\,\varphi + \frac{1}{\gamma^2}\int_{+} ds\,\frac{\partial}{\partial n}\varphi\right)^2}.$$ (6.83)

It may appear more natural to deal with the voltage rather than the current in the search for a proper characteristic impedance definition, since the transverse electric field of an E mode is derived from a potential [(6.32a)].

---

[4] This follows from the fact that $\mathbf{H} = \mathbf{0}$ in the conductor, so that $\oint_C ds \cdot \mathbf{H} = \int_{\sigma} d\sigma \cdot \nabla \times \mathbf{H} = 0$, if the encircling C lies entirely inside the walls of the waveguide, and so encloses both the conduction and displacement current.

The voltage could then be defined as the potential of some fixed point with respect to the wall in a given cross section, thus determining N. In a guide of symmetrical cross section the only natural reference point is the center, which entails the difficulty that there exists an infinite class of modes for which $\varphi = 0$ at the center, and the definition fails. In addition, when this does not occur, as in the lowest E mode, the potential of the center point does not necessarily equal the voltage, if the characteristic impedance is defined on a current basis as we have done. Hence, while significance can always be attached to the E-mode current, no generally valid voltage definition can be offered.

By analogy with the E-mode discussion, we shall base a characteristic admittance definition for H modes on the properties of H_z, which can be said to define a longitudinal magnetic displacement current density, from (6.33d)

$$-i\omega\mu H_z = \gamma^2 \psi(x,y) N^{-1/2} V(z) . \tag{6.84}$$

The total longitudinal magnetic displacement current is zero[5] and we shall identify the total magnetic current flowing in the positive direction with $V(z)$. The voltage thus defined equals the line integral of the electric field intensity taken clockwise around all regions through which positive magnetic displacement flows. Accordingly,

$$N = \gamma^4 \left(\int_+ d\sigma\, \psi \right)^2 = \gamma^2 \frac{\left(\int_+ d\sigma\, \psi \right)^2}{\int_\sigma d\sigma\, \psi^2} \tag{6.85}$$

and [cf. (6.36d)]

$$Y = \eta \frac{\kappa}{k} \frac{1}{\gamma^2} \frac{\int_\sigma d\sigma\, \psi^2}{\left(\int_+ d\sigma\, \psi \right)^2} . \tag{6.86}$$

It would also be possible to base an admittance definition on the identification of the transmission line current with the total longitudinal electric conduction current flowing in the positive direction on the metal walls. However, the characteristic admittance so obtained will not agree in general with that just obtained.

Although we have advanced rather reasonable definitions of characteristic impedance and admittance, it is clear that these choices possess arbitrary features and in no sense can be considered inevitable. This statement may convey the impression that the theory under development is essentially vague and ill-defined, which would be a misunderstanding. Physically observable quantities can in no way depend on the precise definition of a characteristic impedance, but this does not detract from its appearance in a theory which seeks to express its results in conventional circuit language. Indeed, the arbitrariness in definition is a direct expression of the greater complexity of waveguide systems compared with low-frequency transmission lines. For example, in the

---

[5] Because $\mathbf{E} = 0$ in the conductor, so is $\oint_C d\mathbf{s} \cdot \mathbf{E} = \int_\sigma d\boldsymbol{\sigma} \cdot \boldsymbol{\nabla} \times \mathbf{E}$.

junction of two low-frequency transmission lines with different dimensions, the conventional transmission line currents and voltages are continuous to a high degree of approximation and hence the reflection properties of the junction are completely specified by the quantities which relate the current and voltage in each line. In a corresponding waveguide situation, however, physical quantities with such simple continuity properties do not exist in general, and it is therefore not possible to describe the properties of the junction in terms of two quantities which are each characteristic of an individual guide. Armed with this knowledge, which anticipates the results to be obtained in Chap. 13, we are forced to the position that the characteristic impedance is best regarded as a quantity chosen to simplify the electrical representation of a particular situation, and that different definitions may be advantageously employed in different circumstances. In particular, the impedance definition implicitly adopted at the beginning of the chapter, corresponding to $N = 1$, is most convenient for general theoretical discussion since it directly relates the transverse electric and magnetic fields. This choice, which may be termed the field impedance (admittance), will be adhered to in the remainder of this chapter.

6.6 Complex Poynting and Energy Theorems

Before turning to the discussion of particular types of guides, we shall derive a few simple properties of the electric and magnetic energies associated with propagating and nonpropagating modes. The tools for the purpose are provided by the complex Poynting vector theorem (ε, μ real)

$$\boldsymbol{\nabla} \cdot (\mathbf{E} \times \mathbf{H}^*) = i\omega \left(\mu|\mathbf{H}|^2 - \varepsilon|\mathbf{E}|^2 \right) , \tag{6.87}$$

and the energy theorem (ignoring any dependence of ε and μ on the frequency)

$$\boldsymbol{\nabla} \cdot \left(\frac{\partial \mathbf{E}}{\partial \omega} \times \mathbf{H}^* + \mathbf{E}^* \times \frac{\partial \mathbf{H}}{\partial \omega} \right) = i \left(\varepsilon|\mathbf{E}|^2 + \mu|\mathbf{H}|^2 \right) . \tag{6.88}$$

(Proofs and generalizations of these theorems are given in Problems 6.1 and 6.2.) If (6.87) is integrated over a cross section of the guide, only the longitudinal component of the Poynting vector survives (because $\mathbf{n} \times \mathbf{E}$ vanishes on S), and we obtain the transmission line form of the complex Poynting vector theorem, as applied to a single mode [cf. (6.77)]:

$$\frac{\mathrm{d}}{\mathrm{d}z} \left[\frac{1}{2} V(z)I(z)^* \right] = \frac{\mathrm{d}}{\mathrm{d}z} P = 2i\omega(W_H - W_E) , \tag{6.89}$$

where W_E and W_H are the electric and magnetic linear energy densities. A similar operation on (6.88) yields

$$\frac{\mathrm{d}}{\mathrm{d}z} \left\{ \frac{1}{2} \left[\frac{\partial V(z)}{\partial \omega} I(z)^* + V^*(z) \frac{\partial I(z)}{\partial \omega} \right] \right\} = 2i(W_E + W_H) = 2iW , \tag{6.90}$$

since $\partial\mathbf{E}_\perp/\partial\omega$, for example, involves $\partial V(z)/\partial\omega$ in the same way that $\mathbf{E}_\perp$ contains $V(z)$, for the scalar mode functions do not depend upon the frequency. As a first application of these equations we consider a propagating wave progressing in the positive direction, that is,

$$V(z) = V\mathrm{e}^{\mathrm{i}\kappa z}, \quad V = ZI .$$
$$I(z) = I\mathrm{e}^{\mathrm{i}\kappa z}, \qquad \qquad \text{(6.91)}$$

The complex power is real and independent of z:

$$P = \frac{1}{2}VI^* = \frac{1}{2}Z|I|^2 , \qquad (6.92)$$

whence we deduce from (6.89) that $W_E = W_H$; the electric and magnetic linear energy densities are equal in a progressive wave. To apply the energy theorem we observe that

$$\frac{\partial V(z)}{\partial\omega} = \frac{\partial V}{\partial\omega}\mathrm{e}^{\mathrm{i}\kappa z} + \mathrm{i}\frac{\mathrm{d}\kappa}{\mathrm{d}\omega}zV\mathrm{e}^{\mathrm{i}\kappa z} , \qquad (6.93\mathrm{a})$$

$$\frac{\partial I(z)}{\partial\omega} = \frac{\partial I}{\partial\omega}\mathrm{e}^{\mathrm{i}\kappa z} + \mathrm{i}\frac{\mathrm{d}\kappa}{\mathrm{d}\omega}zI\mathrm{e}^{\mathrm{i}\kappa z} , \qquad (6.93\mathrm{b})$$

and that

$$\frac{1}{2}\left[\frac{\partial V(z)}{\partial\omega}I(z)^* + V^*(z)\frac{\partial I(z)}{\partial\omega}\right] = \frac{1}{2}\left(\frac{\partial V}{\partial\omega}I^* + V^*\frac{\partial I}{\partial\omega}\right)$$
$$+ \mathrm{i}\frac{\mathrm{d}\kappa}{\mathrm{d}\omega}z\frac{1}{2}(VI^* + V^*I) . \quad (6.94)$$

Therefore, (6.90) implies

$$\frac{\mathrm{d}\kappa}{\mathrm{d}\omega}\frac{1}{4}(VI^* + V^*I) = \frac{\mathrm{d}\kappa}{\mathrm{d}\omega}P = W , \qquad (6.95)$$

or

$$P = vW , \qquad (6.96)$$

where

$$v = \frac{\mathrm{d}\omega}{\mathrm{d}\kappa} = c\frac{\mathrm{d}k}{\mathrm{d}\kappa} = c\frac{\kappa}{k} , \qquad (6.97)$$

with κ given by (6.36a) and (6.36c). The relation thus obtained expresses a proportionality between the power transported by a progressive wave and the linear energy density. The coefficient v must then be interpreted as the velocity of energy transport. It is consistent with this interpretation that v is always less than c, and vanishes at the cutoff frequency. At frequencies large in comparison with the cutoff frequency, v approaches the intrinsic velocity c of the medium. It is interesting to compare this velocity with the two velocities already introduced in Chap. 1 in discussing the flow of energy and momentum – the phase and group velocities. The phase velocity equals the ratio of the angular frequency and the propagation constant:

$$u = \frac{\omega}{\kappa} = c\frac{k}{\kappa} , \tag{6.98}$$

while the group velocity is the derivative of the angular frequency with respect to the propagation constant:

$$v = \frac{d\omega}{d\kappa} . \tag{6.99}$$

That the group velocity and energy transport velocity are equal is not unexpected. We notice that the phase velocity always exceeds the intrinsic velocity of the medium, and indeed is infinite at the cutoff frequency of the mode. The two velocities are related by

$$uv = c^2 \tag{6.100}$$

which coincides with (1.39). A simple physical picture for the phase and group velocities will be offered in Chap. 7, see Fig. 7.3. See also Problem 6.4, (6.137).

Another derivation of the energy transport velocity, which makes more explicit use of the waveguide fields, is suggested by the defining equation:

$$v = \frac{P}{W} = \frac{\frac{1}{2}\int_\sigma d\sigma \, \mathbf{e} \cdot \mathbf{E} \times \mathbf{H}^*}{\frac{1}{4}\int_\sigma d\sigma \, (\varepsilon|\mathbf{E}|^2 + \mu|\mathbf{H}|^2)} . \tag{6.101}$$

In virtue of the equality of electric and magnetic linear energy densities, and the relation $\mathbf{e} \times \mathbf{E} = Z\mathbf{H}$, which is valid for an E-mode field propagating in the positive direction [(6.45a)], we find using (6.36b)

$$v = \frac{Z}{\mu}\frac{\int_\sigma d\sigma \, |\mathbf{H}_\perp|^2}{\int_\sigma d\sigma \, |\mathbf{H}|^2} = c\frac{\kappa}{k} , \tag{6.102a}$$

since $\mathbf{H}$ has no longitudinal component. Similarly, the energy transport velocity for an H mode is from (6.45b) and (6.36d)

$$v = \frac{Y}{\varepsilon}\frac{\int_\sigma d\sigma \, |\mathbf{E}_\perp|^2}{\int_\sigma d\sigma \, |\mathbf{E}|^2} = c\frac{\kappa}{k} . \tag{6.102b}$$

When the wave motion on the transmission line is not that of a simple progressive wave, but the general superposition of standing waves (or running waves) described by solutions of (6.34a) and (6.34b), or (6.35a) and (6.35b),

$$V(z) = V\cos\kappa z + iZI\sin\kappa z , \tag{6.103a}$$

$$I(z) = I\cos\kappa z + iYV\sin\kappa z , \tag{6.103b}$$

or

$$V(z) = (2Z)^{1/2}\left(Ae^{i\kappa z} + Be^{-i\kappa z}\right) , \tag{6.104a}$$

$$I(z) = (2Y)^{1/2}\left(Ae^{i\kappa z} - Be^{-i\kappa z}\right) , \tag{6.104b}$$

the electric and magnetic linear energy densities are not equal, according to (6.89), because the complex power is a function of position on the line:

$$P(z) = \frac{1}{2}\left[VI^*\cos^2 \kappa z + V^*I\sin^2 \kappa z + i\frac{1}{2}\sin 2\kappa z \left(Z|I|^2 - Y|V|^2\right)\right],$$
(6.105a)

or

$$P(z) = |A|^2 - |B|^2 - AB^* e^{2i\kappa z} + A^* B e^{-2i\kappa z}.$$
(6.105b)

However, equality is obtained for the total electric and magnetic energies stored in any length of line that is an integral multiple of $\frac{1}{2}\lambda_g$. To prove this, we observe that by integrating (6.89) over z from z_1 to z_2,

$$P(z_2) - P(z_1) = 2i\omega(E_H - E_E),$$
(6.106)

where E_E and E_H are the total electric and magnetic energies stored in the length of transmission line between the points z_1 and z_2. Now, the complex power is a periodic function of z with the periodicity interval $\pi/\kappa = \frac{1}{2}\lambda_g$ [(6.39)], from which we conclude that $P(z_2) = P(z_1)$ if the two points are separated by an integral number of half guide wavelengths, which verifies the statement. An equivalent form of this result is that the average electric and magnetic energy densities are equal, providing the averaging process is extended over an integral number of half guide wavelengths, or over a distance large in comparison with $\frac{1}{2}\lambda_g$.

An explicit expression for the average energy density can be obtained from the energy theorem (6.88). The total energy E, stored in the guide between the planes $z = z_1$ and $z = z_2$, is given by the integral of (6.90), or

$$E = \frac{1}{4i}\left[\frac{\partial V(z)}{\partial \omega}I(z)^* + V(z)^*\frac{\partial I(z)}{\partial \omega}\right]_{z=z_1}^{z=z_2}.$$
(6.107)

On differentiating the voltage and current expressions (6.103a) and (6.103b) with respect to the frequency, we find

$$\frac{\partial V(z)}{\partial \omega} = iz\frac{d\kappa}{d\omega}ZI(z) + \left[\cos \kappa z\frac{\partial V}{\partial \omega} + i\sin \kappa z\frac{\partial ZI}{\partial \omega}\right],$$
(6.108a)

$$\frac{\partial I(z)}{\partial \omega} = iz\frac{d\kappa}{d\omega}YV(z) + \left[\cos \kappa z\frac{\partial I}{\partial \omega} + i\sin \kappa z\frac{\partial YV}{\partial \omega}\right].$$
(6.108b)

Hence

$$\frac{\partial V(z)}{\partial \omega}I(z)^* + V(z)^*\frac{\partial I(z)}{\partial \omega} = iz\frac{d\kappa}{d\omega}\left[Z|I(z)|^2 + Y|V(z)|^2\right] + \cdots,$$
(6.109)

where the unwritten part of this equation consists of those terms, arising from the bracketed expressions in (6.108a) and (6.108b), which are periodic functions of z with the period $\frac{1}{2}\lambda_g$. Thus, if the points z_1 and z_2 are separated by a distance that is an integral multiple of $\frac{1}{2}\lambda_g$, these terms make no contribution to the total energy. We also note that the quantity $Z|I(z)|^2 + Y|V(z)|^2$ is independent of z:

$$Z|I(z)|^2 + Y|V(z)|^2 = Z|I|^2 + Y|V|^2 = 4(|A|^2 + |B|^2) , \qquad (6.110)$$

from which we conclude that the total energy (6.107) stored in a length of guide, l, which is an integral number of half guide wavelengths, is, in terms of the energy velocity (6.97)

$$E = l\frac{d\kappa}{d\omega}\frac{1}{4}\left(Z|I|^2 + Y|V|^2\right) = \frac{l}{v}\frac{1}{4}\left(Z|I|^2 + Y|V|^2\right) = \frac{l}{v}\left(|A|^2 + |B|^2\right) . \qquad (6.111)$$

The average total linear energy density is $\overline{W} = E/l$, which has a simple physical significance in terms of running waves, being just the sum of the energy densities associated with each progressive wave component if it alone existed on the transmission line.

The energy relations for a nonpropagating mode are rather different; there is a definite excess of electric or magnetic energy, depending on the type of mode. The propagation constant for a nonpropagating (below cutoff) mode is imaginary:

$$\kappa = i\sqrt{\gamma^2 - k^2} = i|\kappa| , \qquad (6.112)$$

and a field that is attenuating in the positive z-direction is described by

$$\begin{aligned} V(z) &= Ve^{-|\kappa|z} \\ I(z) &= Ie^{-|\kappa|z} \end{aligned} \qquad V = ZI . \qquad (6.113)$$

The imaginary characteristic impedance (admittance) of an E (H) mode is given by

$$\text{E mode:} \quad Z = i\zeta\frac{|\kappa|}{k} = i|Z| , \qquad (6.114a)$$

$$\text{H mode:} \quad Y = i\eta\frac{|\kappa|}{k} = i|Y| . \qquad (6.114b)$$

The energy quantities of interest are the total electric and magnetic energy stored in the positive half of the guide ($z > 0$). The difference of these energies is given by (6.106), where $z_1 = 0$ and $z_2 \to \infty$. Since all field quantities approach zero exponentially for increasing z, $P(z_2) \to 0$, and

$$E_E - E_H = \frac{1}{2i\omega}P(0) = \frac{1}{4i\omega}VI^* . \qquad (6.115)$$

For an E mode

$$VI^* = Z|I|^2 = i|Z||I|^2 , \qquad (6.116)$$

and so

$$\text{E mode:} \quad E_E - E_H = \frac{1}{4\omega}|Z||I|^2 , \qquad (6.117)$$

which is positive. Hence an E mode below cutoff has an excess of electric energy, in agreement with the capacitive reactance form (see below) of the characteristic impedance. Similarly, for an H mode,

$$VI^* = Y^*|V|^2 = -i|Y||V|^2 \,, \tag{6.118}$$

whence

$$\text{H mode:} \quad E_H - E_E = \frac{1}{4\omega}|Y||V|^2 \,, \tag{6.119}$$

implying that an H mode below cutoff preponderantly stores magnetic energy, as the inductive susceptance form (see below) of its characteristic admittance would suggest.

To obtain the total energy stored in a nonpropagating mode, we employ (6.107), again with $z_1 = 0$ and $z_2 \to \infty$:

$$E = \frac{i}{4}\left(\frac{\partial V}{\partial \omega}I^* + V^*\frac{\partial I}{\partial \omega}\right) \,. \tag{6.120}$$

On differentiating the relation $V = ZI$ with respect to ω, and making appropriate substitutions in (6.120), we find

$$E = \frac{i}{4}\left[\frac{dZ}{d\omega}|I|^2 + (Z + Z^*)\frac{\partial I}{\partial \omega}I^*\right] \,. \tag{6.121}$$

The imaginary form of Z ($= i|Z|$ for an E mode) then implies that

$$E = -\frac{1}{4}\frac{d|Z|}{d\omega}|I|^2 \,. \tag{6.122}$$

We may note in passing that the positive nature of the total energy demands that $|Z|$ be a decreasing function of frequency, or better, that the reactance characterizing Z ($= iX$) be a decreasing function of frequency. The requirement is verified by direct differentiation:

$$-\frac{d|Z|}{d\omega} = \frac{1}{\omega}\frac{\gamma^2}{\gamma^2 - k^2}|Z| \,, \tag{6.123}$$

and

$$\text{E mode:} \quad E = \frac{1}{4\omega}\frac{\gamma^2}{\gamma^2 - k^2}|Z||I|^2 \,. \tag{6.124}$$

A comparison of this result with (6.117) shows that

$$\frac{2E_H}{E} = \frac{k^2}{\gamma^2} \,. \tag{6.125}$$

Thus, the electric and magnetic energies are equal just at the cutoff frequency ($k = \gamma$), and as the frequency diminishes, the magnetic energy steadily decreases in comparison with the electric energy. In conformity with the latter remark, the E-mode characteristic impedance approaches $i\zeta\gamma/k = i\gamma/\omega\epsilon$ when $k/\gamma \ll 1$, which implies that a transmission line describing an attenuated E mode at a frequency considerably below the cutoff frequency behaves like a lumped capacitance $C = \varepsilon/\gamma = \varepsilon\lambda_c/(2\pi)$.

The H-mode discussion is completely analogous, with the roles of electric and magnetic fields interchanged. Thus the total stored energy is

$$E = -\frac{1}{4}\frac{d|Y|}{d\omega}|V|^2 , \tag{6.126}$$

implying that the susceptance characterizing Y ($= -iB$) must be an increasing function of frequency. Explicitly,

$$\text{H mode:} \quad E = \frac{1}{4\omega}\frac{\gamma^2}{\gamma^2 - k^2}|Y||V|^2 , \tag{6.127}$$

and

$$\frac{2E_E}{E} = \frac{k^2}{\gamma^2} . \tag{6.128}$$

At frequencies well below the cutoff frequency, the electric energy is negligible in comparison with the magnetic energy, and the characteristic admittance becomes $i\eta\gamma/k = i\gamma/\omega\mu$. Thus, an H-mode transmission line under these circumstances behaves like a lumped inductance $L = \mu/\gamma = \mu\lambda_c/(2\pi)$.

6.7 Problems for Chap. 6

1. Prove the complex Poynting vector theorem, (6.87), and the energy theorem, (6.88), starting from the definitions of the Fourier transforms in time:

$$\mathbf{E}(\omega) = \int_{-\infty}^{\infty} dt\, e^{i\omega t}\mathbf{E}(t) , \tag{6.129a}$$

$$\mathbf{H}^*(\omega) = \int_{-\infty}^{\infty} dt\, e^{-i\omega t}\mathbf{H}(t) . \tag{6.129b}$$

What are the general forms of these theorems if no connection is assumed between $\mathbf{D}(\omega)$ and $\mathbf{E}(\omega)$ and between $\mathbf{B}(\omega)$ and $\mathbf{H}(\omega)$?

2. Show that if dispersion be included, the generalization of (6.88) is

$$\nabla \cdot \left(\frac{\partial \mathbf{E}(\omega)}{\partial \omega} \times \mathbf{H}^*(\omega) + \mathbf{E}^* \times \frac{\partial \mathbf{H}(\omega)}{\partial \omega}\right)$$
$$= i\left[\left(\frac{d}{d\omega}(\omega\varepsilon)\right)|\mathbf{E}|^2 + \left(\frac{d}{d\omega}(\omega\mu)\right)|\mathbf{H}|^2\right] \equiv 4iU . \tag{6.130}$$

3. The energy theorem (6.130) can be used to prove the uniqueness theorem: *the electric and magnetic fields in a region are completely determined by specifying the values of the electric (or magnetic) field tangential to the closed surface bounding that region.* This was demonstrated in another way in Chap. 4. This theorem is true save for sharply defined and isolated values of the frequency. Prove this by noting first that if $\mathbf{E}_1$, $\mathbf{H}_1$ and $\mathbf{E}_2$, $\mathbf{H}_2$ are two solutions of the Maxwell equations in the region, the difference $\mathcal{E} = \mathbf{E}_1 - \mathbf{E}_2$, $\mathcal{H} = \mathbf{H}_1 - \mathbf{H}_2$ must satisfy the homogeneous equations

$$\nabla \times \mathcal{H} = -\mathrm{i}\omega\varepsilon\mathcal{E}\,, \qquad \nabla \times \mathcal{E} = \mathrm{i}\omega\mu\mathcal{H}\,, \qquad (6.131\mathrm{a})$$

as continuous functions of ω, while on the surrounding surface S we have the homogeneous boundary condition

$$\mathbf{n} \times \mathcal{E} = \mathbf{0} \quad \text{on} \quad S\,, \qquad (6.132)$$

since both $\mathbf{E}_1$ and $\mathbf{E}_2$ have the same tangential values on S. Now show that the integral form of the energy theorem, applied to the difference fields, implies that

$$\mathcal{E} = \mathbf{0}, \qquad \mathcal{H} = \mathbf{0} \quad \text{in} \quad V\,. \qquad (6.133)$$

Show that the same result obtains if the tangential magnetic field is prescribed on the boundary, or if mixtures of the two kinds of boundaries conditions are imposed. However, note that this result depends on continuity requirements, so that it may fail to hold at isolated values of ω. Indeed there are infinitely many such exceptional solutions, for physically they correspond to the normal modes of a cavity enclosed by perfectly conducting walls coinciding with the surface S.

4. Consider a unidirectional light pulse, considered in the nondispersive case in Chap. 1. Calculate the corresponding group velocity v, defined as the ratio of the rate of energy flow or power

$$P = \frac{1}{2} \int_\sigma \mathrm{d}\sigma\, \mathbf{E} \times \mathbf{H}^* \cdot \mathbf{e}\,, \qquad (6.134)$$

where $\mathbf{e}$ is the direction of propagation of the electromagnetic disturbance and the integration is over the corresponding perpendicular area σ, to the energy per unit length,

$$W = \int_\sigma \mathrm{d}\sigma\, U\,. \qquad (6.135)$$

Assuming that the time averaged electric and magnetic energies per unit length are equal, show that

$$v = \frac{c}{1 - \frac{\mathrm{d}\ln c}{\mathrm{d}\ln\omega}}\,, \qquad (6.136)$$

where c is the speed of light in the medium. Alternatively, this can be written as

$$v = c\frac{n}{\frac{d}{d\omega}(\omega n)}, \qquad n = \left(\frac{\varepsilon(\omega)\mu(\omega)}{\varepsilon_0\mu_0}\right)^{1/2}. \qquad (6.137)$$

Calculate this in the example of the plasma model, where $\varepsilon = \varepsilon_0(1 - \omega_p^2/\omega^2)$, $\mu = \mu_0$, in terms of the parameter called the plasma frequency ω_p, and show that $v < c_0$.

5. The constant electrostatic field between parallel conducting plates at different potentials can be multiplied by $e^{ikx}e^{-i\omega t}$, $k = \omega/c$, to arrive at a possible electric field E_z for any ω. Check this. What is the associated magnetic field? Find the analogous electromagnetic wave between concentric cylindrical conductors. Interpret these waves in terms of E or H modes, and give the corresponding impedances.

6. Begin with the macroscopic Maxwell equation for frequency ω and transverse propagation vector $\mathbf{k}_\perp$, appropriate to $\varepsilon = \varepsilon(z)$ and $\mu = \mu_0$, and arrive at

$$-\frac{\partial}{\partial z}\mathbf{H}_\perp + i\omega\varepsilon\mathsf{P}\cdot\mathbf{n}\times\mathbf{E}_\perp = \mathbf{n}\times\mathbf{J}_\perp, \qquad (6.138a)$$

$$\frac{\partial}{\partial z}\mathbf{E}_\perp + i\omega\mathsf{P}\cdot\mathbf{n}\times\mathbf{B}_\perp = \mathbf{k}_\perp\frac{1}{\omega\varepsilon}J_z, \qquad (6.138b)$$

where

$$\mathsf{P} = 1_\perp - \frac{c^2}{\omega^2}\mathbf{k}_\perp\mathbf{k}_\perp. \qquad (6.139)$$

(Recall that c is the speed of light in the medium.) Check that another version of these equations is

$$\frac{\partial}{\partial z}\mathbf{n}\times\mathbf{H}_\perp + i\omega\varepsilon\mathsf{Q}\cdot\mathbf{E}_\perp = \mathbf{J}_\perp, \qquad (6.140a)$$

$$-\frac{\partial}{\partial z}\mathbf{n}\times\mathbf{E}_\perp + i\omega\mathsf{Q}\cdot\mathbf{B}_\perp = -\mathbf{n}\times\mathbf{k}_\perp\frac{1}{\omega\varepsilon}J_z, \qquad (6.140b)$$

in which

$$\mathsf{Q} = 1_\perp - \frac{c^2}{\omega^2}\mathbf{n}\times\mathbf{k}_\perp\mathbf{n}\times\mathbf{k}_\perp. \qquad (6.141)$$

7. Show that

$$\mathsf{P}\cdot\mathsf{Q} = \mathsf{Q}\cdot\mathsf{P} = \left(1 - \frac{c^2k_\perp^2}{\omega^2}\right)1_\perp, \qquad (6.142)$$

and that

$$\mathsf{P} + \mathsf{Q} = \left(2 - \frac{c^2k_\perp^2}{\omega^2}\right)1_\perp. \qquad (6.143)$$

Exhibit the quadratic equation that P and Q individually obey. The characteristic (eigen) vectors and values of P and Q and defined by, for example,

$$\mathsf{P} \cdot \mathbf{V} = \lambda \mathbf{V} \ . \tag{6.144}$$

Find the eigenvectors and eigenvalues for P and Q. Do the eigenvalues obey the quadratic equation? What is the geometrical relation between the eigenvalues?

8. Consider a dielectric waveguide with a discontinuous dielectric constant:

$$\varepsilon(z) = \begin{cases} \epsilon_1, & 0 < z < a, \\ \epsilon_2, & z < 0, z > a. \end{cases} \tag{6.145}$$

suppose $\epsilon_2 > \epsilon_1$, and the frequency is so chosen that κ is real for $0 < z < a$ but imaginary outside this range. The longitudinal mode functions for this dielectric waveguide are of the form $\cos(\kappa z - a)$ for $0 < z < a$ and decrease exponentially, as specified by κ, for $z < 0$ and $z > a$. What are the implications of the boundary conditions at $z = 0$ and a? Consider both polarizations.

9. Dielectric 2, a slab of thickness $2a$, is embedded in dielectric 1, which contains sources of waves, with frequency ω and transverse propagation vector $\mathbf{k}_\perp$, that are polarized parallel to the interfaces between the media ($\perp$ polarization). Two symmetrical situations are considered:

   (a) Equal but oppositely signed sources are disposed on the respective sides of the slab, leading to the vanishing of the electric field at the center of the slab.

   (b) Equal sources are used, leading to a maximum of the electric field at the center of the slab.

   By adding the fields and sources of the two circumstances, find the transmitted and reflected amplitudes of waves incident on the slab. Check conservation of energy, both when waves do, and do not, propagate in the slab.

10. Consider a dielectric body in motion with velocity $\mathbf{V}$. Show that it exhibits an intensity of magnetization

$$\mathbf{M} = \mathbf{P} \times \mathbf{V} \ . \tag{6.146}$$

Then recognize that another effect of such motion will be to alter the relation between $\mathbf{P}$ and $\mathbf{E}$ into

$$\mathbf{P} = (\varepsilon - \varepsilon_0)(\mathbf{E} + \mathbf{V} \times \mathbf{B}) \ . \tag{6.147}$$

Now write Maxwell's equations for a wave of frequency ω and propagation vector $\mathbf{k}$ parallel to $\mathbf{V}$ in source-free space, assuming that $|\mathbf{V}|/c \ll 1$. Conclude that, in the moving body, the light speed, ω/k, is

$$c'(\mathbf{V}) = \frac{c}{n} + \left(1 - \frac{1}{n^2}\right)V \equiv c' + \left(1 - \left(\frac{c'}{c}\right)^2\right)V \ . \tag{6.148}$$

Is this familiar?

11. The answer to the previous problem suffices if the medium is nondispersive (n independent of ω). More generally, we must note that the electric field determining the polarization,

$$\mathbf{E}(\mathbf{r}, t) \sim e^{i(\mathbf{k \cdot r} - \omega t)} \, , \tag{6.149}$$

is evaluated at points fixed in the moving body. Infer the correct form of $c'(V)$ for light of frequency ω when the frequency dependence of n is recognized.

12. The speed of energy transport in a dispersive medium is not c' but

$$c'' = \frac{\mathrm{d}\omega}{\mathrm{d}k} \, , \tag{6.150}$$

according to (6.99). Show that

$$c''(V) = c'' + \left(1 - \left(\frac{c''}{c}\right)^2\right) V \, , \tag{6.151}$$

and make explicit the implicit reference to frequency. Should this result have been anticipated?

13. Prove that the lowest E mode of a hollow waveguide is characterized by a function φ which is nowhere negative, and vanishes only on the boundary.

Rectangular and Triangular Waveguides

To illustrate the general waveguide theory thus far developed, we shall determine the mode functions and associated eigenvalues for those few guide shapes that permit exact analytical treatment. In this chapter we will discuss guides constructed from plane surfaces. Circular boundaries will be the subject of Chap. 8.

7.1 Rectangular Waveguide

The cross section of a rectangular guide with dimensions a and b is shown in Fig. 7.1, together with a convenient coordinate system. The positive sense

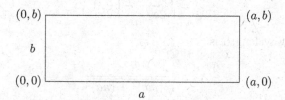

Fig. 7.1. Cross section of a rectangular waveguide, with sides a and b, and coordinates labeled (x, y). The direction of z comes out of the page

of the z-axis is assumed to be out of the plane of the page. The wave equations (6.22a) and (6.23a) can be separated in rectangular coordinates; that is, particular solutions for both E-mode and H-mode functions can be found in the form $X(x)Y(y)$, where the two functions thus introduced satisfy one-dimensional wave equations,

$$\left(\frac{\mathrm{d}^2}{\mathrm{d}x^2} + \gamma_x^2\right) X(x) = 0 \,, \tag{7.1a}$$

$$\left(\frac{d^2}{dy^2} + \gamma_x^2\right) Y(y) = 0 , \tag{7.1b}$$

where

$$\gamma^2 = \gamma_x^2 + \gamma_y^2 . \tag{7.1c}$$

The E-mode functions are characterized by the boundary conditions

$$X(0) = X(a) = 0 , \qquad Y(0) = Y(b) = 0 , \tag{7.2a}$$

while the H-mode functions are determined by

$$\frac{d}{dx}X(0) = \frac{d}{dx}X(a) = 0 , \qquad \frac{d}{dy}Y(0) = \frac{d}{dy}Y(b) = 0 . \tag{7.2b}$$

The scalar mode functions that satisfy these requirements are ($m, n = 0, 1, 2, \ldots$)

$$\text{E mode:} \quad \varphi(x,y) = \sin\frac{m\pi x}{a} \sin\frac{n\pi y}{b} , \tag{7.3a}$$

$$\text{H mode:} \quad \psi(x,y) = \cos\frac{m\pi x}{a} \cos\frac{n\pi y}{b} , \tag{7.3b}$$

and for both types of modes:

$$\gamma^2 = \left(\frac{m\pi}{a}\right)^2 + \left(\frac{n\pi}{b}\right)^2 . \tag{7.4}$$

The latter result can also be written as

$$\frac{1}{\lambda_c^2} = \left(\frac{m}{2a}\right)^2 + \left(\frac{n}{2b}\right)^2 , \tag{7.5}$$

or the cutoff wavelength is

$$\lambda_c = \frac{2ab}{\sqrt{(mb)^2 + (na)^2}} . \tag{7.6}$$

Observe that neither of the integers m or n can be zero for an E mode, or the scalar function vanishes. Although no H mode corresponding to $m = n = 0$ exists, since a constant scalar function generates no electromagnetic field, one of the integers can be zero. Thus, there exists a doubly infinite set of E modes characterized by the integers $m, n = 1, 2, 3, \ldots$; the general member of the set is designated as the E_{mn} mode, the two integers replacing the index a of the general theory. Similarly, a doubly infinite set of H modes exists corresponding to the various combinations of integers $m, n = 0, 1, 2, \ldots$, with $m = n = 0$ excluded. A particular H mode characterized by m and n is called the H_{mn} mode. The double subscript notation is extended to the various quantities describing a mode; thus, the critical wavenumber of an E_{mn} or H_{mn} mode is written γ_{mn}.

The scalar function associated with an E_{mn} mode, normalized in accordance with (6.57), is

$$\varphi_{mn}(x,y) = \frac{2}{\gamma_{mn}\sqrt{ab}} \sin\frac{m\pi x}{a} \sin\frac{n\pi y}{b}$$

$$= \frac{2}{\pi} \frac{1}{\sqrt{m^2\frac{b}{a} + n^2\frac{a}{b}}} \sin\frac{m\pi x}{a} \sin\frac{n\pi y}{b} , \qquad (7.7)$$

and the various field components for the E_{mn} modes are from (6.32a)–(6.32g)

$$E_x = -\frac{2}{a} \frac{m}{\sqrt{m^2\frac{b}{a} + n^2\frac{a}{b}}} \cos\frac{m\pi x}{a} \sin\frac{n\pi y}{b} V(z) , \qquad (7.8a)$$

$$E_y = -\frac{2}{b} \frac{n}{\sqrt{m^2\frac{b}{a} + n^2\frac{a}{b}}} \sin\frac{m\pi x}{a} \cos\frac{n\pi y}{b} V(z) , \qquad (7.8b)$$

$$H_x = \frac{2}{b} \frac{n}{\sqrt{m^2\frac{b}{a} + n^2\frac{a}{b}}} \sin\frac{m\pi x}{a} \cos\frac{n\pi y}{b} I(z) , \qquad (7.8c)$$

$$H_y = -\frac{2}{a} \frac{m}{\sqrt{m^2\frac{b}{a} + n^2\frac{a}{b}}} \cos\frac{m\pi x}{a} \sin\frac{n\pi y}{b} I(z) , \qquad (7.8d)$$

$$E_z = \frac{i\zeta}{k} 2\pi \frac{\sqrt{m^2\frac{b}{a} + n^2\frac{a}{b}}}{ab} \sin\frac{m\pi x}{a} \sin\frac{n\pi y}{b} I(z) , \qquad (7.8e)$$

$$H_z = 0 . \qquad (7.8f)$$

Of this set of waveguide fields, the E_{11} mode has the smallest cutoff wavenumber,

$$\gamma_{11} = \pi\sqrt{\frac{1}{a^2} + \frac{1}{b^2}} = \pi\frac{\sqrt{a^2 + b^2}}{ab} , \qquad (7.9)$$

or, equivalently, the largest cutoff wavelength,

$$(\lambda_c)_{11} = 2\frac{ab}{\sqrt{a^2 + b^2}} . \qquad (7.10)$$

This wavelength may be designated as the absolute E-mode cutoff wavelength, in the sense that if the intrinsic wavelength in the guide exceeds $(\lambda_c)_{11}$, no E modes can be propagated. The E_{11} mode is called the dominant E mode, for in the frequency range between the absolute E-mode cutoff frequency and the next smallest E-mode cutoff frequency (that of the E_{21} mode if a is the larger dimension), E-mode wave propagation in the guide is restricted to the E_{11} mode.

It is convenient to consider separately the H_{mn} modes for which neither integer is zero, and the set of modes for which one integer vanishes, H_{m0} and H_{0n}. The scalar function associated with a member of the former set is

$$\psi_{mn}(x,y) = \frac{2}{\pi} \frac{1}{\sqrt{m^2 \frac{b}{a} + n^2 \frac{a}{b}}} \cos \frac{m\pi x}{a} \cos \frac{n\pi y}{b} \,, \quad m,n \neq 0\,, \qquad (7.11)$$

normalized according to (6.59); the field components are, according to (6.33a)–(6.33g)

$$E_x = \frac{2}{b} \frac{n}{\sqrt{m^2 \frac{b}{a} + n^2 \frac{a}{b}}} \cos \frac{m\pi x}{a} \sin \frac{n\pi y}{b} V(z)\,, \qquad (7.12a)$$

$$E_y = -\frac{2}{a} \frac{m}{\sqrt{m^2 \frac{b}{a} + n^2 \frac{a}{b}}} \sin \frac{m\pi x}{a} \cos \frac{n\pi y}{b} V(z)\,, \qquad (7.12b)$$

$$H_x = \frac{2}{a} \frac{m}{\sqrt{m^2 \frac{b}{a} + n^2 \frac{a}{b}}} \sin \frac{m\pi x}{a} \cos \frac{n\pi y}{b} I(z)\,, \qquad (7.12c)$$

$$H_y = \frac{2}{b} \frac{n}{\sqrt{m^2 \frac{b}{a} + n^2 \frac{a}{b}}} \cos \frac{m\pi x}{a} \sin \frac{n\pi y}{b} I(z)\,, \qquad (7.12d)$$

$$H_z = \frac{i\eta}{k} 2\pi \frac{\sqrt{m^2 \frac{b}{a} + n^2 \frac{a}{b}}}{ab} \cos \frac{m\pi x}{a} \cos \frac{n\pi y}{b} V(z)\,, \qquad (7.12e)$$

$$E_z = 0\,. \qquad (7.12f)$$

If $n = 0$, the appropriately normalized scalar function is

$$\psi_{m0}(x,y) = \frac{1}{m\pi} \sqrt{\frac{2a}{b}} \cos \frac{m\pi x}{a}\,, \qquad (7.13)$$

which differs by a factor of $1/\sqrt{2}$ from the result obtained on placing $n = 0$ in (7.11). The field components of the H_{m0} mode are

$$E_x = 0\,, \qquad (7.14a)$$

$$E_y = -\sqrt{\frac{2}{ab}} \sin \frac{m\pi x}{a} V(z)\,, \qquad (7.14b)$$

$$H_x = \sqrt{\frac{2}{ab}} \sin \frac{m\pi x}{a} I(z)\,, \qquad (7.14c)$$

$$H_y = 0\,, \qquad (7.14d)$$

$$E_z = 0\,, \qquad (7.14e)$$

$$H_z = \frac{i\eta}{ka} m\pi \sqrt{\frac{2}{ab}} \cos \frac{m\pi x}{a} V(z)\,, \qquad (7.14f)$$

of which the most notable feature is the absence of all electric field components save E_y. We also observe that $H_y = 0$, whence the H_{m0} modes behave like E modes with respect to the y-axis. The field structure of the H_{0n} modes is analogous, with the x-axis as the preferred direction. The smallest H-mode cutoff wavenumber is that of the H_{10} mode if a is the larger dimension:

$$\gamma_{10} = \frac{\pi}{a} \; . \tag{7.15}$$

Thus, the cutoff wavelength is simply

$$(\lambda_c)_{10} = 2a \; , \tag{7.16}$$

and is independent of the dimension b, the height of the guide. We may designate $(\lambda_c)_{10}$ both as the absolute H-mode cutoff wavelength, and as the absolute cutoff wavelength of the guide, for the H_{10} has the smallest critical frequency of all the waveguide modes. Thus the H_{10} mode is called the dominant H mode, and the dominant mode of the guide, for in the frequency range between the absolute cutoff frequency of the guide and the next smallest cutoff frequency (which mode this represents depends upon the ratio b/a), only H_{10} wave propagation can occur. It is evident from the result (7.6) that a hollow waveguide is intrinsically suited to microwave frequencies, if metal tubes of convenient dimensions are to be employed. Furthermore, the frequency range over which a rectangular guide can be operated as a simple transmission line (only dominant mode propagation) is necessarily less than an octave, for the restrictions thereby imposed on the wavelength are $2a > \lambda > a$ (expressing the gap between the $m = 1$, $n = 0$ and the $m = 2$, $n = 0$ modes) and $\lambda > 2b$ (which marks the beginning of the $m = 0$, $n = 1$ mode). (If $b > a/\sqrt{3}$ the $m = 1$, $n = 1$ mode sets in before λ gets as small as a.)

The field components of the H_{10} mode contained in (7.14a)–(7.14f) involve voltages and currents defined with respect to the field impedance choice of characteristic impedance. (Recall Sect. 6.5.) If the definition discussed in (6.85) is adopted, $N = 2b/a$ (because the + subscript on the integral in (6.85) means that x ranges only from 0 to $a/2$), and the nonvanishing H_{10} mode field components obtained by $V \to N^{-1/2}V$, $I \to N^{1/2}I$ read

$$E_y = -\frac{1}{b} \sin \frac{\pi x}{a} V(z) \; , \tag{7.17a}$$

$$H_x = \frac{2}{a} \sin \frac{\pi x}{a} I(z) \; , \tag{7.17b}$$

$$H_z = \mathrm{i}\frac{\pi}{\kappa a}\frac{2}{a} \cos \frac{\pi x}{a} YV(z) \; , \tag{7.17c}$$

where the characteristic admittance is obtained by dividing (6.36d) by N, or

$$Y = \eta\frac{\kappa}{k}\frac{a}{2b} \; . \tag{7.18}$$

Thus the voltage at a given cross section is defined as the (negative of the) line integral of the electric field between the top and bottom faces of the guide taken along the line of maximum field intensity, $x = a/2$. Some important properties of the dominant mode emerge from a study of the surface current flowing in the various guide walls. On the side walls, $x = 0, a$, the only tangential magnetic field component is H_z and therefore from (4.64b) the surface current flows entirely in the y-direction. The surface current density on the $x = 0$ face is $K_y = -H_z$, while that on the $x = a$ face is $K_y = H_z$. However, since H_z is of opposite sign on the two surfaces, the current flows in the same direction on both side surfaces, with the density

$$x = 0, a : \quad K_y = -\mathrm{i}\frac{\pi}{\kappa a}\frac{2}{a}YV(z) . \tag{7.19}$$

On the top and bottom surfaces, $y = b, 0$, there are two tangential field components, H_x and H_z, and correspondingly two components of surface current, $K_z = \pm H_x$ and $K_x = \mp H_z$. The upper and lower signs refer to the top and bottom surface, respectively. On the top surface ($y = b$), the lateral and longitudinal components of surface current are

$$K_x = -\mathrm{i}\frac{\pi}{\kappa a}\frac{2}{a}\cos\frac{\pi x}{a}YV(z) , \tag{7.20a}$$

$$K_z = \frac{2}{a}\sin\frac{\pi x}{a}I(z) . \tag{7.20b}$$

Equal and opposite currents flow on the bottom surface $y = 0$. Since $\cos\frac{\pi x}{a}$ has reversed signs on opposite sides of the center line $x = a/2$, and vanishes at the latter point, it is clear that the transverse current flowing up the side walls continues to move toward the center of the top surface, but with diminishing magnitude, and vanishes at $x = a/2$. Thus the lines of current flow must turn as the center is approached, the flow becoming entirely longitudinal at $x = a/2$. In agreement with this, we observe that K_z vanishes at the boundaries of the top surface $x = 0, a$, and is a maximum at the center line. As a function of position along the guide, the direction of current flow reverses every half guide wavelength in consequence of the corresponding behavior of the current and voltage. The essential character of the current distribution can also be seen by noting that the surface charge density $\tau = -\mathbf{n} \cdot \mathbf{D} = -\varepsilon \mathbf{n} \cdot \mathbf{E}$ is confined to the top and bottom surfaces where the charge densities are $\tau = \mp \varepsilon E_y$, respectively. Thus, at $y = b$

$$\tau = \frac{\varepsilon}{b}\sin\frac{\pi x}{a}V(z) . \tag{7.21}$$

Hence, current flows around the circumference of the guide between the charges of opposite sign residing on the top and bottom surfaces, and current flows longitudinally on each surface between the oppositely charged regions separated by half the guide wavelength. A contour plot of the intensity of the current illustrating these general remarks is given in Fig. 7.2. The most immediate practical conclusion to be drawn from this analysis follows from the fact that no current crosses the center line of the top and bottom surfaces. Thus, if the metal were cut along the center line no disturbance of the current flow or of the field in the guide would result. In practice, a slot of appreciable dimensions can be cut in the guide wall without appreciable effect, which permits the insertion of a probe to determine the state of the field, without thereby markedly altering the field to be measured.

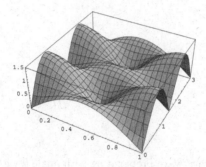

Fig. 7.2. Contour plot of the current intensity flowing on the top surface of the waveguide in the dominant mode H_{10}, in units of $2I/a$. Here we have chosen $\lambda_g = 3a$. The coordinates shown are in units of a

The field of a propagating mode in a rectangular waveguide can be regarded as a superposition of elementary plane waves arising from successive reflections at the various inner guide surfaces. This is illustrated most simply for a progressive H_{10} wave, where the single component of the electric field has the form, from (7.17a)

$$E_y = -\frac{V}{b}\sin\frac{\pi x}{a}e^{i\kappa z} = \frac{i}{2}\frac{V}{b}\left[e^{i(\kappa z+\pi x/a)} - e^{i(\kappa z-\pi x/a)}\right], \qquad (7.22)$$

which is simply a superposition of the two plane waves $\exp[ik(\cos\theta z \pm \sin\theta x)]$ that travel in the x–z-plane at an angle θ with respect to the z-axis. Here

$$\cos\theta = \frac{\kappa}{k}, \qquad \sin\theta = \frac{1}{k}\frac{\pi}{a} \qquad (7.23)$$

which defines a real angle since $\kappa < k$, and in fact

$$k^2 - \kappa^2 = \gamma_{10}^2 = \left(\frac{\pi}{a}\right)^2. \qquad (7.24)$$

Each plane wave results from the other on reflection at the two surfaces $x = 0$ and a, and each component is an elementary free-space wave with its electric vector directed along the y-axis and its magnetic vector contained in the plane of propagation. As an obvious generalization to be drawn from this simple situation, we may regard the various propagating modes in a waveguide as the result of the coherent interference of the secondary plane waves produced by the successive reflection of an elementary wave at the guide walls. From this point of view, the difference between E and H modes is just that of the polarizations of the plane wave components. The plane wave point of view also affords a simple picture of the phase and group velocities associated with a guide mode. A study of Fig. 7.3, which depicts a plane wave moving with speed c at an angle θ relative to the z-axis, shows that during the time interval dt the z-projection of a point on an equiphase surface advances a distance

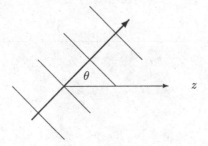

Fig. 7.3. Elementary wave propagating with angle θ with respect to z-axis

$c \cos \theta \, dt$, while the z-intercept of the equiphase surface advances a distance $\frac{c}{\cos \theta} dt$. Thus the group velocity is

$$v = c \cos \theta = c \frac{\kappa}{k} \, , \tag{7.25a}$$

while the phase velocity is

$$u = \frac{c}{\cos \theta} = c \frac{k}{\kappa} \, , \tag{7.25b}$$

in agreement with our previous results (6.97) and (6.98). The different nature of these velocities in particularly apparent at the cutoff frequency, where the group velocity is zero and the phase velocity infinite. From the plane wave viewpoint, the field then consists of elementary wave moving perpendicularly to the guide axis. There is then no progression of waves along the guide – zero group velocity – while all points on a line parallel to the z-axis have the same phase – infinite phase velocity. Thus the essential difference between the two velocities is contained in the statement that the former is a physical velocity, the latter a geometrical velocity.

It has been remarked above that the H_{m0} modes behave like E modes with respect to the y-axis. A complete set of waveguide fields with E or H character relative to the y-axis can be constructed from those already obtained. The E_{mn} and H_{mn} modes associated with the same nonvanishing integers are degenerate, since they possess equal critical frequencies and propagation constants. Furthermore, the respective transverse field components manifest the same dependence upon x and y, cf. (7.8a)–(7.8d) and (7.12a)–(7.12d). Hence, by a suitable linear combination of these modes, it is possible to construct fields for which either E_y or H_y vanishes. Similar remarks apply to the x-axis. Thus, a decomposition into E and H modes can be performed with any of the three axes as preferred directions. The result is to be anticipated from the general analysis of Sect. 6.1, since a rectangular waveguide has cylindrical symmetry with respect to all three axes.

7.2 Isosceles Right Triangular Waveguide

A square waveguide, $a = b$, has a further type of degeneracy, since the E_{mn} and E_{nm} modes have the same critical frequency, as do the H_{mn} and H_{nm} modes. By suitable linear combinations of these degenerate modes, it is possible to construct the mode functions appropriate to a guide with a cross section in the form of an isosceles right triangle. The (unnormalized) mode function,

$$\varphi_{mn}(x, y) = \sin \frac{m\pi x}{a} \sin \frac{n\pi y}{a} - \sin \frac{n\pi x}{a} \sin \frac{m\pi y}{a} \,, \tag{7.26}$$

describes a possible E mode in a square guide, with the cutoff wavenumber

$$\gamma_{mn} = \frac{\pi}{a}\sqrt{m^2 + n^2} \,. \tag{7.27}$$

The function thus constructed vanishes on the line $y = x$ as well as on the boundaries $y = 0$, $x = a$, and therefore satisfies all the boundary conditions for an E mode in an isosceles right triangular guide, as shown in Fig. 7.4. The

Fig. 7.4. Isosceles right triangular waveguide (shown in cross section) obtained by bisecting a square waveguide by a plane diagonal to the square

linearly independent square guide E-mode function

$$\sin \frac{m\pi x}{a} \sin \frac{n\pi y}{a} + \sin \frac{n\pi x}{a} \sin \frac{m\pi y}{a} \,, \tag{7.28}$$

does not vanish on the line $y = x$, and therefore does not describe a possible triangular mode. Note that the function (7.26) vanishes if $m = n$, and that therefore an interchange of the integers produces a trivial change in sign of the function. Hence the possible E modes of an isosceles right triangular guide are obtained from (7.26) with the integers restricted by $0 < m < n$. Thus the dominant E mode corresponds to $m = 1$, $n = 2$, and has the cutoff wavelength $\lambda_c = \frac{2}{\sqrt{5}}a$. The mode function

$$\psi_{mn}(x, y) = \cos \frac{m\pi x}{a} \cos \frac{n\pi y}{a} + \cos \frac{n\pi x}{a} \cos \frac{m\pi y}{a} \,, \tag{7.29}$$

describing an H mode in the square guide, has a vanishing derivative normal to the line $y = x$:

$$\frac{\partial}{\partial n}\psi_{mn} = \frac{1}{\sqrt{2}}\left(-\frac{\partial}{\partial x} + \frac{\partial}{\partial y}\right)\psi_{mn} = 0 \,, \quad y = x \,, \tag{7.30}$$

and therefore satisfies all boundary conditions for an H mode in the triangular guide under consideration. The linearly independent function

$$\cos \frac{m\pi x}{a} \cos \frac{n\pi y}{a} - \cos \frac{n\pi x}{a} \cos \frac{m\pi y}{a} , \qquad (7.31)$$

is not acceptable for this purpose. The function (7.29) is symmetrical in the integers m and n, and therefore the possible H modes of an isosceles right triangular guide are derived from (7.29) with the integers restricted by $0 \leq m \leq n$, but with $m = n = 0$ excluded. Thus the dominant H mode, and the dominant mode of the guide, corresponds to $m = 0$, $n = 1$, and has the cutoff wavelength $\lambda_c = 2a$, which is identical with the dominant mode cutoff wavelength of a rectangular guide with the maximum dimension a.

The discussion just presented may appear incomplete, for although we have constructed a set of triangular modes from the modes of the square guide, the possibility remains that there exist other modes not so obtainable. We shall demonstrate that all the modes of the isosceles right triangle are contained among those of the square, thereby introducing a method that will prove fruitful in the derivation of the modes of an equilateral triangle. It will be convenient to use a coordinate system in which the x-axis coincides with the diagonal side of the triangle, the latter then occupying a space below the x-axis. Let $\varphi(x, y)$ be an E-mode function of the triangle, which therefore satisfies the wave equation (6.22a) and vanishes on the three triangular boundaries. We now *define* the function $\varphi(x, y)$ for positive y, in terms of its known values within the triangle for negative y, by

$$\varphi(x, y) = -\varphi(x, -y) , \quad y > 0 . \qquad (7.32)$$

The definition of the function is thereby extended to the triangle obtained from the original triangle by reflection in the x-axis, the two regions together forming a square, as seen in Fig. 7.4. The two parts of the extended function are continuous and have continuous normal derivatives across the line $y = 0$, since $\varphi(x, +0) = 0$, and

$$\varphi_y(x, y) \equiv \frac{\partial}{\partial y} \varphi(x, y) = \frac{\partial}{\partial(-y)} \varphi(x, -y) \equiv \varphi_y(x, -y) , \qquad (7.33)$$

whence $\varphi_y(x, +0) = \varphi_y(x, -0)$. Furthermore, $\varphi(x, y)$, $y > 0$, satisfies the wave equation:

$$\left(\frac{\partial^2}{\partial x^2} + \frac{\partial^2}{\partial y^2} + \gamma^2 \right) \varphi(x, y) = -\left(\frac{\partial^2}{\partial x^2} + \frac{\partial^2}{\partial(-y)^2} + \gamma^2 \right) \varphi(x, -y) = 0 , \qquad (7.34)$$

and clearly vanishes on the orthogonal sides of the triangle produced by reflection. Hence, the extended function satisfies all the requirements for an E-mode function of the square, and thus our contention is proved, since every triangular E-mode function generates an E mode of the square. The analogous H-mode discussion employs the reflection

$$\psi(x,y) = \psi(x,-y) \,, \quad y > 0 \tag{7.35}$$

to extend the definition of the function. All the necessary continuity and boundary conditions are easily verified.

7.3 Equilateral Triangular Waveguide

The method just discussed may be used to great advantage in the construction of a complete set of E and H modes associated with a guide that has a cross section in the form of an equilateral triangle. A mode function within the triangle can be extended to three neighboring equilateral triangles by reflection about the three lines, parallel to the sides of the triangle, that intersect the opposite vertices, as shown in Fig. 7.5. We now suppose that the function thus defined in these new regions is further extended by similar reflection processes, and so on, indefinitely, as sketched in Fig. 7.6. It is apparent from Fig. 7.6 that the infinite system of equilateral triangles so formed uniformly

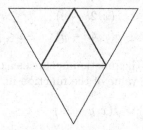

Fig. 7.5. Reflected equilateral triangle

cover the entire x–y plane. Hence, a function has been defined which is finite, continuous, has continuous derivatives, and satisfies the wave equation at every point of the x–y plane. Such a function must be composed of uniform plane waves, and therefore the modes of the equilateral triangle must be constructible from such plane waves. To proceed further, we observe that the extended mode function has special periodicity properties. Consider a row of triangles parallel to one of the three sides of the triangle. The value of the function at a point within a second neighboring parallel row is obtained by two successive reflections from the value at a corresponding point in the original row, and is therefore identical for both E and H modes. Hence the extended mode function must be periodic with respect to the three dimensions normal to the sides of the triangle, the periodicity interval being $2h$, where h is the length of the perpendicular from an apex to the opposite side. In terms of a, the base length of the triangle, $h = \frac{\sqrt{3}}{2}a$. To supply a mathematical proof for these statements we need merely write the equations that provide the successive definitions of the extended function in one of the directions normal to the sides of the triangle. Thus, for an E mode,

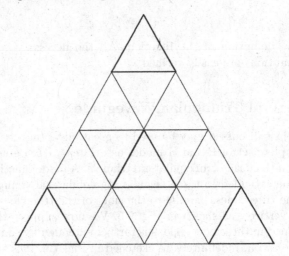

Fig. 7.6. Repeated reflection of equilateral triangle

$$\varphi(x,y) = -\varphi(x, 2h - y) , \quad h \leq y \leq 2h , \tag{7.36a}$$

$$\varphi(x,y) = -\varphi(x, 4h - y) , \quad 2h \leq y \leq 3h , \tag{7.36b}$$

which relates the values of the function in three successive rows parallel to the x-axis. On eliminating the value of the function in the center row, we obtain

$$\varphi(x,y) = \varphi(x, y + 2h) , \quad 0 \leq y \leq h , \tag{7.37}$$

which establishes the stated periodicity. The same proof is immediately applicable to translation in the other two directions, if a corresponding coordinate system is employed. The verification for H modes is similar. To construct a function having the desired periodicity properties, we introduce the three unit vectors normal to the sides of the triangle, and oriented in an inward sense (see Fig. 7.7)

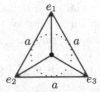

Fig. 7.7. Cross section of equilateral triangular cylinder

$$\mathbf{e}_1 = \mathbf{j} , \quad \mathbf{e}_2 = -\frac{\sqrt{3}}{2}\mathbf{i} - \frac{1}{2}\mathbf{j} , \quad \mathbf{e}_3 = \frac{\sqrt{3}}{2}\mathbf{i} - \frac{1}{2}\mathbf{j} , \tag{7.38}$$

and remark that the function of vectorial position in the plane, $\mathbf{r}$

$$F(\mathbf{r}) = \sum_{\lambda,\mu,\nu} f(\mathbf{r} - \lambda 2h\mathbf{e}_1 - \mu 2h\mathbf{e}_2 - \nu 2h\mathbf{e}_3) \tag{7.39}$$

is unaltered by a translation of magnitude $2h$ in any of the directions specified by $\mathbf{e}_1$, $\mathbf{e}_2$, and $\mathbf{e}_3$. The summation is to be extended over all integral values of λ, μ, ν. If $f(\mathbf{r})$ is a solution of the wave equation everywhere, $F(\mathbf{r})$ will also be a solution, with the proper periodicity. Furthermore, if $f(\mathbf{r})$ is a uniform plane wave, $F(\mathbf{r})$ will also be of this form, and by a proper combination of such elementary fields, we can construct the modes of the equilateral triangle. Thus, assuming that

$$f(\mathbf{r}) = e^{i\boldsymbol{\gamma}\cdot\mathbf{r}} , \tag{7.40}$$

where the magnitude of the real vector $\boldsymbol{\gamma}$ is the cutoff wavenumber of the mode, we find

$$F(\mathbf{r}) = e^{i\boldsymbol{\gamma}\cdot\mathbf{r}} \sum_{\lambda=-\infty}^{\infty} e^{-2ih\lambda\mathbf{e}_1\cdot\boldsymbol{\gamma}} \sum_{\mu=-\infty}^{\infty} e^{-2ih\mu\mathbf{e}_2\cdot\boldsymbol{\gamma}} \sum_{\nu=-\infty}^{\infty} e^{-2ih\nu\mathbf{e}_3\cdot\boldsymbol{\gamma}} . \tag{7.41}$$

Each summation is of the form

$$\sum_{\mu=-\infty}^{\infty} e^{i\mu x} , \tag{7.42}$$

which is a periodic function of x, with the periodicity 2π. Within the range $-\pi \leq x < \pi$, the series may be recognized as the Fourier series expansion of $2\pi\delta(x)$:

$$\delta(x) = \frac{1}{2\pi} \sum_{\mu=-\infty}^{\infty} e^{i\mu x} \int_{-\pi}^{\pi} dx' e^{-i\mu x'} \delta(x') = \frac{1}{2\pi} \sum_{\mu=-\infty}^{\infty} e^{i\mu x} . \tag{7.43}$$

We can therefore write, for all x,

$$\sum_{\mu=-\infty}^{\infty} e^{i\mu x} = 2\pi \sum_{m=-\infty}^{\infty} \delta(x - 2\pi m) , \tag{7.44}$$

since this equation reduces to (7.43) in the range $-\pi \leq x < \pi$, all delta function terms but the $m = 0$ one being nowhere different from zero in this range, and since both sides of the equation are periodic functions with periodicity 2π. The relation (7.44) is known as the Poisson sum formula. Applying this result to the three summations contained in (7.41), we obtain

$$F(\mathbf{r}) = e^{i\boldsymbol{\gamma}\cdot\mathbf{r}}(2\pi)^3 \sum_{l=-\infty}^{\infty} \sum_{m=-\infty}^{\infty} \sum_{n=-\infty}^{\infty} \delta(2h\mathbf{e}_1 \cdot \boldsymbol{\gamma} - 2\pi l)$$
$$\times \delta(2h\mathbf{e}_2 \cdot \boldsymbol{\gamma} - 2\pi m)\delta(2h\mathbf{e}_3 \cdot \boldsymbol{\gamma} - 2\pi n) . \tag{7.45}$$

Hence, the function vanishes entirely unless the arguments of all three delta functions vanish simultaneously, which requires that

$$\mathbf{e}_1 \cdot \boldsymbol{\gamma} = \frac{\pi l}{h}, \quad \mathbf{e}_2 \cdot \boldsymbol{\gamma} = \frac{\pi m}{h}, \quad \mathbf{e}_3 \cdot \boldsymbol{\gamma} = \frac{\pi n}{h}, \tag{7.46}$$

where l, m, and n are integers, positive, negative, or zero. The three integers may not be assigned independently, however, for

$$(\mathbf{e}_1 + \mathbf{e}_2 + \mathbf{e}_3) \cdot \boldsymbol{\gamma} = \frac{\pi}{h}(l + m + n) = 0, \tag{7.47}$$

since

$$\mathbf{e}_1 + \mathbf{e}_2 + \mathbf{e}_3 = \mathbf{0}, \tag{7.48}$$

which is evident from the geometry in Fig. 7.7, or from the explicit form of the vectors, (7.38). Therefore,

$$l + m + n = 0. \tag{7.49}$$

With the relations (7.46) and (7.49) we have determined the cutoff wavenumbers of all E and H modes in the triangular guide. Thus

$$(\mathbf{e}_1 \cdot \boldsymbol{\gamma})^2 + (\mathbf{e}_2 \cdot \boldsymbol{\gamma})^2 + (\mathbf{e}_3 \cdot \boldsymbol{\gamma})^2 = \boldsymbol{\gamma} \cdot (\mathbf{e}_1\mathbf{e}_1 + \mathbf{e}_2\mathbf{e}_2 + \mathbf{e}_3\mathbf{e}_3) \cdot \boldsymbol{\gamma}$$
$$= \frac{\pi^2}{h^2}(l^2 + m^2 + n^2). \tag{7.50}$$

However, it follows from the explicit form of the vectors that the dyadic

$$\mathbf{e}_1\mathbf{e}_1 + \mathbf{e}_2\mathbf{e}_2 + \mathbf{e}_3\mathbf{e}_3 = \frac{3}{2}(\mathbf{ii} + \mathbf{jj}) \tag{7.51}$$

a multiple of the unit dyadic in two dimensions,[1] whence

$$\gamma^2 = \frac{2}{3}\frac{\pi^2}{h^2}(l^2 + m^2 + n^2) = \frac{8}{9}\frac{\pi^2}{a^2}(l^2 + m^2 + n^2). \tag{7.52}$$

If we wish, l can be eliminated by means of (7.49), with the result

$$\gamma = \frac{4}{3}\frac{\pi}{a}\sqrt{m^2 + mn + n^2}. \tag{7.53}$$

We have shown that the elementary functions from which the triangular mode functions are to be constructed have the form

$$e^{i\boldsymbol{\gamma}\cdot\mathbf{r}}, \tag{7.54}$$

with the components of $\boldsymbol{\gamma}$ determined by (7.46). In view of the latter relations, and of (7.51), the quantity $\boldsymbol{\gamma} \cdot \mathbf{r}$ is conveniently rewritten as follows:

---

[1] The fact that the multiple is not unity signifies the overcompleteness of the vectors $\mathbf{e}_a$.

$$\boldsymbol{\gamma} \cdot \mathbf{r} = \boldsymbol{\gamma} \cdot (\mathbf{ii} + \mathbf{jj}) \cdot \mathbf{r} = \frac{2}{3} \boldsymbol{\gamma} \cdot (\mathbf{e}_1 \mathbf{e}_1 + \mathbf{e}_2 \mathbf{e}_2 + \mathbf{e}_3 \mathbf{e}_3) \cdot \mathbf{r}$$

$$= \frac{2}{3} \frac{\pi}{h} (l \mathbf{e}_1 \cdot \mathbf{r} + m \mathbf{e}_2 \cdot \mathbf{r} + n \mathbf{e}_3 \cdot \mathbf{r}) = \frac{2}{3} \frac{\pi}{h} (lu + mv + nw) . \quad (7.55)$$

The three new variables,

$$u = \mathbf{e}_1 \cdot \mathbf{r} = y , \tag{7.56a}$$

$$v = \mathbf{e}_2 \cdot \mathbf{r} = -\frac{\sqrt{3}}{2} x - \frac{1}{2} y , \tag{7.56b}$$

$$w = \mathbf{e}_3 \cdot \mathbf{r} = \frac{\sqrt{3}}{2} x - \frac{1}{2} y , \tag{7.56c}$$

known as trilinear coordinates, are the projections of the position vector on the three directions specified by the unit vectors $\mathbf{e}_1$, $\mathbf{e}_2$, $\mathbf{e}_3$, and are related by

$$u + v + w = 0 . \tag{7.57}$$

If the origin of coordinates is placed at the intersection of the three perpendiculars from each vertex to the opposite side, the trilinear coordinates of a point are the perpendicular distances from the origin to the three lines drawn through the point parallel to the sides of the triangle. A coordinate is considered negative if the line lies between the origin and the corresponding side; thus, the equations for the sides of the triangle are $u = -r$, $v = -r$, $w = -r$, where r is the radius of the inscribed circle (see Fig. 7.7)

$$r = \frac{1}{3} h = \frac{a}{2\sqrt{3}} . \tag{7.58}$$

To construct the mode functions, we must attempt to satisfy the boundary conditions by combining all functions of the form (7.45) that correspond to the same value of $\boldsymbol{\gamma}$. There are 12 such functions, of which six are obtained from (7.54) and (7.55) by permutation of the integers l, m, and n, and another six from these by a common sign reversal of l, m, and n, or equivalently, by taking the complex conjugate, cf. (7.41). It would not be possible to reverse the sign of l alone, for example, since the condition (7.49) would be violated. We first consider E modes and construct pairs of functions that vanish on the boundary $u = -r$. Thus

$$e^{i \frac{2\pi}{3h}(lu + mv + nw - lh)} - e^{-i \frac{2\pi}{3h}(lu + nv + mw - lh)} \tag{7.59}$$

has this property, for in consequence of the relations (7.49) and (7.57) it can be written

$$e^{i \frac{2\pi}{3h} \left[l\left(\frac{3}{2}u - h\right) + (m-n)\frac{1}{2}(v-w) \right]} - e^{i \frac{2\pi}{3h} \left[-l\left(\frac{3}{2}u - h\right) + (m-n)\frac{1}{2}(v-w) \right]}$$

$$= 2i \sin \frac{l\pi}{h} \left(u - \frac{2}{3} h \right) e^{i \frac{2\pi}{3h}(m-n)\frac{1}{2}(v-w)} ,$$

$$\tag{7.60}$$

which clearly vanishes when $u = -\frac{1}{3}h$. The two additional functions obtained by a cyclic permutation of l, m, n also vanish on the boundary $u = -r$, as do the complex conjugates of the three pairs of functions. Of course, the six functions and their conjugates could also be grouped in pairs so that they vanish on the boundary $v = -r$, or $w = -r$. It is now our task to construct one or more functions that have the proper grouping relative to all three boundaries, and are thus the desired general E-mode functions. It is easy to verify that the required functions are

$$\varphi(x,y) = e^{i\frac{2\pi}{3h}(lu+mv+nw-lh)} - e^{-i\frac{2\pi}{3h}(lu+nv+mw-lh)}$$
$$+ e^{i\frac{2\pi}{3h}(mu+nv+lw-mh)} - e^{-i\frac{2\pi}{3h}(mu+lv+nw-mh)}$$
$$+ e^{i\frac{2\pi}{3h}(nu+lv+mw-nh)} - e^{-i\frac{2\pi}{3h}(nu+mv+lw-nh)}, \qquad (7.61)$$

and its complex conjugate. Each E mode is therefore twofold degenerate, and the real and imaginary parts of (7.61) are equally admissible mode functions. Note that permutations of the integers l, m, n yield no new functions, for (7.61) is unchanged under the three cyclic permutations, and the other three permutations convert (7.61) into its negative complex conjugate. On combining the pairs of functions in the manner of (7.60), and taking real and imaginary parts, we obtain the two E-mode functions associated with the integers l, m, n in the form

$$\varphi(x,y) = \sin\frac{l\pi}{h}\left(y - \frac{2h}{3}\right)\frac{\cos}{\sin}\frac{\pi}{\sqrt{3}h}(m-n)x$$
$$+ \sin\frac{m\pi}{h}\left(y - \frac{2h}{3}\right)\frac{\cos}{\sin}\frac{\pi}{\sqrt{3}h}(n-l)x$$
$$+ \sin\frac{n\pi}{h}\left(y - \frac{2h}{3}\right)\frac{\cos}{\sin}\frac{\pi}{\sqrt{3}h}(l-m)x, \qquad (7.62)$$

where we have replaced u and $\frac{v-w}{2}$ by y and $-\frac{\sqrt{3}}{2}x$, respectively, according to (7.56a)–(7.56c).

To find the dominant E mode, we note that (7.62) vanishes identically if any integer is zero, and therefore the E mode of lowest cutoff wavenumber corresponds to $m = n = 1$, $l = -2$, and thus from (7.53) has the cutoff wavelength $\lambda_c = \frac{\sqrt{3}}{2}a = h$. It is important to observe that the E-mode function constructed from the sine functions of x vanishes if two integers are equal. Hence the dominant E mode is nondegenerate.

The H-mode function analogous to (7.61) is

$$\psi(x,y) = e^{i\frac{2\pi}{3h}(lu+mv+nw-lh)} + e^{-i\frac{2\pi}{3h}(lu+nv+mw-lh)}$$
$$+ e^{i\frac{2\pi}{3h}(mu+nv+lw-mh)} + e^{-i\frac{2\pi}{3h}(mu+lv+nw-mh)}$$
$$+ e^{i\frac{2\pi}{3h}(nu+lv+mw-nh)} + e^{-i\frac{2\pi}{3h}(nu+mv+lw-mh)}. \qquad (7.63)$$

To verify this statement, it must be shown that $\psi(x,y)$ has vanishing derivatives normal to each of the triangular sides. Now

$$\mathbf{e}_1 \cdot \boldsymbol{\nabla} = \frac{\partial}{\partial y} = \frac{\partial}{\partial u} - \frac{1}{2}\frac{\partial}{\partial v} - \frac{1}{2}\frac{\partial}{\partial w} = \frac{3}{2}\frac{\partial}{\partial u} , \tag{7.64}$$

since the operator $\frac{\partial}{\partial u} + \frac{\partial}{\partial v} + \frac{\partial}{\partial w}$ annihilates all of the 12 plane waves in consequence of $l + m + n = 0$. The symmetry of this relation is sufficient assurance for the validity of

$$\mathbf{e}_2 \cdot \boldsymbol{\nabla} = \frac{3}{2}\frac{\partial}{\partial v} , \qquad \mathbf{e}_3 \cdot \boldsymbol{\nabla} = \frac{3}{2}\frac{\partial}{\partial w} . \tag{7.65}$$

Thus, each normal derivative is proportional to the derivative with respect to the corresponding trilinear coordinate, and the application of any of these operators to (7.63) produces a form in which the same cancellation occurs on the boundaries as for the E modes. The twofold degenerate mode functions obtained as the real and imaginary parts of (7.63) are

$$\begin{aligned} \psi(x, y) = \cos\frac{l\pi}{h}\left(y - \frac{2h}{3}\right)\frac{\cos}{\sin}\frac{\pi}{\sqrt{3}h}(m - n)x \\ + \cos\frac{m\pi}{h}\left(y - \frac{2h}{3}\right)\frac{\cos}{\sin}\frac{\pi}{\sqrt{3}h}(n - l)x \\ + \cos\frac{n\pi}{h}\left(y - \frac{2h}{3}\right)\frac{\cos}{\sin}\frac{\pi}{\sqrt{3}h}(l - m)x . \end{aligned} \tag{7.66}$$

It should again be noted that a permutation of the integers produces no new mode, the function (7.66) either remaining invariant or reversing sign under such an operation. It is permissible for one integer to be zero, and therefore the dominant H mode, which is also the dominant mode of the guide, corresponds to $m = 1$, $n = -1$, $l = 0$, and has the cutoff wavelength $\lambda_c = \frac{3}{2}a$. Unlike the dominant E mode, the dominant H mode is doubly degenerate.

The mode functions of the equilateral triangle have interesting properties relative to coordinate transformations that leave the triangle invariant. If, for example, the coordinate system indicated in Fig. 7.7 is rotated counterclockwise through 120°, the triangle has the same aspect relative to the new coordinate system that it had in the original system. The relationship between the unit vectors and trilinear coordinates in the two coordinate systems is expressed by

$$\mathbf{e}_1 = \mathbf{e}_3' , \qquad \mathbf{e}_2 = \mathbf{e}_1' , \qquad \mathbf{e}_3 = \mathbf{e}_2' , \tag{7.67a}$$
$$u = w' , \qquad v = u' , \qquad w = v' , \tag{7.67b}$$

where primes indicate that the new reference system is involved. Now, the wave equation is invariant in form under a rotation of coordinates and the boundary conditions have the same form in the two systems. If therefore a mode function $f(u, v, w)$ is expressed in the new coordinates:

$$f(u, v, w) = f'(u', v', w') , \tag{7.68}$$

the new *function* $f'(u, v, w)$ must also be a possible mode function corresponding to the same eigenvalue, and can therefore be expressed as a linear combination of the degenerate mode functions associated with that eigenvalue. This conclusion is most easily verified by considering the complex E- and H-mode functions (7.61) and (7.63). It is not difficult to show that the effect of the substitution (7.67b) is to produce a function of the primed coordinates that has exactly the same form, save that each function is multiplied by the constant

$$C = e^{i\frac{2\pi}{3}(n-m)} = e^{i\frac{2\pi}{3}(l-n)} = e^{i\frac{2\pi}{3}(m-l)} . \tag{7.69}$$

The equivalence of these expressions follows from

$$(n - m) - (l - n) = 3n , \tag{7.70a}$$
$$(m - l) - (n - m) = 3m , \tag{7.70b}$$
$$(l - n) - (m - l) = 3l . \tag{7.70c}$$

Of course, the complex conjugate of a mode function is multiplied by $C^* = e^{i2\pi(m-n)/3}$. If two such rotations are applied in succession, implying that the coordinate system is rotated through 240°, each mode function is multiplied by C^2. If a rotation through 360° is performed, we must expect that the mode function preserve their original form, which requires that

$$C^3 = 1 , \tag{7.71}$$

as indeed it is. Hence all modes can be divided into three classes, depending on whether the mode function is multiplied by 1, $e^{2\pi i/3}$, or $e^{4\pi i/3} = e^{-2\pi i/3}$, under the influence of a rotation through 120°. The corresponding classification of $n - m$ is that it is divisible by 3 with a remainder that is either 0, 1, or 2. Thus, the dominant E-mode functions ($m = n = 1$) are invariant under rotation, while the dominant H-mode function ($m = -n = 1$) and its complex conjugate are multiplied by $e^{-2\pi i/3}$ and $e^{2\pi i/3}$, respectively, under a rotation through 120°. It should be noted that although the complex functions (7.61) and (7.63) are transformed into multiples of themselves by a rotation, the real and imaginary parts will not have this simple behavior, save for the class that is invariant under rotation.

Another coordinate transformation that leaves the triangle unaltered is reflection. If, in the coordinate system of Fig. 7.7, the positive sense of the x-axis is reversed, the triangle has the same aspect relative to the new system. Again the wave equation is unaltered in form by the reflection transformation

$$x = -x' , \quad y = y' , \tag{7.72a}$$

or

$$u = u' , \quad v = w' , \quad w = v' , \tag{7.72b}$$

and we conclude that a mode function expressed in the new coordinates must be a linear combination of the mode functions associated with the same eigenvalue. Indeed, the H-mode function (7.63) is converted into its complex conjugate, while the E-mode function (7.61) becomes its negative complex conjugate. However, it should be observed that now the real functions (7.62) and (7.66) change into multiples of themselves under the transformation, either remaining unaltered in form or reversing sign. It follows that two successive reflections produces no change in the mode function, as must be required. Two other reflection operations exist, associated with the remaining perpendiculars from the vertices to opposite sides. They are described by

$$u = w', \qquad v = v', \qquad w = u', \qquad (7.73a)$$

and

$$u = v', \qquad v = u', \qquad w = w'. \qquad (7.73b)$$

Thus, the entire set of symmetry operations of the triangle, consisting of three rotations and three reflections, are described by the six permutations of the trilinear coordinates u, v, and w. Only two of these transformations are independent, in the sense that successive application of them will generate all the other transformations. Thus, a rotation through 120°, $u = w'$, $v = u'$, $w = v'$, (7.67b), followed by the reflection (7.72b), $u' = u''$, $v' = w''$, $w' = v''$, is equivalent to $u = v''$, $v = u''$, $w = w''$, the reflection (7.73b). Similarly, a rotation through 240°, which is the transformation (7.67b) applied twice, followed by the reflection (7.72b), generates the reflection (7.73a). As a particular consequence of these statements, note that a mode function which is invariant with respect to rotations and one of the reflections, is also invariant under the other two reflections.

In constructing the complete set of E and H modes for a guide with a cross section in the form of an equilateral triangle, we have also obtained a complete solution for the modes of a right angle triangle with angles of 30° and 60°. The connection between the two problems is the same as that between the square and the isosceles right angle triangle. The set of equilateral E modes (7.62) that involve sine functions of x vanish on the line $x = 0$ and therefore satisfy all the boundary conditions for an E mode of the 30° and 60° triangles thus obtained. To find the dominant E mode we must recall that no integer can be zero and that no two integers can be equal for the mode that is an odd function of x. Hence the dominant mode corresponds to $m = 1$, $n = 2$, $l = -3$, and has from (7.53) the cutoff wavelength $\lambda_c = \frac{3}{2\sqrt{7}}a$, where a is now the length of the diagonal side. The set of equilateral triangular H modes (7.66) that contain cosine functions of x have vanishing derivatives normal to the y-axis and therefore yield the H modes of a 30° and 60° triangle. The dominant H mode of the latter triangle is the same as that of the equilateral triangle ($m = 1$, $n = -1$, $l = 0$) and has the same cutoff wavelength $\lambda_c = \frac{3}{2}a$. The modes of the equilateral triangle also furnish us with modes for a guide that has a cross section formed by any closed curve drawn along the lines of Fig. 7.6. In particular, we obtain modes for a cross section in the form of a regular hexagon, but only some of the modes are obtained in this way.

7.4 Problems for Chap. 7

1. Verify the analysis concerning the current flow in the guide walls for the H_{10} mode. In particular, show that the charge density (7.21) is consistent with the conservation of surface current (4.59).
2. Verify that the H modes that satisfy (7.35) fulfill all the necessary requirements to be H modes for the square, and that therefore all the H modes of the isosceles right triangle are contained within those for the square.

8

Electromagnetic Fields in Waveguides with Circular Cross Sections

Waveguides with cross-sectional boundaries composed wholly, or in part, of circular arcs are most conveniently discussed with the aid of polar coordinates r and ϕ,

$$x = r \cos \phi, \qquad y = r \sin \phi . \tag{8.1}$$

In terms of these coordinates, the wave equation satisfied by an E- or H-mode function reads

$$\left(\frac{\partial^2}{\partial r^2} + \frac{1}{r} \frac{\partial}{\partial r} + \frac{1}{r^2} \frac{\partial^2}{\partial \phi^2} + \gamma^2 \right) F(r, \phi) = 0 . \tag{8.2}$$

This wave equation can be separated; particular solutions exist in the form

$$F(r, \phi) = \Phi(\phi) Z(\xi) , \tag{8.3}$$

where

$$\left(\frac{\mathrm{d}^2}{\mathrm{d}\phi^2} + \mu^2 \right) \Phi(\phi) = 0 , \tag{8.4a}$$

$$\left(\frac{\mathrm{d}^2}{\mathrm{d}\xi^2} + \frac{1}{\xi} \frac{\mathrm{d}}{\mathrm{d}\xi} + 1 - \frac{\mu^2}{\xi^2} \right) Z(\xi) = 0 , \tag{8.4b}$$

and

$$\xi = \gamma r . \tag{8.5}$$

The solutions of (8.4a) are

$$\Phi(\phi) = \frac{\cos}{\sin} \mu\phi \quad \text{or} \quad \Phi(\phi) = \mathrm{e}^{\pm \mathrm{i}\mu\phi} , \tag{8.6}$$

where μ is as yet unrestricted. Equation (8.4b) will be recognized as Bessel's equation and its general solutions $Z_\mu(\xi)$ are known as circular cylinder functions of order μ. The particular solution $J_\mu(\xi)$, as defined below, will be specifically called a Bessel function of order μ as distinguished from the other linearly independent solutions of the equation.

8.1 Cylinder Functions

Cylinder functions of the complex variable ξ and the complex order μ may be defined as solutions of the pair of recurrence relations

$$Z_{\mu-1}(\xi) + Z_{\mu+1}(\xi) = \frac{2\mu}{\xi} Z_\mu(\xi) , \tag{8.7a}$$

$$Z_{\mu-1}(\xi) - Z_{\mu+1}(\xi) = 2\frac{\mathrm{d}}{\mathrm{d}\xi} Z_\mu(\xi) , \tag{8.7b}$$

or, equivalently, of

$$\frac{\mathrm{d}}{\mathrm{d}\xi} Z_\mu(\xi) + \frac{\mu}{\xi} Z_\mu(\xi) = Z_{\mu-1}(\xi) , \tag{8.8a}$$

$$-\frac{\mathrm{d}}{\mathrm{d}\xi} Z_\mu(\xi) + \frac{\mu}{\xi} Z_\mu(\xi) = Z_{\mu+1}(\xi) . \tag{8.8b}$$

It is an immediate consequence of these relations that

$$\left(\frac{\mathrm{d}^2}{\mathrm{d}\xi^2} + \frac{1}{\xi}\frac{\mathrm{d}}{\mathrm{d}\xi} + 1 - \frac{\mu^2}{\xi^2} \right) Z_\mu(\xi) = 0 . \tag{8.9}$$

A particular solution of (8.9), in the form of an ascending power series, is the Bessel function

$$J_\mu(\xi) = \sum_{n=0}^{\infty} (-1)^n \frac{\left(\frac{\xi}{2}\right)^{\mu+2n}}{n!\,\Gamma(\mu+n+1)} . \tag{8.10}$$

If μ is not an integer, the function $J_{-\mu}(\xi)$ provides the second independent solution of (8.9), although it must be multiplied by an odd periodic function of μ, with period unity, in order to satisfy the recurrence relations (8.8a) and (8.8b). When $\mu = m$ is integral, however,

$$J_m(\xi) = (-1)^m J_{-m}(\xi) , \tag{8.11}$$

and an additional function is required. For this purpose, a second solution of Bessel's equation, called Neumann's function, is defined by

$$N_\mu(\xi) = \frac{1}{\sin \mu\pi} \left[\cos \mu\pi J_\mu(\xi) - J_{-\mu}(\xi) \right] . \tag{8.12}$$

The two independent cylinder functions J_μ, N_μ satisfy the Wronskian relation

$$\xi \left[J_\mu(\xi)\frac{\mathrm{d}}{\mathrm{d}\xi} N_\mu(\xi) - N_\mu(\xi)\frac{\mathrm{d}}{\mathrm{d}\xi} J_\mu(\xi) \right] = \frac{2}{\pi} . \tag{8.13}$$

When μ is integral, the expression (8.12) is indeterminate, and the definition of the Neumann function of integral order, obtained from (8.12) by a limiting process, is

$$N_m(\xi) = \frac{1}{\pi}\left[\frac{\partial}{\partial\mu}J_\mu(\xi) - (-1)^m\frac{\partial}{\partial\mu}J_{-\mu}(\xi)\right]_{\mu=m} = (-1)^m N_{-m}(\xi) \, . \quad (8.14)$$

In particular

$$N_0(\xi) = \frac{2}{\pi}\left[\frac{\partial}{\partial\mu}J_\mu(\xi)\right]_{\mu=0} \, . \quad (8.15)$$

Expansions for the integral order Neumann functions, obtained in this manner, are

$$N_0(\xi) = \frac{2}{\pi}\log\frac{\gamma\xi}{2}J_0(\xi) + \frac{2}{\pi}\sum_{n=1}^\infty (-1)^{n+1}\frac{\left(\frac{\xi}{2}\right)^{2n}}{(n!)^2}\left(1 + \frac{1}{2} + \cdots + \frac{1}{n}\right) \, , \quad (8.16a)$$

$$N_m(\xi) = \frac{2}{\pi}\log\frac{\gamma\xi}{2}J_m(\xi) - \frac{1}{\pi}\sum_{n=0}^{m-1}\frac{(m-n-1)!}{n!}\left(\frac{2}{\xi}\right)^{m-2n}$$

$$-\frac{1}{\pi}\sum_{n=0}^\infty (-1)^n\frac{\left(\frac{\xi}{2}\right)^{m+2n}}{n!\,(m+n)!}\left(1 + \frac{1}{2} + \cdots + \frac{1}{n} + 1 + \frac{1}{2} + \cdots + \frac{1}{n+m}\right) \, .$$

$$(8.16b)$$

In the last summation the $n = 0$ term is to be understood as $1 + \frac{1}{2} + \cdots + \frac{1}{m}$. The quantity $\gamma = 1.78107\ldots$ is related to the Eulerian constant $C = \log\gamma = 0.577216\ldots$.

Bessel functions of integral order can also be defined with aid of the generating function

$$e^{\frac{1}{2}\xi(t-1/t)} = \sum_{m=-\infty}^\infty J_m(\xi)t^m \, . \quad (8.17)$$

On placing $t = ie^{i\phi}$, we obtain the expansion

$$e^{i\xi\cos\phi} = \sum_{m=-\infty}^\infty i^m J_m(\xi)e^{im\phi} = J_0(\xi) + 2\sum_{m=1}^\infty i^m J_m(\xi)\cos m\phi \, . \quad (8.18)$$

By regarding this as a Fourier expansion in ϕ, we get the integral representations

$$i^m J_m(\xi) = \frac{1}{2\pi}\int_0^{2\pi} d\phi\, e^{i\xi\cos\phi}e^{-im\phi} = \frac{1}{\pi}\int_0^\pi d\phi\, e^{i\xi\cos\phi}\cos m\phi \, , \quad (8.19)$$

or

$$(-1)^{m/2}J_m(\xi) = \frac{1}{\pi}\int_0^\pi d\phi\,\cos(\xi\cos\phi)\cos m\phi \, , \quad m \text{ even}\, , \quad (8.20a)$$

$$(-1)^{(m-1)/2}J_m(\xi) = \frac{1}{\pi}\int_0^\pi d\phi\,\sin(\xi\cos\phi)\cos m\phi \, , \quad m \text{ odd}\, . \quad (8.20b)$$

An equivalent integral representation is

$$J_m(\xi) = \frac{\left(\frac{1}{2}\xi\right)^m}{\Gamma\left(m + \frac{1}{2}\right)\Gamma\left(\frac{1}{2}\right)} \int_0^\pi d\phi \, \cos(\xi\cos\phi) \sin^{2m}\phi \, , \qquad (8.21)$$

which is also valid for a complex order, provided $\mathrm{Re}\,\mu > -\frac{1}{2}$.

In addition to the Bessel and Neumann functions, it is convenient to introduce two new cylinder functions

$$H_\mu^{(1)}(\xi) = J_\mu(\xi) + iN_\mu(\xi) \, , \qquad (8.22a)$$

$$H_\mu^{(2)}(\xi) = J_\mu(\xi) - iN_\mu(\xi) \, , \qquad (8.22b)$$

known as Hankel's functions of the first and second kind, respectively. They may be defined directly in terms of Bessel functions by

$$H_\mu^{(1)}(\xi) = \frac{i}{\sin\mu\pi} \left[e^{-\mu\pi i} J_\mu(\xi) - J_{-\mu}(\xi) \right] \, , \qquad (8.23a)$$

$$H_\mu^{(2)}(\xi) = -\frac{i}{\sin\mu\pi} \left[e^{\mu\pi i} J_\mu(\xi) - J_{-\mu}(\xi) \right] \, . \qquad (8.23b)$$

It follows from the latter forms that

$$H_{-\mu}^{(1)}(\xi) = e^{\mu\pi i} H_\mu^{(1)}(\xi) \, , \qquad H_{-\mu}^{(2)}(\xi) = e^{-\mu\pi i} H_\mu^{(2)}(\xi) \, , \qquad (8.24)$$

whereas similar relations only exist for Bessel and Neumann functions of integral order.

All cylinder functions, other than Bessel functions of integral order, are multiple valued. It is apparent from the series (8.10) that $J_\mu(\xi)/\xi^\mu$ is a single-valued, even function of ξ, and therefore, the multiple-valued nature of $J_\mu(\xi)$ is a consequence of the branch point possessed by ξ^μ at $\xi = 0$, when μ is not an integer. The principal branch of $\xi^\mu = e^{\mu\log\xi}$, and correspondingly that of $J_\mu(\xi)$, is defined by restricting the phase, or argument, of ξ to its principal value:

$$-\pi < \arg\xi \le \pi \, . \qquad (8.25)$$

It is convenient to define $J_\mu(\xi)$ when the argument of ξ is unrestricted by giving the values of $J_\mu\left(\xi e^{n\pi i}\right)$ where the argument of ξ has its principal value and n is any integer. Since $J_\mu(\xi)/\xi^\mu$ is an even single-valued function,

$$J_\mu\left(\xi e^{n\pi i}\right) = e^{n\mu\pi i} J_\mu(\xi) \qquad (8.26a)$$

and corresponding formulae for the Neumann and Hankel functions are obtained from the defining equations (8.14) and (8.22a)–(8.22b):

$$N_\mu\left(\xi e^{n\pi i}\right) = e^{-n\mu\pi i} N_\mu(\xi) + 2i\sin n\mu\pi \cot\mu\pi J_\mu(\xi) \, , \qquad (8.26b)$$

$$H_\mu^{(1)}\left(\xi e^{n\pi i}\right) = e^{-n\mu\pi i} H_\mu^{(1)}(\xi) - 2e^{-\mu\pi i} \sin n\mu\pi \csc\mu\pi J_\mu(\xi) \, , \qquad (8.26c)$$

$$H_\mu^{(2)}\left(\xi e^{n\pi i}\right) = e^{-n\mu\pi i} H_\mu^{(2)}(\xi) + 2e^{\mu\pi i} \sin n\mu\pi \csc\mu\pi J_\mu(\xi) \, . \qquad (8.26d)$$

An important consequence of these formulae is that

$$J_\mu(\alpha\xi)N_\mu(\beta\xi) - J_\mu(\beta\xi)N_\mu(\alpha\xi) \tag{8.27}$$

is an even single-valued function of ξ.

The behavior of $J_\mu(\xi)$ near the origin is indicated by the first term in (8.10):

$$J_\mu(\xi) \sim \frac{1}{\Gamma(\mu+1)} \left(\frac{\xi}{2}\right)^\mu , \quad |\xi| \ll 1 . \tag{8.28}$$

Thus all Bessel functions remain finite at the origin, provided Re μ is not negative. The form of the Neumann functions near the origin can be inferred from the definition (8.12):

$$N_\mu(\xi) \sim -\frac{\Gamma(\mu)}{\pi} \left(\frac{2}{\xi}\right)^\mu , \qquad \text{Re}\,\mu > 0 , \quad |\xi| \ll 1 , \tag{8.29a}$$

$$N_0(\xi) \sim \frac{2}{\pi} \log \frac{\gamma\xi}{2}, \quad |\xi| \ll 1 . \tag{8.29b}$$

Hence no Neumann function remains finite at the origin. The latter statement is also applicable to the Hankel functions:

$$H_\mu^{(1)}(\xi) \sim -\mathrm{i}\frac{\Gamma(\mu)}{\pi} \left(\frac{2}{\xi}\right)^\mu , \quad H_\mu^{(2)}(\xi) \sim \mathrm{i}\frac{\Gamma(\mu)}{\pi} \left(\frac{2}{\xi}\right)^\mu , \quad \text{Re}\,\mu > 0 ,$$
$$|\xi| \ll 1 \tag{8.30a}$$

$$H_0^{(1)}(\xi) \sim \frac{2\mathrm{i}}{\pi} \log \frac{\gamma\xi}{2\mathrm{i}} , \quad H_0^{(2)}(\xi) \sim -\frac{2\mathrm{i}}{\pi} \log \frac{\mathrm{i}\gamma\xi}{2} . \tag{8.30b}$$

The values of the Hankel functions for sufficiently large magnitudes of ξ can be obtained from the asymptotic expansions

$$H_\mu^{(1)}(\xi) \sim \sqrt{\frac{2}{\pi\xi}} e^{\mathrm{i}(\xi-\mu\pi/2-\pi/4)} [P_\mu(\xi) + \mathrm{i}Q_\mu(\xi)] , \tag{8.31a}$$

$$|\xi| \gg 1 , \quad |\xi| \gg \mu$$

$$H_\mu^{(2)}(\xi) \sim \sqrt{\frac{2}{\pi\xi}} e^{-\mathrm{i}(\xi-\mu\pi/2-\pi/4)} [P_\mu(\xi) - \mathrm{i}Q_\mu(\xi)] , \tag{8.31b}$$

where

$$P_\mu(\xi) = 1 - \frac{(4\mu^2 - 1)(4\mu^2 - 9)}{2!\,(8\xi)^2} + \frac{(4\mu^2 - 1)(4\mu^2 - 9)(4\mu^2 - 25)(4\mu^2 - 49)}{4!\,(8\xi)^4}$$
$$- \cdots , \tag{8.32a}$$

$$Q_\mu(\xi) = \frac{(4\mu^2 - 1)}{8\xi} - \frac{(4\mu^2 - 1)(4\mu^2 - 9)(4\mu^2 - 25)}{3!\,(8\xi)^3} + \cdots . \tag{8.32b}$$

These expansions are valid when the phase of ξ is restricted to the range $-\pi < \arg\xi < 2\pi$ for $H_\mu^{(1)}(\xi)$, and $-2\pi < \arg\xi < \pi$ for $H_\mu^{(2)}(\xi)$. The analogous expansions of the Bessel and Neumann functions are [cf. (8.22a)–(8.22b)]:

$$J_\mu(\xi) \sim \sqrt{\frac{2}{\pi\xi}} \left[\cos\left(\xi - \mu\frac{\pi}{2} - \frac{\pi}{4}\right) P_\mu(\xi) - \sin\left(\xi - \mu\frac{\pi}{2} - \frac{\pi}{4}\right) Q_\mu(\xi) \right] ,$$

$$(8.33a)$$

$$N_\mu(\xi) \sim \sqrt{\frac{2}{\pi\xi}} \left[\sin\left(\xi - \mu\frac{\pi}{2} - \frac{\pi}{4}\right) P_\mu(\xi) + \cos\left(\xi - \mu\frac{\pi}{2} - \frac{\pi}{4}\right) Q_\mu(\xi) \right] ,$$

$$|\xi| \gg 1 , \quad |\arg\xi| < \pi . \qquad (8.33b)$$

The latter formulae are applicable in the phase interval common to the expansions of the two Hankel functions, $-\pi < \arg\xi < \pi$. When μ is real and nonnegative and ξ real and positive, the remainder after p terms of the expansion of $P_\mu(\xi)$ is of the same sign and numerically less than the $(p+1)$th term, provided that $2p > \mu - \frac{1}{2}$; a corresponding statement applies to $Q_\mu(\xi)$ if $2p > \mu - \frac{3}{2}$. Notice that the series for $P_\mu(\xi)$ and $Q_\mu(\xi)$ terminate when 2μ is an odd integer and therefore provide exact rather than asymptotic expansions. Hence cylinder functions of half-integral order can be expressed in terms of elementary functions. For example,

$$H^{(1)}_{1/2}(\xi) = -i\sqrt{\frac{2}{\pi\xi}}e^{i\xi} , \qquad H^{(2)}_{1/2}(\xi) = i\sqrt{\frac{2}{\pi\xi}}e^{-i\xi} , \qquad (8.34a)$$

$$J_{1/2}(\xi) = \sqrt{\frac{2}{\pi\xi}}\sin\xi , \qquad N_{1/2}(\xi) = -\sqrt{\frac{2}{\pi\xi}}\cos\xi . \qquad (8.34b)$$

Finally, we record two useful indefinite integrals:

$$\int \xi\,d\xi\, Z_\mu(\alpha\xi)\overline{Z}_\mu(\beta\xi) = \frac{\xi}{\alpha^2 - \beta^2} \left[Z_\mu(\alpha\xi)\frac{d}{d\xi}\overline{Z}_\mu(\beta\xi) - \overline{Z}_\mu(\beta\xi)\frac{d}{d\xi}Z_\mu(\alpha\xi) \right] ,$$

$$(8.35a)$$

$$\int \xi\,d\xi\, Z_\mu^2(\xi) = \frac{1}{2}\xi^2 \left[\left(1 - \frac{\mu^2}{\xi^2}\right) Z_\mu^2(\xi) + \left[\frac{d}{d\xi}Z_\mu(\xi)\right]^2 \right] , \qquad (8.35b)$$

where Z_μ and $\overline{Z}_\mu$ denote any two cylinder functions of order μ.

8.2 Circular Guide

The simplest example of a guide to which the solutions (8.3) are applicable is one with a cross section in the form of a circle. Here, all values of ϕ are permissible, and the angular coordinates ϕ and $\phi + 2\pi$ designate the same physical point. In order that the mode functions be single valued, the quantity μ must be an integer $m = 0, 1, 2, \ldots$. Furthermore, the cylinder functions must be restricted to the Bessel function solutions, for the Neumann functions become infinite at the origin, implying that the wave equation is not satisfied at the latter point. Hence the mode functions of a hollow circular guide are of the form

$$J_m(\gamma r) \, {\textstyle{\cos \atop \sin}} \, m\phi \, , \quad m = 0, 1, 2, \ldots \, . \tag{8.36}$$

The E modes are obtained by the condition that the mode functions vanish on the circular boundary of radius a:

$$\text{E mode:} \quad J_m(\gamma a) = 0 \, , \tag{8.37a}$$

while the H modes are derived from the requirement that the radial derivative vanish at $r = a$:

$$\text{H mode:} \quad J_m'(\gamma a) = 0 \, . \tag{8.37b}$$

Thus the E and H cutoff wavenumbers are found from the zeros and extrema of the Bessel functions, respectively. In view of the oscillatory character of these functions, as indicated by the asymptotic form (8.33a), it is clear that for each m there exists an infinite number of modes of both types. The E mode associated with the nth nonvanishing root of J_m will be designated as the E_{mn} mode, and the cutoff wavenumber correspondingly written γ_{mn}'; the H mode associated with the nth nonvanishing root of J_m' will be called the H_{mn} mode and the cutoff wavenumber written γ_{mn}''. Unlike the rectangular guide, the cutoff wavenumbers of the E and H modes do not coincide, and the notation must be complicated accordingly. Explicit formulae for the two types of roots can be obtained from the Bessel function asymptotic formulae:

$$\gamma_{mn}' a \sim \left(m + 2n - \frac{1}{2}\right)\frac{\pi}{2} - \frac{4m^2 - 1}{4\pi\left(m + 2n - \frac{1}{2}\right)} - \frac{(4m^2 - 1)(28m^2 - 31)}{48\pi^3\left(m + 2n - \frac{1}{2}\right)^3}$$
$$- \cdots \, , \tag{8.38a}$$

$$\gamma_{mn}'' a \sim \left(m + 2n - \frac{3}{2}\right)\frac{\pi}{2} - \frac{4m^2 + 3}{4\pi\left(m + 2n - \frac{3}{2}\right)} - \frac{112m^4 + 328m^2 - 9}{48\pi^3\left(m + 2n - \frac{3}{2}\right)^3}$$
$$- \cdots \, . \tag{8.38b}$$

For each value of m, the successive roots are obtained by placing $n = 1, 2, \ldots$, with the exception of the H_{0n} modes, where the first nonvanishing root, 3.8317, corresponds to $n = 2$ in (8.38b). To avoid this slight difficulty, we restrict the applicability of (8.38b) to $m > 0$ and remark that the cutoff wavenumber of the H_{0n} mode equals that of the E_{1n} mode, for the roots of J_1 and $J_0' = -J_1$ coincide. Numerical values for both types or roots are listed in Tables 8.1 and 8.2 for small values of m and n. It is apparent from Table 8.1 that the dominant E mode is E_{01}, with the cutoff wavelength $(\lambda_c)_{01}' = 2.6127a$, while Table 8.2 assures us that the dominant H mode, and the dominant mode of the guide, is H_{11}, with the cutoff wavelength $(\lambda_c)_{11}'' = 3.4126a$.

The circular guide E-mode function $\varphi_{mn}(r, \phi)$, normalized in accordance with (6.57), or

$$(\gamma_{mn}')^2 \int_\sigma d\sigma \, \varphi_{mn}^2 = 1 \tag{8.39}$$

is

Table 8.1. $\gamma'_{mn}a$: nth zero of J_m. Only those zeroes below 25 are given

$m \setminus n$	1	2	3	4	5	6	7	8
0	2.40483	5.52008	8.65373	11.79153	14.93092	18.07106	21.21164	24.35247
1	3.83171	7.01559	10.17347	13.32369	16.47063	19.61586	22.76008	
2	5.13562	8.41724	11.61984	14.79595	17.95982	21.11700	24.27011	
3	6.38016	9.76102	13.01520	16.22347	19.40942	22.58273		
4	7.58834	11.06471	14.37254	17.61597	20.82693	24.01902		
5	8.77148	12.33860	15.70017	18.98013	22.21780			
6	9.93611	13.58929	17.00382	20.32079	23.58608			
7	11.08637	14.82127	18.28758	21.64154	24.93493			
8	12.22509	16.03778	19.55454	22.94517				
9	13.35430	17.2412	20.8070	24.2339				
10	14.47550	18.4335	22.0470					
11	15.58985	19.6160	23.2759					
12	16.6983	20.7899	24.4949					
13	17.8014	21.9562						
14	18.9000	21.1158						
15	19.9944	24.2692						
16	21.0851							
17	22.1725							
18	23.2568							
19	24.3383							

$$\varphi_{mn}(r,\phi) = \sqrt{\frac{2}{\pi}} \frac{1}{\gamma'_{mn}a} \frac{J_m(\gamma'_{mn}r)}{J_{m+1}(\gamma'_{mn}a)} \frac{\cos}{\sin} m\phi \,, \quad m > 0 \,, \tag{8.40a}$$

and

$$\varphi_{0n}(r,\phi) = \frac{1}{\sqrt{\pi}} \frac{1}{\gamma'_{0n}a} \frac{J_0(\gamma'_{0n}r)}{J_1(\gamma'_{0n}a)} \,, \tag{8.40b}$$

for the E_{0n} modes. The verification of the normalization involves the integral

$$\int_0^a r\,dr\, J_m^2(\gamma'_{mn}r) = \frac{1}{2}a^2 \left[J'_m(\gamma'_{mn}a)\right]^2 = \frac{1}{2}a^2 J_{m+1}^2(\gamma'_{mn}a) \,, \tag{8.41}$$

deduced from (8.35b) and the recurrence relation (8.8b). In stating the fields derived from these mode functions, it is convenient to resolve the transverse parts of the electric and magnetic fields relative to the unit vectors $\mathbf{e}_r$ and $\mathbf{e}_\phi$, which are drawn at each point in the direction of increasing r and increasing ϕ, respectively. The three vectors $\mathbf{e}_r$, $\mathbf{e}_\phi$, $\mathbf{e}_z$ form a right-handed reference system which differs from the usual reference system in that the directions of $\mathbf{e}_r$ and $\mathbf{e}_\phi$ vary from point to point. The latter variation is expressed by the relations

$$\frac{\partial \mathbf{e}_r}{\partial \phi} = \mathbf{e}_\phi \,, \qquad \frac{\partial \mathbf{e}_\phi}{\partial \phi} = -\mathbf{e}_r \,. \tag{8.42}$$

The representation of the transverse part of the gradient operator, in this coordinate system, is

Table 8.2. $\gamma''_{mn}a$: nth zero of J'_m. Except for the first row, only zeroes below 25 are listed

$m \setminus n$	1	2	3	4	5	6	7	8
0	3.8317	7.0156	10.1735	13.3237	16.4706	19.6159	22.7601	25.9037
1	1.8412	5.3314	8.5363	11.7060	14.8636	18.0155	21.1644	24.3113
2	3.0542	6.7061	9.9695	13.1704	16.3475	19.5129	22.6721	
3	4.2012	8.0152	11.3459	14.5859	17.7888	20.9724	24.1469	
4	5.3175	9.2824	12.6819	15.9641	19.1960	22.4010		
5	6.4156	10.5199	13.9872	17.3128	20.5755	23.8033		
6	7.5013	11.7349	15.2682	18.6374	21.9318			
7	8.5778	12.9324	16.5294	19.9419	23.2681			
8	9.6474	14.1156	17.7740	21.2291	24.5872			
9	10.7114	15.2868	19.0045	22.5014				
10	11.7709	16.4479	20.2230	23.7608				
11	12.8265	17.6003	21.4309					
12	13.8788	18.7451	22.6293					
13	14.9284	19.8832	23.8194					
14	15.9754	21.0154						
15	17.0203	22.1423						
16	18.0633	23.2644						
17	19.1045	24.3819						
18	20.1441							
19	21.1823							
20	22.2191							
21	23.2548							
22	24.2894							

$$\boldsymbol{\nabla}_\perp = \mathbf{e}_r \frac{\partial}{\partial r} + \mathbf{e}_\phi \frac{1}{r} \frac{\partial}{\partial \phi} \tag{8.43a}$$

and

$$\mathbf{e}_z \times \boldsymbol{\nabla} = \mathbf{e}_\phi \frac{\partial}{\partial r} - \mathbf{e}_r \frac{1}{r} \frac{\partial}{\partial \phi} \, . \tag{8.43b}$$

Hence the various field components of an E_{mn} mode with $m > 0$ are (omitting indices on γ for simplicity) are from (6.32a) to (6.32d)

$$E_r = -\sqrt{\frac{2}{\pi}} \frac{1}{a} \frac{J'_m(\gamma r)}{J_{m+1}(\gamma a)} \frac{\cos}{\sin} m\phi \, V(z) \, , \tag{8.44a}$$

$$E_\phi = \pm\sqrt{\frac{2}{\pi}} \frac{m}{\gamma a} \frac{1}{r} \frac{J_m(\gamma r)}{J_{m+1}(\gamma a)} \frac{\sin}{\cos} m\phi \, V(z) \, , \tag{8.44b}$$

$$H_r = \mp\sqrt{\frac{2}{\pi}} \frac{m}{\gamma a} \frac{1}{r} \frac{J_m(\gamma r)}{J_{m+1}(\gamma a)} \frac{\sin}{\cos} m\phi \, I(z) \, , \tag{8.44c}$$

$$H_\phi = -\sqrt{\frac{2}{\pi}} \frac{1}{a} \frac{J'_m(\gamma r)}{J_{m+1}(\gamma a)} \frac{\cos}{\sin} m\phi \, I(z) \, , \tag{8.44d}$$

$$E_z = i\zeta \frac{\gamma}{ka} \sqrt{\frac{2}{\pi}} \frac{J_m(\gamma r)}{J_{m+1}(\gamma a)} {\cos \atop \sin} m\phi \, I(z) \,, \tag{8.44e}$$

$$H_z = 0 \,. \tag{8.44f}$$

All E_{mn} modes with $m > 0$ are doubly degenerate; either of the two trigonometric functions that describe the mode function's dependence upon ϕ generates a possible mode. However, the E_{0n} modes are nondegenerate since one of the two possibilities now vanishes. The field components of an E_{0n} mode are:

$$E_r = \frac{1}{\sqrt{\pi}} \frac{1}{a} \frac{J_1(\gamma r)}{J_1(\gamma a)} V(z) \,, \tag{8.45a}$$

$$E_\phi = 0 \,, \tag{8.45b}$$

$$H_r = 0 \,, \tag{8.45c}$$

$$H_\phi = \frac{1}{\sqrt{\pi}} \frac{1}{a} \frac{J_1(\gamma r)}{J_1(\gamma a)} I(z) \,, \tag{8.45d}$$

$$E_z = i\zeta \frac{\gamma}{ka} \frac{1}{\sqrt{\pi}} \frac{J_0(\gamma r)}{J_1(\gamma a)} I(z) \,, \tag{8.45e}$$

$$H_z = 0 \,. \tag{8.45f}$$

It will be noticed that the only nonvanishing magnetic field component is H_ϕ, and that $E_\phi = 0$.

The normalization for the H-mode function is given by (6.59), or

$$(\gamma''_{mn})^2 \int_\sigma d\sigma \, \psi^2_{mn} = 1 \,, \tag{8.46}$$

so the normalized H-mode functions for $m > 0$ are

$$\psi_{mn} = \sqrt{\frac{2}{\pi}} \frac{1}{\sqrt{(\gamma''_{mn}a)^2 - m^2}} \frac{J_m(\gamma''_{mn}r)}{J_m(\gamma''_{mn}a)} {\cos \atop \sin} m\phi \,, \tag{8.47a}$$

which now uses

$$\int_0^a dr \, r \, J_m^2(\gamma''_{mn}r) = \frac{1}{2}a^2 \left(1 - \frac{m^2}{\gamma''_{mn}a^2}\right) J_m^2(\gamma''_{mn}a) \,, \tag{8.47b}$$

derived from (8.35b), while, for $m = 0$, we write

$$\psi_{0n} = \frac{1}{\sqrt{\pi}} \frac{1}{\gamma''_{mn}a} \frac{J_0(\gamma''_{0n}r)}{J_0(\gamma''_{0n}a)} \,. \tag{8.47c}$$

The field components of an H_{mn} mode, $m > 0$, are obtained from (6.33a) to (6.33d) and (8.43a), (8.43b):

$$E_r = \pm\sqrt{\frac{2}{\pi}} \frac{m}{\sqrt{(\gamma a)^2 - m^2}} \frac{1}{r} \frac{J_m(\gamma r)}{J_m(\gamma a)} {\sin \atop \cos} m\phi \, V(z) \,, \tag{8.48a}$$

$$E_\phi = \sqrt{\frac{2}{\pi}} \frac{\gamma}{\sqrt{(\gamma a)^2 - m^2}} \frac{J'_m(\gamma r)}{J_m(\gamma a)} \frac{\cos}{\sin} m\phi\, V(z)\,, \qquad (8.48b)$$

$$H_r = -\sqrt{\frac{2}{\pi}} \frac{\gamma}{\sqrt{(\gamma a)^2 - m^2}} \frac{J'_m(\gamma r)}{J_m(\gamma a)} \frac{\cos}{\sin} m\phi\, I(z)\,, \qquad (8.48c)$$

$$H_\phi = \pm\sqrt{\frac{2}{\pi}} \frac{m}{\sqrt{(\gamma a)^2 - m^2}} \frac{1}{r} \frac{J_m(\gamma r)}{J_m(\gamma a)} \frac{\sin}{\cos} m\phi\, I(z)\,, \qquad (8.48d)$$

$$E_z = 0\,, \qquad (8.48e)$$

$$H_z = i\eta\frac{\gamma}{k} \frac{\gamma}{(\sqrt{(\gamma a)^2 - m^2}} \sqrt{\frac{2}{\pi}} \frac{J_m(\gamma r)}{J_m(\gamma a)} \frac{\cos}{\sin} m\phi\, V(z)\,, \qquad (8.48f)$$

while those for an H_{0n} mode are simply

$$E_r = 0\,, \qquad (8.49a)$$

$$E_\phi = -\frac{1}{\sqrt{\pi}} \frac{1}{a} \frac{J_1(\gamma r)}{J_0(\gamma a)}\, V(z)\,, \qquad (8.49b)$$

$$H_r = \frac{1}{\sqrt{\pi}} \frac{1}{a} \frac{J_1(\gamma r)}{J_0(\gamma a)}\, I(z)\,, \qquad (8.49c)$$

$$H_\phi = 0\,, \qquad (8.49d)$$

$$E_z = 0\,, \qquad (8.49e)$$

$$H_z = i\eta\frac{\gamma}{ka} \frac{1}{\sqrt{\pi}} \frac{J_0(\gamma r)}{J_0(\gamma a)}\, V(z)\,. \qquad (8.49f)$$

Thus, an H mode with $m = 0$ is characterized by the single electric field component E_ϕ, and $H_\phi = 0$.

8.3 Circular Guide with Metallic Cylindrical Wedge

Another type of waveguide, which can be rigorously described with the aid of Bessel functions of nonintegral order, is constructed from the hollow circular guide by the insertion of a metallic cylindrical wedge with its apex at the center of the circle, as illustrated in Fig. 8.1. The external angle of the wedge

Fig. 8.1. Cross section of circular guide with metal wedge of angle α inserted

is denoted by α, and it is supposed that the boundaries of the wedge coincide with the radial planes $\phi = 0$ and $\phi = \alpha$. The mode functions are of the form

(8.3), and the function $\Phi(\phi)$ is determined by the requirement that it vanish at $\phi = 0$ and $\phi = \alpha$ for E modes, and that the derivative with respect to ϕ vanish at $\phi = 0$ and $\phi = \alpha$ for H modes. Hence

$$\text{E mode:} \quad \Phi(\phi) = \sin \frac{m\pi}{\alpha}\phi \,, \quad m = 1, 2, \ldots \,, \tag{8.50a}$$

$$\text{H mode:} \quad \Phi(\phi) = \cos \frac{m\pi}{\alpha}\phi \,, \quad m = 0, 1, \ldots \,, \tag{8.50b}$$

and the allowed values of μ are of the form $m\pi/\alpha$ with m a nonnegative integer that may assume the value zero for H modes. The Neumann functions must again be excluded in consequence of their lack of finiteness at $r = 0$, whence the E- and H-mode scalar functions take the form

$$\text{E mode:} \quad \varphi = J_{m\pi/\alpha}(\gamma r) \sin \frac{m\pi}{\alpha}\phi \,, \tag{8.51a}$$

$$\text{H mode:} \quad \psi = J_{m\pi/\alpha}(\gamma r) \cos \frac{m\pi}{\alpha}\phi \,. \tag{8.51b}$$

The cutoff wavenumbers follow from the imposition of the boundary conditions at $r = a$:

$$\text{E mode:} \quad J_{m\pi/\alpha}(\gamma a) = 0 \,, \tag{8.52a}$$

$$\text{H mode:} \quad J'_{m\pi/\alpha}(\gamma a) = 0 \,, \tag{8.52b}$$

and are obtained from the zeros and extrema of the Bessel functions of order $m\pi/\alpha$, $m = 0, 1, \ldots$. The order is generally nonintegral save when α equals π or a fraction thereof.

A particularly simple situation is $\alpha = \pi$, which corresponds to a guide in the form of a semicircle of radius a. The cutoff wavenumbers are found from the zeros and extrema of Bessel functions of integral order and are therefore identical with those of the original circular guide. The only exception to the latter statement is that no E mode with $m = 0$ exists in the semicircular guide. It is clear that the connection between the modes of the circular and semicircular guides is identical with that already encountered in isosceles triangular guides, see Sect. 7.2; modes of a given cross section possessing proper behavior relative to a line of reflection symmetry form the modes of the semi-cross section. The dominant E mode of the semicicular guide is an E_{11} mode of the circular guide and has the cutoff wavelength $(\lambda_c)'_{11} = 1.6398a$. The dominant H mode, and the dominant mode, of the semicircular guide is an H_{11} mode of the circular guide and has the same cutoff wavelength $(\lambda_c)''_{11} = 3.4126a$.

If $\alpha = 2\pi$, the physical situation is that of a circular guide with a wall of negligible thickness inserted along a radial plane from the center to the circumference. The two kinds of cutoff wavenumbers are obtained from the zeros and extrema of the Bessel functions of order $m/2$. If m is even, the order is integral and the modes are those of the semicircle whose base coincides with the wall. When m is odd, however, the order is half-integral, and a new set of modes is obtained. In particular, if the order is 1/2, the radial function that enters into the mode function is [cf. (8.34b)]

$$J_{1/2}(\gamma r) = \sqrt{\frac{2}{\pi}} \frac{\sin \gamma r}{\sqrt{\gamma r}} , \qquad (8.53)$$

and therefore, the cutoff wavenumbers for the E modes of order 1/2 are determined by

$$J_{1/2}(\gamma a) = 0 \Rightarrow \sin \gamma a = 0 , \qquad (8.54a)$$

while the H-mode cutoff wavenumbers are given by

$$J'_{1/2}(\gamma a) = 0 \Rightarrow \tan \gamma a = 2\gamma a . \qquad (8.54b)$$

The minimum E-mode cutoff wavenumber of this type is found from the smallest nonvanishing root of (8.54a), $\gamma a = \pi$, which is less than the smallest root of the Bessel function of order unity, and hence represents the absolute E-mode cutoff wavenumber of the guide under discussion. The dominant E mode so obtained is designated as $E_{\frac{1}{2}1}$, and has the cutoff wavelength $(\lambda_c)'_{\frac{1}{2}1} = 2a$. The smallest nonvanishing solution of (8.54b) is $\gamma a = 1.1655$ which gives the cutoff wavenumber of the $H_{\frac{1}{2}1}$ mode. On comparison with the cutoff wavenumber of the H_{11} mode, it is apparent that the dominant H mode, and the dominant mode, of the circular guide with a radial wall is $H_{\frac{1}{2}1}$. The absolute cutoff wavelength is $(\lambda_c)''_{\frac{1}{2}1} = 5.3910a$. It will be seen that the insertion of the radial wall in the circular guide depresses the absolute H-mode cutoff wavenumber, and raises the absolute E-mode cutoff wavenumber.

8.4 Coaxial Guide

The cylinder function solutions (8.3) can be used to describe the modes of a guide that has a cross section in the form of two concentric circles, a coaxial line. The dependence upon ϕ is identical with that of a circular guide, and, in particular, μ is restricted to integral values. The radial dependence of the mode functions must be such as to satisfy appropriate boundary conditions at the outer and inner radii, a and b, respectively. In view of the double boundary condition, a Bessel function alone will not suffice in general, and a linear combination of Bessel and Neumann functions must be used. Furthermore, the Neumann function can no longer be excluded by virtue of its singularity at the origin, for the latter point is not within the region occupied by the field. To construct a suitable E-mode function, we observe that

$$\varphi = [J_m(\gamma r)N_m(\gamma b) - N_m(\gamma r)J_m(\gamma b)] \frac{\cos}{\sin} m\phi \qquad (8.55)$$

vanishes at $r = b$, and will also vanish at $r = a$ if

$$J_m(\gamma a)N_m(\gamma b) = N_m(\gamma a)J_m(\gamma b) \,, \tag{8.56a}$$

or

$$\frac{J_m(\gamma a)}{N_m(\gamma a)} = \frac{J_m(\gamma b)}{N_m(\gamma b)} \,, \tag{8.56b}$$

which is the equation that determines the cutoff wavenumbers of the various E modes. Similarly,

$$\psi = [J_m(\gamma r)N_m'(\gamma b) - N_m(\gamma r)J_m'(\gamma b)] \, {\cos \atop \sin} m\phi \tag{8.57}$$

has vanishing radial derivative at $r = b$, and will show similar behavior at $r = a$ if

$$J_m'(\gamma a)N_m'(\gamma b) = N_m'(\gamma a)J_m'(\gamma b) \,, \tag{8.58a}$$

or

$$\frac{J_m'(\gamma a)}{N_m'(\gamma a)} = \frac{J_m'(\gamma b)}{N_m'(\gamma b)} \,, \tag{8.58b}$$

thus providing the eigenvalue equation for the H-mode cutoff wavenumbers.

A general idea of the nature of the mode functions and their eigenvalues can be obtained by considering the situation in which the difference between the radii, a and b, is small compared to the average radius, $a - b \ll \frac{a+b}{2}$. In this event, Bessel's equation (8.4b) can be simplified to

$$\left(\frac{d^2}{dr^2} + \gamma^2 - \frac{m^2}{\bar{r}^2}\right) Z = 0 \,, \tag{8.59}$$

where $\bar{r}$ is some average radius, such as $\frac{a+b}{2}$, and the term $\frac{1}{r}\frac{d}{dr}Z$ has been neglected, which will be justified by the ensuing results. The appropriate solutions of (8.59) are

$$\text{E mode:} \quad Z = \sin\frac{n\pi}{a-b}(r-b) \,, \qquad n = 1, 2, \dots \,, \tag{8.60a}$$

$$\text{H mode:} \quad Z = \cos\frac{n\pi}{a-b}(r-b) \,, \qquad n = 0, 1, \dots \,, \tag{8.60b}$$

and

$$\gamma^2 = \left(\frac{n\pi}{a-b}\right)^2 + \left(\frac{m}{\bar{r}}\right)^2 . \tag{8.60c}$$

Returning to the approximations involved in (8.59), we notice that Z is either a constant, and $\frac{1}{r}\frac{d}{dr}Z$ is zero, or Z is a trigonometric function of the argument $\frac{n\pi}{a-b}(r-b)$, whence $\frac{1}{r}\frac{d}{dr}Z$ has the order of magnitude of $\frac{n\pi}{a-b}\frac{1}{\bar{r}}Z$, which is negligible compared with $\frac{d^2}{dr^2}Z = -\left(\frac{n\pi}{a-b}\right)^2 Z$, in consequence of the assumption $a - b \ll \bar{r}$. The modes obtained in this manner can be classified by the integers m and n, forming the E_{mn} amd H_{mn} modes. The dominant E mode of the coaxial line (excluding the T mode, of course) is obtained by placing

$n = 1$ and $m = 0$ and the approximate formula for the cutoff wavelength of the dominant E_{01} modes is $(\lambda_c)'_{01} = 2(a - b)$. The dominant H mode is H_{10}, corresponding to $m = 1$, $n = 0$. No H mode exists with both integers equal to zero since the mode function is a constant. The cutoff wavelength of the dominant H mode is $(\lambda_c)''_{10} = 2\pi\bar{r} = \pi(a + b)$, or the average circumference of the concentric circles. With the exception of the H_{m0} modes, more precise expressions for the cutoff wavenumbers can be obtained from the asymptotic forms (8.38a) and (8.38b) employed in conjunction with (8.56b) and (8.58b). The cutoff wavenumber of the E_{mn} mode is given by

$$
\gamma'_{mn} \sim \frac{n\pi}{a - b} + \frac{a - b}{2n\pi}\frac{m^2 - 1/4}{ab} + \left(\frac{a - b}{2n\pi}\right)^3 \frac{m^2 - 1/4}{(ab)^3}
$$
$$
\times \left[\frac{1}{3}\left(m^2 - \frac{25}{4}\right)(a^2 + ab + b^2) - 2\left(m^2 - \frac{1}{4}\right)ab\right] + \cdots ,
$$

(8.61a)

and that of the H_{mn} mode is $(n \neq 0)$

$$
\gamma''_{mn} \sim \frac{n\pi}{a - b} + \frac{a - b}{2n\pi}\frac{m^2 + 3/4}{ab} + \left(\frac{a - b}{2n\pi}\right)^3 \frac{1}{(ab)^3}
$$
$$
\times \left[\frac{1}{3}\left(m^4 + \frac{23}{2}m^2 - \frac{63}{16}\right)(a^2 + ab + b^2) - 2\left(m^2 + \frac{3}{4}\right)^2 ab\right] + \cdots .
$$

(8.61b)

Figure 8.2 contains graphs of $\frac{a-b}{n\pi}\gamma'_{0n}$, as a function of a/b, for the first few E_{0n} modes. The E_{0n} coaxial mode approaches the E_{0n} circular guide mode as $b/a \to 0$. The cutoff wavenumber of the coaxial H_{10} mode, multiplied by the average radius $\frac{a+b}{2}$, is plotted in Fig. 8.3 as a function of b/a. As the latter quantity approaches zero, the H_{10} coaxial mode becomes the H_{11} circular guide mode, and correspondingly, the intercept in Fig. 8.3 is 1.8412/2. Finally, we may note that, as in a circular guide, all modes with $m > 0$ are doubly degenerate, and the cutoff wavenumbers of the E modes with $m = 1$ coincide with those of the H modes with $m = 0$.

8.5 Coaxial Guide with Metallic Cylindrical Wedge

The insertion in the coaxial line of a metallic wedge formed by two radial planes can be rigorously treated by cylinder functions of fractional order, as in the corresponding circular guide case. In particular, if the external angle of the wedge is $\alpha = 2\pi$, we deal with a coaxial line that has inner and outer cylinders connected by a radial plane of negligible thickness, as in Fig. 8.4. Of course, no T mode exists in such circumstances. The dominant E mode is derived from cylinder functions of order 1/2, and it is evident from (8.56b),

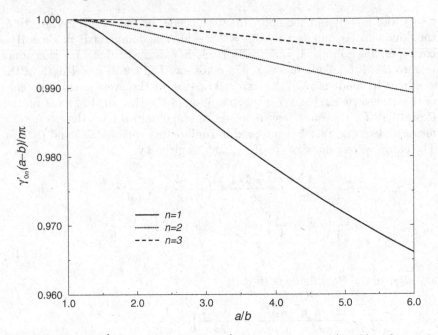

Fig. 8.2. Plot of $\frac{a-b}{n\pi}\gamma'_{0n}$ versus a/b where γ'_{0n} is the nth root of (8.56b) with $m = 0$, a being the outer radius of the coaxial guide, and b the inner radius

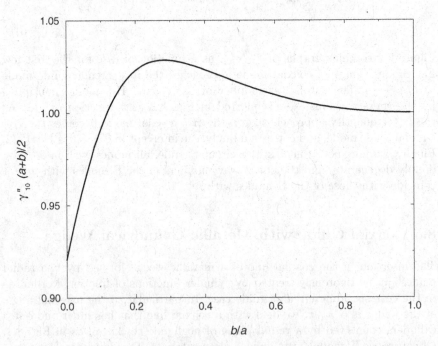

Fig. 8.3. Plot of $\frac{a+b}{2}\gamma''_{10}$ versus b/a where γ''_{10} is the first root of (8.58b) with $m = 1$

Fig. 8.4. Waveguide consisting of two concentric cylinders, connected by a radial plane

which is also applicable to noninteger orders, that, according to (8.34b), the cutoff wavenumbers of such modes are rigorously given by

$$\gamma'_{\frac{1}{2}n} = \frac{n\pi}{a-b} \,.$$

(8.62)

Thus the cutoff wavelength of the dominant $E_{\frac{1}{2}n}$ mode is $(\lambda_c)'_{\frac{1}{2}1} = 2(a-b)$. The H modes of order $1/2$, a member of which is the dominant H mode, have cutoff wavenumbers that are solutions of

$$\frac{\tan \gamma(a-b)}{\gamma(a-b)} = \frac{2}{4\gamma^2 ab + 1} \,.$$

(8.63)

If $a - b \ll \frac{a+b}{2}$, an approximate expression for the smallest solution of (8.63) is $\gamma = \frac{1}{2\bar{r}}$ or $(\lambda_c)''_{\frac{1}{2}0} = 4\pi\bar{r} = 2\pi(a+b)$; the cutoff wavenumber of the dominant H mode is decreased by a factor of 2 owing to the insertion of the wall. As $b/a \to 0$, the $H_{\frac{1}{2}0}$ mode of the connected coaxial cylinders approaches the dominant mode of the circular guide with a radial wall, the mode that has been called $H_{\frac{1}{2}1}$ in Sect. 8.3 [see (8.54b)].

8.6 Elliptic and Parabolic Cylinder Coordinates

Thus far, two coordinate systems have been introduced that permit the separation of the wave equation – rectangular and circular cylindrical coordinates. It will now be shown that only two other coordinate systems exist with this property, namely elliptic and parabolic cylinder coordinates. Further, since parabolic and circular cylinder coordinates are, as we shall demonstrate, limiting forms of elliptical cylinder coordinates, only two general types of coordinate systems exist that permit the separation of the wave equation in two dimensions – rectangular and elliptical cylinder coordinates. If the two-dimensional wave equation

$$(\nabla^2 + \gamma^2)f = 0$$

(8.64)

is separable in the coordinates ξ and η, that is, if a solution can be found in the form

$$f(\xi, \eta) = X(\xi)Y(\eta) \,,$$

(8.65)

the separate functions must satisfy second-order differential equations of the type

$$\left(\frac{d^2}{d\xi^2} + F(\xi)\right) X(\xi) = 0 ,$$ (8.66a)

$$\left(\frac{d^2}{d\eta^2} + G(\eta)\right) Y(\eta) = 0 .$$ (8.66b)

First derivatives can be omitted without loss of generality since they can always be eliminated by a suitable transformation of the wave functions. In consequence of (8.66a) and (8.66b) the differential equation satisfied by (8.65) is of the form

$$\left(\frac{\partial^2}{\partial\xi^2} + \frac{\partial^2}{\partial\eta^2} + F(\xi) + G(\xi)\right) f(\xi,\eta) = 0 ,$$ (8.67)

which must be obtained from the wave equation (8.64) on expressing x and y as a function ξ and η. Hence the relation between the coordinates must be such that the Laplacian in the x, y coordinates transforms into a multiple of the Laplacian in the ξ, η coordinates. Now, by direct transformation of the Laplacian, we find

$$\nabla^2 = (\nabla\xi)^2 \frac{\partial^2}{\partial\xi^2} + (\nabla\eta)^2 \frac{\partial^2}{\partial\eta^2} + 2\nabla\xi \cdot \nabla\eta \frac{\partial}{\partial\xi}\frac{\partial}{\partial\eta} + \nabla^2\xi \frac{\partial}{\partial\xi} + \nabla^2\eta \frac{\partial}{\partial\eta} ,$$ (8.68)

and therefore, the coordinates ξ and η as functions of x and y are restricted by

$$\nabla^2\xi = 0 , \qquad \nabla^2\eta = 0 ,$$ (8.69a)

$$\nabla\xi \cdot \nabla\eta = 0 ,$$ (8.69b)

$$(\nabla\xi)^2 = (\nabla\eta)^2 .$$ (8.69c)

The equation (8.69b) informs us, incidentally, that the contours of constant ξ and constant η intersect orthogonally, for the vectors $\nabla\xi$ and $\nabla\eta$ are respectively orthogonal to the lines of constant ξ and η. In rewriting the relation (8.69b) in the form

$$\frac{\frac{\partial\xi}{\partial x}}{\frac{\partial\eta}{\partial y}} = -\frac{\frac{\partial\xi}{\partial y}}{\frac{\partial\eta}{\partial x}} \equiv \mu ,$$ (8.70)

and employing this equation to replace the ξ derivatives by η derivatives in (8.69c) we discover that

$$\mu^2 = 1 .$$ (8.71)

Without loss of generality, we may choose $\mu = 1$, for the opposite choice would be equivalent to reversing the sign of η, say. Hence (8.69b) and (8.69c) are combined in

$$\frac{\partial \xi}{\partial x} = \frac{\partial \eta}{\partial y}, \qquad \frac{\partial \xi}{\partial y} = -\frac{\partial \eta}{\partial x}, \tag{8.72}$$

from which we may immediately deduce (8.69a). Hence the restriction on ξ and η as a function of x and y are entirely contained in (8.72), which will be recognized as the Cauchy–Riemann equations, the conditions that the complex variable $\zeta = \xi + i\eta$ be an analytic function of the complex variable $z = x + iy$:

$$\zeta = F(z). \tag{8.73}$$

It is well known that an analytic function, regarded as defining a mapping of the x–y plane into the ξ–η plane, has the conformal property; an infinitesimal neighborhood of a point in the former plane is mapped into a geometrically similar neighborhood of a point in the latter plane.

Now that we have established a necessary relation between the coordinates ξ, η and x, y, a functional relation between the complex variables ζ and z, it is convenient to re-express the wave equation (8.64) in terms of the independent coordinates z and $z^* = x - iy$, transform to the variables ζ and ζ^*, and compare the result with the form (8.67). Thus

$$\frac{\partial^2}{\partial x^2} + \frac{\partial^2}{\partial y^2} = 4\frac{\partial}{\partial z}\frac{\partial}{\partial z^*} = 4\frac{\mathrm{d}\zeta}{\mathrm{d}z}\frac{\mathrm{d}\zeta^*}{\mathrm{d}z^*}\frac{\partial}{\partial \zeta}\frac{\partial}{\partial \zeta^*} = \left|\frac{\mathrm{d}\zeta}{\mathrm{d}z}\right|^2 \left(\frac{\partial^2}{\partial \xi^2} + \frac{\partial^2}{\partial \eta^2}\right), \tag{8.74}$$

whence

$$\left(\frac{\partial^2}{\partial \xi^2} + \frac{\partial^2}{\partial \eta^2} + \gamma^2 \left|\frac{\mathrm{d}z}{\mathrm{d}\zeta}\right|^2\right) f = 0. \tag{8.75}$$

In order that the wave equation be separable in the coordinates ξ and η, it is necessary that $|\mathrm{d}z/\mathrm{d}\zeta|^2$ be additively composed of two functions, each containing a single variable. Hence, the quantity $|\mathrm{d}z/\mathrm{d}\zeta|^2$ must be annihilated by the differential operator

$$\frac{\partial}{\partial \xi}\frac{\partial}{\partial \eta} = i\left(\frac{\partial^2}{\partial \zeta^2} - \frac{\partial^2}{\partial \zeta^{*2}}\right), \tag{8.76}$$

which requirement can be written as

$$\frac{\mathrm{d}z^*}{\mathrm{d}\zeta^*}\frac{\mathrm{d}^2}{\mathrm{d}\zeta^2}\left(\frac{\mathrm{d}z}{\mathrm{d}\zeta}\right) = \frac{\mathrm{d}z}{\mathrm{d}\zeta}\frac{\mathrm{d}^2}{\mathrm{d}\zeta^{*2}}\left(\frac{\mathrm{d}z^*}{\mathrm{d}\zeta^*}\right), \tag{8.77}$$

or

$$\frac{\frac{\mathrm{d}^2}{\mathrm{d}\zeta^2}\left(\frac{\mathrm{d}z}{\mathrm{d}\zeta}\right)}{\frac{\mathrm{d}z}{\mathrm{d}\zeta}} = \frac{\frac{\mathrm{d}^2}{\mathrm{d}\zeta^{*2}}\left(\frac{\mathrm{d}z^*}{\mathrm{d}\zeta^*}\right)}{\frac{\mathrm{d}z^*}{\mathrm{d}\zeta^*}} = \nu^2, \tag{8.78}$$

where ν^2 must a real constant, since ζ and ζ^* are independent variables. Two possibilities must be distinguished, $\nu \neq 0$ and $\nu = 0$. In the former eventuality,

$$\frac{\mathrm{d}z}{\mathrm{d}\zeta} = c\sinh(\nu\zeta + c'), \tag{8.79}$$

where c and c' are arbitrary constants. The constant c' may be placed equal to zero without loss of generality, for it can be reinstated by a suitable translation of the ξ–η reference system. Similarly, ν may be chosen as unity, since any other value corresponds to a changed scale in the ξ–η plane. Finally,

$$z = c \cosh \zeta \,, \tag{8.80}$$

again discarding a constant of integration. The transform (8.80) is equivalent to the real equations, assuming c is real and positive

$$x = c \cosh \xi \cos \eta \,, \qquad y = c \sinh \xi \sin \eta \,, \tag{8.81}$$

which exhibits a one-to-one correspondence between the entire x–y plane and a strip in the ξ–η plane, $\xi \geq 0,\ -\pi \leq \eta \leq \pi$. The restriction to positive ξ is needed to avoid the duplicity in the x–y plane of the points ξ, η and $-\xi$, $-\eta$. The curves of constant ξ,

$$\frac{x^2}{c^2 \cosh^2 \xi} + \frac{y^2}{c^2 \sinh^2 \xi} = 1 \,, \tag{8.82}$$

form a family of confocal ellipses, with semi-major and minor axes equal to $c \cosh \xi$ and $c \sinh \xi$, respectively, semi-focal distance c, and eccentricity $1/\cosh \xi$. The contours of constant η,

$$\frac{x^2}{c^2 \cos^2 \eta} - \frac{y^2}{c^2 \sin^2 \eta} = 1 \,, \tag{8.83}$$

form confocal hyperbolas, with η equal to the angle of the asymptotes. The quantities ξ and η are called elliptic cylinder coordinates. In the limit as the semi-focal distance c and the eccentricity of the ellipse $1/\cosh \xi$ become zero, the semi-major and minor axes approach equality

$$c \cosh \xi = c \sinh \xi = r \,, \tag{8.84}$$

and it is evident from (8.81) that the elliptic cylinder coordinates degenerate into circular cylindrical coordinates (8.1), with η equal to the polar angle ϕ.

Returning to the second possibility $\nu = 0$, which is already a limiting form of the situation just discussed, we deduce that

$$\frac{\mathrm{d}z}{\mathrm{d}\zeta} = \alpha + \beta \zeta \,. \tag{8.85}$$

Here α may be placed equal to zero and β equal to unity with no loss in generality, whence

$$z = \frac{1}{2} \zeta^2 \,, \tag{8.86}$$

or

$$x = \frac{1}{2}(\xi^2 - \eta^2) \,, \qquad y = \xi \eta \,. \tag{8.87}$$

This transformation defines a one-to-one correspondence between the entire x–y plane and the semi-infinite ξ–η plane $\xi \geq 0$, $-\infty < \eta < \infty$. The curves of constant ξ are

$$x = \frac{1}{2}\xi^2 - \frac{y^2}{2\xi^2} , \tag{8.88}$$

and constant η

$$x = -\frac{1}{2}\eta^2 + \frac{y^2}{2\eta^2} , \tag{8.89}$$

form families of parabolas that are symmetrical relative to the x-axis. In particular the parabola $\xi = 0$ coincides with the negative x-axis, while $\eta = 0$ describes the positive x-axis.

The wave equation, in elliptical cylinder coordinates, from (8.75), and $dz/d\zeta = c \sinh \zeta$, reads

$$\left[\frac{\partial^2}{\partial \xi^2} + \frac{\partial^2}{\partial \eta^2} + (\gamma c)^2 (\cosh^2 \xi - \cos^2 \eta) \right] f = 0 , \tag{8.90}$$

which separates into

$$\frac{d^2}{d\xi^2} X(\xi) - \left(A - (\gamma c)^2 \cosh^2 \xi \right) X(\xi) = 0 , \tag{8.91a}$$

$$\frac{d^2}{d\eta^2} Y(\eta) + \left(A - (\gamma c)^2 \cos^2 \eta \right) Y(\eta) = 0 , \tag{8.91b}$$

where A is an arbitrary constant. The two equations are of essentially the same form, (8.91a) being obtained from (8.91b) on replacing η by $i\xi$. The solution of (8.91a) and (8.91b) is known as the elliptic cylinder function, or Mathieu function.[1] Expressed in parabolic cylinder coordinates, the wave equation becomes, because $dz/d\zeta = \zeta$,

$$\left[\frac{\partial^2}{\partial \xi^2} + \frac{\partial^2}{\partial \eta^2} + \gamma^2 (\xi^2 + \eta^2) \right] f = 0 , \tag{8.92}$$

or in separated form

$$\frac{d^2}{d\xi^2} X(\xi) + (A + \gamma^2 \xi^2) X(\xi) = 0 , \tag{8.93a}$$

$$\frac{d^2}{d\eta^2} Y(\eta) + (-A + \gamma^2 \eta^2) Y(\eta) = 0 , \tag{8.93b}$$

where A again denotes an arbitrary constant. Solutions of (8.93a) and (8.93b) are known as parabolic cylinder functions or Weber–Hermite functions.[2] Elliptic cylinder functions are suitable for treating variously shaped guides

---

[1] For example, see Whittaker and Watson [17], Chap. XIX.
[2] See Whittaker and Watson [17], p. 347 et seq.

bounded by elliptical and hyperbolic arcs, while parabolic cylinder functions permit a discussion of guides with boundaries consisting of intersecting parabolas. The importance of these problems is insufficient, however, to warrant the inclusion of the rather elaborate analysis necessary to obtain complete numerical results. Parabolic shaped guides are of little interest, while elliptical guides are of importance only insofar as they indicate the effects on circular guides of a slight deviation from perfect symmetry. An adequate answer to the latter problem is readily obtained by the approximation methods to which attention will be turned in Chap. 10.

8.7 Problems for Chap. 8

1. Prove the following theorems for summing Bessel functions of integer order:

$$\sum_{m=-\infty}^{\infty} [J_m(t)]^2 = 1 , \tag{8.94a}$$

and, for $n \neq 0$ an integer,

$$\sum_{m=-\infty}^{\infty} J_{m+n}(t) J_m(t) = 0 . \tag{8.94b}$$

2. Prove the following statements, for t and a real numbers:

$$|a| < \pi : \quad \sum_{m=-\infty}^{\infty} (-1)^m J_0(t+ma) = 0 , \tag{8.95a}$$

$$|a| < 2\pi : \quad \sum_{m=-\infty}^{\infty} J_0(t+ma) = \frac{2}{a} , \tag{8.95b}$$

$$2\pi < |a| < 4\pi : \quad \sum_{m=-\infty}^{\infty} J_0(t+ma) = \frac{2}{a} \left[1 + \frac{2\cos\frac{2\pi t}{a}}{\sqrt{1 - \left(\frac{2\pi}{a}\right)^2}} \right] . \tag{8.95c}$$

3. Demonstrate that any two solutions of the Bessel differential equation, φ and ψ, satisfy

$$t \left[\varphi \frac{d\psi}{dt} - \psi \frac{d\varphi}{dt} \right] = \text{const.} \tag{8.96}$$

Evaluate the constant for $\varphi = J_m$, $\psi = N_m$, by using their asymptotic forms. Show that at a value of t such that $J_m(t) = 0$,

$$N_m(t) = -\frac{2}{\pi t} \frac{1}{J'_m(t)} . \tag{8.97}$$

Apply the known behavior of $J_m(t)$, $t \ll 1$, given in (8.28), to learn that

$$t \ll 1: \quad m > 0: \ N_m(t) \sim -\frac{1}{\pi}(m-1)! \left(\frac{2}{t}\right)^m, \qquad (8.98a)$$

$$N_0(t) \sim -\frac{2}{\pi} \log \frac{2}{t} + \text{const.}, \qquad (8.98b)$$

which are the behaviors shown in (8.29a) and (8.29b).

4. To find the value of the latter constant, begin with

$$N_0(t) = -\frac{2}{\pi} \int_0^\infty ds \frac{\cos \sqrt{s^2 + t^2}}{\sqrt{s^2 + t^2}} \qquad (8.99)$$

[this is the imaginary part of the representation (16.3)], divide the integral into two parts at $t \ll s_1 \ll 1$, and get

$$t \ll 1: \quad N_0(t) \sim -\frac{2}{\pi}\left(\log \frac{2}{t} - C\right), \qquad (8.100)$$

where

$$C = -\left[\int_{s_1}^\infty \frac{ds}{s} \cos s + \log s_1\right] = -\int_0^\infty \frac{ds}{s}\left(\cos s - \frac{1}{1+s}\right) \qquad (8.101)$$

is Euler's constant, $C = 0.5772\ldots$.

5. Use the information gained in Problem 8.3 to show that

$$\frac{3}{2} > m > 0: \quad \int_0^\infty ds\, s^{m-1} J_m(s) = 2^{m-1}\Gamma(m). \qquad (8.102)$$

Produce another proof of this by applying a recurrence relation.

6. Combine the differential equation obeyed by

$$G_0(\mathbf{r}, \mathbf{r}') = \frac{e^{ik|\mathbf{r}-\mathbf{r}'|}}{|\mathbf{r}-\mathbf{r}'|}, \qquad (8.103)$$

with the analogous equation for Green's function, appropriate to the longitudinal electric field in a circular pipe with perfectly conducting walls, and arrive at an integral equation. What is the physical interpretation of the surface charge density?

7. The modified Bessel function K_0 may be taken to be defined by

$$K_0(\lambda P) = 2\pi \int \frac{(d\mathbf{k}_\perp)}{(2\pi)^2} \frac{e^{i\mathbf{k}_\perp \cdot (\mathbf{r}-\mathbf{r}')_\perp}}{k^2 + \lambda^2}, \qquad (8.104)$$

where P (capital rho) is the distance between the two points in the plane,

$$P = |(\mathbf{r} - \mathbf{r}')_\perp|. \qquad (8.105)$$

Choose $(\mathbf{r} - \mathbf{r}')_\perp$ to lie in the x-direction, integrate over k_y, and get

$$K_0(t) = \int_0^\infty \frac{d\lambda}{\sqrt{\lambda^2 + 1}} \cos \lambda t \,. \tag{8.106}$$

Now think complex, and derive

$$K_0(t) = \int_1^\infty \frac{d\mu}{\sqrt{\mu^2 - 1}} e^{-\mu t} \,, \tag{8.107}$$

from which you can infer the asymptotic form

$$t \gg 1: \quad K_0(t) \sim \sqrt{\frac{\pi}{2t}} \, e^{-t} \,. \tag{8.108}$$

[Another way of writing the form (8.107) of K_0 is

$$K_0(t) = \int_0^\infty d\theta \, e^{-t \cosh \theta} \,, \tag{8.109}$$

which is reminiscent of J_0. Verify that this construction of K_0 obeys the appropriate differential equation.]

8. The modified Bessel functions are also referred to as Bessel functions of imaginary argument because they may be defined in terms of ordinary Bessel functions by, for $-\pi < \arg z \leq \pi/2$,

$$I_\nu(z) = e^{-\nu \pi i/2} J_\nu(z e^{\pi i/2}) \,, \tag{8.110a}$$

$$K_\nu(z) = \frac{1}{2} \pi i e^{\nu \pi i/2} H_\nu^{(1)}(z e^{\pi i/2}) \,. \tag{8.110b}$$

What differential equation do I_ν and K_ν satisfy? What are the behaviors of these functions for large and small values of z? What is the corresponding Wronskian?

9. This problem explores the toroidal coordinate system. Consider two points A, B, separated by a distance $2c$. If $\overline{AP}$ and $\overline{BP}$ are line segments from A and B to an arbitrary point, respectively, define the toroidal coordinate η by

$$\eta = \ln \frac{|\overline{AP}|}{|\overline{BP}|} \,. \tag{8.111}$$

The interior angle between the two lines locating P from A and B, respectively, is the other toroidal coordinate ϑ (see Fig. 8.5). Show that the contours of constant η define two circles, as shown in the figure, which may be thought of as the cross section of a torus in a plane of constant ϕ. Because $|\overline{AB}|/2 = c$, show that the center of each circle is at distance ρ from the vertical symmetry line of

$$\rho = R = c \coth \eta \,, \tag{8.112}$$

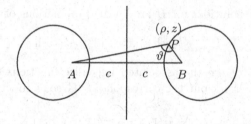

Fig. 8.5. Diagram defining the toroidal coordinates η and ϑ of a point P. Here η is defined by (8.111), and ϑ is the angle between the lines $\overline{PA}$ and $\overline{PB}$. The circles represent lines of constant η. Point P is also located by the coordinates from the horizontal and vertical symmetry axes, z and ρ, two of the three cylindrical coordinates

and the radius of each circle is

$$r_0 = c \operatorname{csch} \eta , \tag{8.113}$$

so we have

$$\frac{R}{r_0} = \cosh \eta , \qquad \eta = \ln \left(\frac{R}{r_0} + \sqrt{\left(\frac{R}{r_0} \right)^2 - 1} \right) . \tag{8.114}$$

Express the cylindrical coordinates ρ and z in terms of toroidal coordinates η and ϑ, and show that the three-dimensional line element has the form

$$\begin{aligned}
\mathrm{d}s^2 &= \mathrm{d}z^2 + \mathrm{d}\rho^2 + \rho^2 \mathrm{d}\phi^2 \\
&= \frac{c^2}{(\cosh \eta - \cos \vartheta)^2} \left(\mathrm{d}\eta^2 + \mathrm{d}\vartheta^2 + \sinh^2 \eta \, \mathrm{d}\varphi^2 \right) .
\end{aligned} \tag{8.115}$$

Consider the action of the Laplacian on a scalar potential V, and show that if the latter is redefined according to

$$V(\rho, \phi, z) = (\cosh \eta - \cos \vartheta)^{1/2} U(\eta, \vartheta, \phi) , \tag{8.116}$$

we have

$$\begin{aligned}
\nabla^2 V = \frac{1}{c^2} &(\cosh \eta - \cos \vartheta)^{5/2} \\
&\times \left(\frac{\partial^2 U}{\partial \eta^2} + \coth \eta \frac{\partial U}{\partial \eta} + \frac{\partial^2 U}{\partial \vartheta^2} + \frac{1}{4} U + \frac{1}{\sinh^2 \eta} \frac{\partial^2 U}{\partial \phi^2} \right) .
\end{aligned} \tag{8.117}$$

We can separate variables in Laplace's equation,

$$U = \mathrm{e}^{\mathrm{i}m\phi} \mathrm{e}^{\mathrm{i}n\vartheta} H(\eta) ; \tag{8.118}$$

show that the solutions for H are linear combinations of

$$P^m_{n-1/2}(\cosh \eta) , \qquad Q^m_{n-1/2}(\cosh \eta) , \qquad (8.119)$$

where P and Q are the associated Legendre functions of the first and second kinds. Work out the corresponding Green's function, that is, the solution of

$$\nabla^2 G(\mathbf{r}, \mathbf{r}') = -\delta(\mathbf{r} - \mathbf{r}') , \qquad (8.120)$$

in toroidal coordinates.

9

Reflection and Refraction

In this chapter we take a brief digression to illustrate a familiar application of the mode decomposition discussed in Chap. 6. The notions discussed here will be significantly generalized in Chap. 13, when junctions between waveguides will be considered.

Using the formalism developed in Chap. 6, it is easy to derive the laws of reflection and refraction of a plane electromagnetic wave by a plane interface separating two dielectric media. Consider the geometry shown in Fig. 9.1. Let

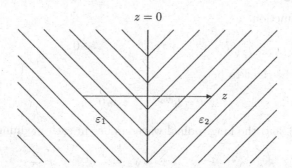

Fig. 9.1. Plane interface between two dielectric media

us consider an E mode, propagating in a purely dielectric medium. Then from (6.32a) to (6.32c) we have

$$\mathbf{E}_\perp = -\boldsymbol{\nabla}_\perp \varphi V \,, \tag{9.1a}$$

$$\mathbf{H}_\perp = -\mathbf{e} \times \boldsymbol{\nabla} \varphi I \,, \tag{9.1b}$$

$$E_z = \mathrm{i}\frac{\gamma^2}{\omega\varepsilon}\varphi I \,. \tag{9.1c}$$

Here, because there are no boundaries in the transverse $(x\text{--}y)$ directions, we may take φ to be a transverse plane wave:

$$\varphi(x,y) = e^{i\mathbf{k}_\perp \cdot \mathbf{r}_\perp} \, , \tag{9.2a}$$

and the corresponding eigenvalue of the transverse Laplacian is

$$\gamma^2 = k_\perp^2 \, . \tag{9.2b}$$

Let us consider a plane wave incident on the interface from the left; there will be a wave reflected by the interface, and one refracted into the region on the right (see Fig. 9.2). Thus we take the wave on the left to be described by

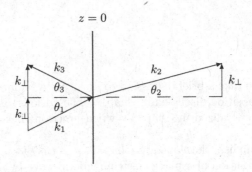

Fig. 9.2. Reflection and refraction of a plane wave by a plane interface

the current function

$$I = e^{i\kappa_1 z} + r e^{-i\kappa_1 z} \, , \quad z < 0 \, , \tag{9.3a}$$

and that on the right to be

$$I = t e^{i\kappa_2 z} \, , \quad z > 0 \, . \tag{9.3b}$$

Here, from (6.36a) the longitudinal wavenumber in each medium is given by

$$\kappa^2 = k^2 - k_\perp^2 \, , \quad k = \frac{\omega}{c}\sqrt{\tilde{\varepsilon}} \equiv \frac{\omega}{c} n \, , \quad \tilde{\varepsilon} = \frac{\varepsilon}{\varepsilon_0} \, , \tag{9.4}$$

where we have introduced the index of refraction $n = \sqrt{\tilde{\varepsilon}}$, and c is the speed of light in vacuum (deviating from our practice elsewhere). The subscripts in (9.3a) and (9.3b) refer to the media 1 and 2.

The laws of reflection,

$$\theta_1 = \theta_3 \, , \tag{9.5}$$

and of refraction (Snell's law),

$$k_1 \sin \theta_1 = k_2 \sin \theta_2 = k_\perp \, , \tag{9.6}$$

or

$$n_1 \sin \theta_1 = n_2 \sin \theta_2 \, , \tag{9.7}$$

are geometrically obvious from Fig. 9.2.

Now we impose the boundary conditions on the electric field at the interface. Recall from (4.42a) and (4.46) that

$$\mathbf{n} \times \mathbf{E} \quad \text{is continuous},$$ (9.8a)

$$\mathbf{n} \cdot \varepsilon\mathbf{E} \quad \text{is continuous}.$$ (9.8b)

Then from (9.1a) and (9.1c) we learn that both I and V must be continuous. From the construction (9.3a) and (9.3b) we learn the continuity condition at $z = 0$:

$$1 + r = t.$$ (9.9)

The relation between the I and V functions, (6.32f), reads

$$V = \frac{1}{i\omega\varepsilon}\frac{d}{dz}I,$$ (9.10)

then implies from (9.3a) and (9.3b) at $z = 0$

$$\frac{1}{\varepsilon_1}\kappa_1(1 - r) = \frac{1}{\varepsilon_2}\kappa_2 t,$$ (9.11)

which when multiplied by (9.9) yields

$$\frac{1}{\varepsilon_1}\kappa_1(1 - r^2) = \frac{1}{\varepsilon_2}\kappa_2 t^2.$$ (9.12)

Equations (9.9) and (9.12) are the usual equations for the reflection and transmission coefficients, as for example given in [9], (41.96) and (41.97), for the case of $\parallel$ polarization ($\mathbf{E}_\perp$ lies in the plane of incidence, defined by $\mathbf{k}_\perp$ and $\mathbf{n}$, hence the name). These equations may be easily solved:

$$\parallel \text{polarization:} \quad r = \frac{\kappa_1/\varepsilon_1 - \kappa_2/\varepsilon_2}{\kappa_1/\varepsilon_1 + \kappa_2/\varepsilon_2}, \quad t = \frac{2\kappa_1/\varepsilon_1}{\kappa_1/\varepsilon_1 + \kappa_2/\varepsilon_2}.$$ (9.13)

Equation (9.12) actually is a statement of energy conservation. If we compute the power (6.77) just to the left, and just to the right, of the interface at $z = 0$, we obtain from (9.10)

$$\frac{1}{2}VI^* = \frac{1}{2}\frac{\kappa_1}{\omega\varepsilon_1}(1 - r^2) \quad (\text{at } z = 0-)$$

$$= \frac{1}{2}\frac{\kappa_2}{\omega\varepsilon_2}t^2 \quad (\text{at } z = 0+),$$ (9.14)

which is the desired equality.

Now we repeat this calculation for the H mode ($\perp$ polarization, since $\mathbf{H}_\perp$ lies in the plane of incidence). The fields are now given by

$$H_z = \frac{i}{\omega\mu} k_\perp^2 e^{ik_\perp \cdot r_\perp} V \,, \tag{9.15a}$$

$$\mathbf{H}_\perp = -i\mathbf{k}_\perp e^{ik_\perp \cdot r_\perp} I \,, \tag{9.15b}$$

$$\mathbf{E}_\perp = \mathbf{e} \times i\mathbf{k}_\perp e^{ik_\perp \cdot r_\perp} V \,, \tag{9.15c}$$

which follow from (6.33a) to (6.33c) when we write

$$\psi = e^{ik_\perp \cdot r_\perp} \,. \tag{9.16}$$

The continuity of $\mathbf{E}_\perp$ implies that of V, while the continuity of $\mathbf{H}_\perp$ [no surface current – see (4.42b)] implies that I is continuous. Now we write[1]

$$V = e^{i\kappa_1 z} + re^{-i\kappa_1 z} \,, \quad z < 0 \,, \tag{9.17a}$$

$$V = te^{i\kappa_2 z} \,, \quad z > 0 \,, \tag{9.17b}$$

so the continuity of V at $z = 0$ implies

$$1 + r = t \,, \tag{9.18}$$

of the same form as (9.9). Using (6.33g), or

$$I = \frac{1}{i} \frac{1}{\omega\mu} \frac{\mathrm{d}}{\mathrm{d}z} V, \tag{9.19}$$

we see that $\mathrm{d}V/\mathrm{d}z$ must be continuous, or

$$\kappa_1(1 - r) = \kappa_2 t \,. \tag{9.20}$$

Multiplying (9.20) by (9.18) yields the equation of energy conservation,

$$\kappa_1(1 - r^2) = \kappa_2 t^2 \,, \tag{9.21}$$

which follows from the continuity of the complex power, $\frac{1}{2}IV^*$. This time the solution for the reflection and transmission coefficients is

$$\perp \text{ polarization:} \quad r = \frac{\kappa_1 - \kappa_2}{\kappa_1 + \kappa_2} \,, \quad t = \frac{2\kappa_1}{\kappa_1 + \kappa_2}. \tag{9.22}$$

For further discussion of these phenomena, the reader is referred to standard textbooks, in particular [9].

9.1 Problems for Chap. 9

1. A wave of frequency ω, moving in the vacuum along the z-axis, is normally incident on a plane mirror (perfect conductor) that is advancing with

---

[1] Note that the way we have defined the reflection and transmission coefficients are such that they refer to $\mathbf{H}_\perp$ for ∥ polarization, and to $\mathbf{E}_\perp$ for ⊥ polarization.

constant speed v perpendicular to its plane. What is ω', the frequency of the reflected wave? (Here, it is enough to say that there is some linear field condition at the conducting surface.) Extend this to a wave in the z–x plane that is incident at angle θ; what relations specify ω' and θ' for the reflected wave? For definiteness consider $\perp$ polarization. Give the explicit forms of θ' and ω' for $|\beta| \ll 1$, and for arbitrary β when both θ and θ' are small.

2. Derive the results found in the previous problem by applying a Lorentz transformation to a known situation.

3. Now let the mirror move with speed v along the x-axis, that is, parallel to its plane. What are ω', θ' for given ω and θ?

4. Return to the wave normally incident on the mirror, moving perpendicular to its plane. What is the boundary condition on the electric field vector? Compute the ratio of the reflected to the incident electric field amplitudes in terms of the velocity of the mirror, or in terms of ω, ω'.

5. Two homogeneous media with refractive indices n_1 and n_3 are separated by a plane sheet of thickness d and refractive index n_2. All substances have unit permeability. Prove that a plane wave normally incident upon the boundary surfaces is completely transmitted provided $n_2 = \sqrt{n_1 n_3}$ and d is an odd integral multiple of a quarter wavelength in the sheet.

10

Variational Methods

10.1 Variational Principles

In view of the very limited number of guide shapes that are amenable to exact analysis in terms of known functions, further progress requires the development of approximation methods applicable to boundaries of more general form. Although there are several special methods that serve this purpose, it is convenient to discuss here a number of general principles that are particularly relevant to the determination of guide cutoff wavenumbers and mode functions. These principles are fundamentally based on the variational formulation of the field equations, as set forth in Chap. 4. It was there shown that if one of the Maxwell equations be admitted as a definition of one field variable, the other equation may be derived as the requirement that a certain volume integral, the Lagrangian, be stationary with respect to variations of the other field variable. For a dissipationless region enclosed by perfectly conducting walls and devoid of internal currents, the variational principles (4.28) and (4.26) state that

$$\delta \int_V (d\mathbf{r})(\varepsilon \mathbf{E}^2 + \mu \mathbf{H}^2) = 0 \,, \tag{10.1}$$

where either the field quantity $\mathbf{H}$ is varied and $\mathbf{E}$ is defined by

$$\nabla \times \mathbf{H} = -ik\eta \mathbf{E} \,, \tag{10.2}$$

or $\mathbf{E}$ is varied and $\mathbf{H}$ is defined by

$$\nabla \times \mathbf{E} = ik\zeta \mathbf{H} \,. \tag{10.3}$$

Furthermore, when $\mathbf{H}$ is the fundamental variable, the stationary requirement, applied to surface variations, automatically yields the boundary conditions (6.19) appropriate to metallic walls, while if $\mathbf{E}$ is the independent field quantity subject to variation it is necessary to restrict the variation to fields that satisfy the boundary condition in order that the surface integral contribution to the variation be removed [cf. (4.26)]. We shall apply these results to construct two variational formulations of the wave equation and boundary conditions satisfied by waveguide mode functions.

For this purpose, we consider the field components of a mode with the intrinsic wavenumber equal to the cutoff wavenumber, $k = \gamma$, so that the guide wavelength is infinite and all field quantities are independent of z. On referring to (6.32g) and (6.33f), we see that an E-mode voltage and an H-mode current vanish under these circumstances (zero characteristic impedance and admittance, respectively) and that, from (6.32c), (6.32b), (6.33d), and (6.33a), the only nonvanishing field components are proportional to

$$\text{E mode:} \quad E_z = \mathrm{i}\gamma\zeta\varphi(x,y)\,, \quad \mathbf{H} = -\mathbf{e} \times \boldsymbol{\nabla}\varphi(x,y)\,, \qquad (10.4a)$$

$$\text{H mode:} \quad H_z = \mathrm{i}\gamma\eta\psi(x,y)\,, \quad \mathbf{E} = \mathbf{e} \times \boldsymbol{\nabla}\psi(x,y)\,. \qquad (10.4b)$$

In this form, the E-mode field satisfies (6.1a) identically, while the H-mode field satisfies (6.1b). To reverse this situation, we may, with equal validity, write

$$\text{E mode:} \quad E_z = -\mathrm{i}\frac{\zeta}{\gamma}\nabla^2\varphi(x,y)\,, \quad \mathbf{H} = -\mathbf{e} \times \boldsymbol{\nabla}\varphi(x,y)\,, \qquad (10.5a)$$

$$\text{H mode:} \quad H_z = -\mathrm{i}\frac{\eta}{\gamma}\nabla^2\psi(x,y)\,, \quad \mathbf{E} = \mathbf{e} \times \boldsymbol{\nabla}\psi(x,y)\,. \qquad (10.5b)$$

Since all field quantities are independent of z, the volume integral in (10.1) degenerates into a surface integral extended across a section of the guide. On substituting the representations (10.4a)–(10.4b) in (10.1), we obtain

$$\delta \int_\sigma \mathrm{d}\sigma\,[(\boldsymbol{\nabla}f)^2 - \gamma^2 f^2] = 0 \qquad (10.6)$$

for both E ($f = \varphi$) and H ($f = \psi$) mode functions. However, since this representation employs the electric and magnetic fields, respectively, as the fundamental field variables for the E and H modes, the variational principle (10.6), applied to the E modes, excludes any variation that violates the boundary condition $\varphi = 0$ on the boundary curve C, while automatically yielding the H-mode boundary conditions, $\frac{\partial}{\partial n}\psi = 0$ on C, for unrestricted variations on the boundary. Although we have derived this result from the variational formulation of the field equations, in the interests of generality, it is immediately confirmed on performing the variation,

$$-\int_\sigma \mathrm{d}\sigma\,\delta f\,(\nabla^2 f + \gamma^2 f) + \oint_C \mathrm{d}s\,\delta f\frac{\partial}{\partial n}f = 0\,, \qquad (10.7)$$

that the stationary requirement yields the wave equation and the H-mode boundary condition, for unrestricted variations, and requires the imposition of the boundary condition on the varied function, $\delta f = 0$, when dealing with

E modes. Another variational formulation is provided by the representation (10.5a)–(10.5b), which when substituted into (10.1) yields

$$\delta \int_\sigma d\sigma \left[(\nabla^2 f)^2 - \gamma^2 (\nabla f)^2 \right] = 0 , \tag{10.8}$$

where now the variational principle yields the E-mode boundary condition for unrestricted variations, and when applied to H modes, the variation must not violate the boundary conditions. To confirm these statements directly, we perform the indicated variations in (10.8), and obtain

$$-\int_\sigma d\sigma \, \nabla (\nabla^2 f + \gamma^2 f) \cdot \delta (\nabla f) + \oint_C ds \, \delta \left(\frac{\partial f}{\partial n} \right) \nabla^2 f = 0 . \tag{10.9}$$

The stationary requirement for arbitrary variations of ∇f in the interior of the region yields

$$\nabla^2 f + \gamma^2 f = \text{const.} , \tag{10.10}$$

which is fully equivalent to the wave equation, for the constant of integration can be absorbed into the function f if γ is not zero. Indeed, it is apparent that the variational principle (10.8) is unaffected by the addition of a constant to f. The requirement that the line integral vanish for arbitrary variations of $\frac{\partial f}{\partial n}$ on the boundary implies that $\nabla^2 f = 0$ on the boundary C, which is equivalent to the E-mode condition if $\gamma \neq 0$. In order that the variational principle be applicable to H modes, the boundary condition, $\frac{\partial}{\partial n} f = 0$ on C, must not be violated by the variation.

We shall later find it advantageous to consider more general boundary conditions of the type

$$\frac{\partial}{\partial n} f + \rho f = 0 \text{ on } C , \tag{10.11}$$

where ρ is a prescribed function of position on the boundary curve. The E-mode and H-mode boundary conditions are particular forms of (10.11) corresponding to $\rho = \infty$ and $\rho = 0$, respectively. [The general linear boundary condition was considered in (4.19) and (4.23).] This boundary condition can be derived from the two variational principles by the addition of suitable line integrals to the surface integrals of (10.6) and (10.8). Thus, the stationary requirement expressed by the vanishing variation of

$$\int_\sigma d\sigma \left[(\nabla f)^2 - \gamma^2 f^2 \right] + \oint_C ds \, \rho f^2 , \tag{10.12}$$

namely

$$-\int_\sigma d\sigma \, \delta f \, (\nabla^2 + \gamma^2) f + \oint_C ds \, \delta f \left(\frac{\partial}{\partial n} f + \rho f \right) = 0 , \tag{10.13}$$

yields the wave equation and the boundary condition, for arbitrary variations within the region and on the boundary. Similarly, the stationary property of

$$\int_\sigma d\sigma \left[(\nabla^2 f)^2 - \gamma^2 (\nabla f)^2 \right] - \gamma^2 \oint_C ds \frac{1}{\rho} \left(\frac{\partial}{\partial n} f \right)^2 \tag{10.14}$$

as expressed by

$$-\int_\sigma d\sigma \, \nabla (\nabla^2 f + \gamma^2 f) \cdot \delta(\nabla f) + \oint_C ds \, \delta \left(\frac{\partial}{\partial n} f \right) \left(\nabla^2 f - \frac{\gamma^2}{\rho} \frac{\partial}{\partial n} f \right) = 0 \,, \tag{10.15}$$

implies the wave equation and a boundary condition that is equivalent to (10.11), if $\gamma \neq 0$. In both these more general variational principles, no restrictions are imposed upon the nature of the variation. If it be desired, however, to let ρ approach ∞ in (10.12), in order to obtain the E-mode boundary condition, the line integral can only be omitted if the boundary condition $f = 0$ is imposed on the varied function. Similarly, if the expression (10.14) is restricted to the H-mode boundary condition by letting ρ approach zero, it is necessary to rigidly impose the boundary condition $\frac{\partial}{\partial n} f = 0$ before omitting the line integral. It is in this way that the restrictions arise of the allowable variations in (10.6) and (10.8) for E and H modes, respectively.

10.2 Rayleigh's Principle

Before indicating the practical utility of these variational principles, it is necessary to remark that the various quantities that are stationary in value relative to variations about certain functions f have the value zero for such functions. To demonstrate this, we need only observe that the expressions (10.12) and (10.14), of which (10.6) and (10.8) are special variations, have the property of being homogeneous in f. If, therefore, we consider a variation in f that is proportional to itself, say $\delta f = \epsilon f$ with ϵ an infinitesimal quantity, the variations of the expressions (10.12) and (10.14) are a multiple of the original form, which on the other hand must be zero, since the stationary property is independent of the particular form of the variation. Thus the assertion is proved. Hence, if f is a solution to the wave equation and satisfies the boundary condition (10.11),

$$\int_\sigma d\sigma \left[(\nabla f)^2 - \gamma^2 f^2 \right] + \oint_C ds \, \rho f^2 = 0 \,, \tag{10.16a}$$

$$\int_\sigma d\sigma \left[(\nabla^2 f)^2 - \gamma^2 (\nabla f)^2 \right] - \gamma^2 \oint_C ds \frac{1}{\rho} \left(\frac{\partial}{\partial n} f \right)^2 = 0 \,, \tag{10.16b}$$

or

$$\gamma^2 = \frac{\int_\sigma d\sigma \, (\nabla f)^2 + \oint_C ds \, \rho f^2}{\int_\sigma d\sigma \, f^2} \,, \tag{10.17a}$$

$$\gamma^2 = \frac{\int_\sigma d\sigma \, (\nabla^2 f)^2}{\int_\sigma d\sigma \, (\nabla f)^2 + \oint_C ds \frac{1}{\rho} \left(\frac{\partial}{\partial n} f \right)^2} \,. \tag{10.17b}$$

When E- or H-mode boundary conditions obtain, the latter expressions reduce
to

$$\gamma^2 = \frac{\int_\sigma d\sigma \, (\nabla f)^2}{\int_\sigma d\sigma \, f^2} \, , \tag{10.18a}$$

$$\gamma^2 = \frac{\int_\sigma d\sigma \, (\nabla^2 f)^2}{\int_\sigma d\sigma \, (\nabla f)^2} \, , \tag{10.18b}$$

of which the first has already been stated in (6.43). Let us now choose an
arbitrary function f, subject to certain continuity restrictions, and define a
quantity γ associated with it by means of (10.17a) and (10.17b). Of the class
of arbitrary functions, the eigenfunctions f, which satisfy the wave equation
and the boundary condition, are unique in possessing the property that the
quantity γ is stationary with respect to arbitrary variations about f. Thus a
function that deviates from a true eigenfunction to the first-order of a small
quantity will enable a value of γ to be calculated that deviates from the correct
value only in the second-order of the small quantity. To prove this statement
we consider an infinitesimal deviation of f from an eigenfunction and calculate the concomitant change in γ, in accordance with (10.17a) or (10.17b).
Now the expressions contained in the latter equations have the property of
being stationary relative to variations of the function f alone, and therefore
to preserve the equality the quantity γ must also be stationary. Conversely, in
order that γ be stationary, the right-hand members of (10.17a) and (10.17b)
must be stationary with respect to variations of f alone, which are the two
variational principles that yield the wave equation and the boundary condition. As a further property of (10.17a) and (10.17b) we remark that if ρ is
positive, or more particularly if E- or H-mode boundary conditions are opera-
tive, the expressions for γ^2 are never negative and, therefore, of the functions
that render γ^2 stationary there must exist one that makes γ^2 an absolute
minimum. The eigenfunction that possesses this property is the mode function of minimum cutoff wavenumber, the dominant E or H mode of the guide.
(This statement is subject to important qualifications. See Sect. 10.4.) Any
other function must yield a larger value of γ^2, that is, a function f that differs from the dominant mode function produces a value of γ^2 in excess of
the true value. The stationary property of the cutoff wavenumbers, and the
minimum property of the absolute cutoff wavenumber, are aspects of what is
known as Rayleigh's principle, which is generally applicable to the frequencies of systems with dynamical equations that are obtained from a variational
principle.

The two expressions for the cutoff wavenumber can be given another significance in the following manner. The quantity[1]

$$\int d\sigma \, (\nabla^2 f + \gamma^2 f)^2 \tag{10.19}$$

---

[1] Henceforward, in this chapter, when only surface integrals appear, we will understand that they extend over the cross-sectional area σ.

is evidently greater than zero unless f is a solution of the wave equation and γ^2 the corresponding eigenvalue. The eigenfunctions are therefore characterized by the vanishing of (10.19). If f is not an eigenfunction, the best approximation to γ^2 is obtained by minimizing (10.19) with respect to γ. This yields

$$\gamma^2 = -\frac{\int d\sigma\, f\nabla^2 f}{\int d\sigma\, f^2} = \frac{\int d\sigma\, (\nabla f)^2}{\int d\sigma\, f^2}\,, \tag{10.20}$$

the latter form obtaining if E- or H-mode boundary conditions are imposed on f. Furthermore, for this choice of γ^2,

$$\int d\sigma\, (\nabla^2 f + \gamma^2 f)^2 = \int d\sigma\, (\nabla^2 f)^2 - \frac{\left[\int d\sigma\, (\nabla f)^2\right]^2}{\int d\sigma\, f^2}\,, \tag{10.21}$$

which shows that[2]

$$\frac{\int d\sigma\, (\nabla^2 f)^2}{\int d\sigma\, (\nabla f)^2} \geq \frac{\int d\sigma\, (\nabla f)^2}{\int d\sigma\, f^2}\,. \tag{10.23}$$

Hence the value of γ computed from an assumed function f by means of (10.18a) never exceeds that obtained from (10.18b), and the two values of γ^2 are only equal if f is an eigenfunction with the common value of γ^2 as eigenvalue. Thus a measure of the extent to which a function approximates an eigenfunction is provided by the degree of agreement between (10.18a) and (10.18b). Similarly,

$$\int d\sigma\, \left[\nabla(\nabla^2 f + \gamma^2 f)\right]^2 \tag{10.24}$$

is greater than zero unless f is a solution of (10.10). For an assumed function f, the minimum of (10.24) as a function of γ^2 is obtained when

$$\gamma^2 = -\frac{\int d\sigma\, \nabla(\nabla^2 f) \cdot \nabla f}{\int d\sigma\, (\nabla f)^2} = \frac{\int d\sigma\, (\nabla^2 f)^2}{\int d\sigma\, (\nabla f)^2}\,. \tag{10.25}$$

In writing the latter form it is assumed that f satisfies H-mode boundary conditions or that $\nabla^2 f$ obeys E-mode boundary conditions. With this value of γ^2,

$$\int d\sigma\, \left[\nabla(\nabla^2 f + \gamma^2 f)\right]^2 = \int d\sigma\, \left[\nabla(\nabla^2 f)\right]^2 - \frac{\left[\int d\sigma\, (\nabla^2 f)^2\right]^2}{\int d\sigma\, (\nabla f)^2}\,, \tag{10.26}$$

---

[2] Mathematically, this is called the Cauchy–Schwarz–Bunyakovskii inequality, which generally states that

$$\int d\sigma\, |f|^2 \int d\sigma\, |g|^2 \geq \left|\int d\sigma\, f^* g\right|^2\,. \tag{10.22}$$

whence

$$\frac{\int d\sigma \left[\boldsymbol{\nabla}(\nabla^2 f)\right]^2}{\int d\sigma \, (\nabla^2 f)^2} \geq \frac{\int d\sigma \, (\nabla^2 f)^2}{\int d\sigma \, (\boldsymbol{\nabla} f)^2} \geq \frac{\int d\sigma \, (\boldsymbol{\nabla} f)^2}{\int d\sigma \, f^2} \, , \qquad (10.27)$$

provided

$$\frac{\partial}{\partial n} f = 0 \text{ on } C \, , \qquad (10.28a)$$

or

$$f = 0 \, , \qquad \nabla^2 f = 0 \text{ on } C \, . \qquad (10.28b)$$

The equality signs hold only if f is a solution of the wave equation obeying either E- or H-mode boundary conditions.

The eigenfunctions can be independently defined and, in principle, constructed by means of the stationary property of the eigenvalues. Thus, of the class of admissible functions, those that are continuous and have sectionally continuous derivatives, the function f_1 that makes (10.17a) an absolute minimum is a solution of the wave equation obeying the boundary condition (10.11). The minimum value of (10.17a), γ_1^2, is the eigenvalue associated with the function f_1. If, now, the minimum of (10.17a) is sought among the class of admissible functions that are orthogonal to f_1:

$$\int d\sigma \, f f_1 = 0 \, , \qquad (10.29)$$

the function f_2 that meets these requirements is a solution of the wave equation, obeying the boundary conditions, and associated with the eigenvalue $\gamma_2^2 \geq \gamma_1^2$. More generally, if (10.17a) is minimized among the class of admissible functions orthogonal to the first n eigenfunctions,

$$\int d\sigma \, f f_i = 0 \, , \qquad i = 1, 2, \ldots, n \, , \qquad (10.30)$$

the minimizing function f_{n+1} is a solution of the wave equation and the concomitant boundary condition, associated with an eigenvalue γ_{n+1}^2 which exceeds or possibly equals the nth eigenvalue obtained by the minimizing process. In this manner, an infinite set of eigenfunctions and eigenvalues is obtained, arranged in ascending magnitude of the eigenvalues. That the set of eigenfunctions formed in this way constitute the entire set of eigenfunctions will be demonstrated shortly. To prove that the minimization process indicated above generates successive eigenfunctions, we proceed by induction, assuming that the first n functions, f_i, $i = 1, \ldots, n$ are each solutions of the wave equation and the boundary condition and that each function is orthogonal to the previous members of the set. Furthermore, it is assumed that the eigenvalues form a monotonically increasing sequence, $\gamma_{i+1}^2 \geq \gamma_i^2$. The variation equation (10.13) is a necessary condition that (10.17a) be a minimum, which must be supplemented by the conditions deduced from (10.30):

$$\int d\sigma \, \delta f f_i = 0 \, , \qquad i = 1, 2, \ldots, n \, , \qquad (10.31)$$

n equations of constraint on the arbitrary functions δf. On introducing the Lagrangian multipliers α_i, $i = 1, \ldots, n$, we deduce that the conditions imposed on f in order that (10.17a) be a minimum are

$$\frac{\partial}{\partial n}f + \rho f = 0 \text{ on } C , \tag{10.32a}$$

$$(\nabla^2 + \gamma^2)f = \sum_{j=1}^{n} \alpha_j f_j , \tag{10.32b}$$

where the α_j are to be determined by the supplementary conditions (10.31). The result of multiplying (10.32b) by f_i, a member of the set of mutually orthogonal functions, and integrating over the guide cross section, is

$$\alpha_i \int_\sigma \mathrm{d}\sigma\, f_i^2 = \int_\sigma \mathrm{d}\sigma\, f(\nabla^2 + \gamma^2)f_i + \oint_C \mathrm{d}s \left(f_i \frac{\partial}{\partial n}f - f \frac{\partial}{\partial n}f_i \right)$$

$$= (\gamma^2 - \gamma_i^2) \int_\sigma \mathrm{d}\sigma\, f f_i$$

$$+ \oint_C \mathrm{d}s \left[f_i \left(\frac{\partial}{\partial n}f + \rho f \right) - f \left(\frac{\partial}{\partial n}f_i + \rho f_i \right) \right] . \tag{10.33}$$

Now the line integral is zero when both f and f_i satisfy the boundary condition (10.32a), and the surface integral is also zero, in consequence of (10.30). Hence all the Lagrangian multipliers vanish, and the minimizing function f_{n+1} is a solution of the wave equation and the boundary condition. Furthermore, it is evident that if a quantity involving a function f is to be minimized among a class of admissible functions, any restriction imposed on that class cannot decrease the minimum value of the quantity, but either increases it or leaves in unaffected. Since the functions f_{n+1} that yields the minimum of γ^2, subject to orthogonality with the first n eigenfunctions, is to be found among a more restricted class of functions than f_n, which is orthogonal to the first $n-1$ eigenfunctions, we conclude that

$$\gamma_{n+1}^2 \geq \gamma_n^2 , \tag{10.34}$$

and thus our assertions are verified.

The same process can be used with the expression (10.17b) for γ^2. For simplicity, we consider only E- or H-mode boundary conditions, so (10.18b) applies. The class of admissible functions are those that have continuous first derivatives and sectionally continuous second derivatives. In addition H-mode functions must satisfy the proper boundary condition. The function that minimizes (10.18b) is the eigenfunction f_1, to within an additive constant. The second eigenfunction f_2 is obtained by minimizing (10.18b) with f restricted by the orthogonality condition,

$$\int \mathrm{d}\sigma\, \boldsymbol{\nabla} f \cdot \boldsymbol{\nabla} f_1 = 0 , \tag{10.35}$$

and, more generally, the eigenfunction f_{n+1} is found by minimizing (10.18b) among the class of functions that satisfy

$$\int d\sigma \, \boldsymbol{\nabla} f \cdot \boldsymbol{\nabla} f_i = 0 \, , \quad i = 1, \ldots, n \, . \tag{10.36}$$

To prove the latter statement we combine (10.9), the minimum conditions for γ^2, with the supplementary conditions deduced from (10.36):

$$\int d\sigma \, \delta(\boldsymbol{\nabla} f) \cdot (\boldsymbol{\nabla} f_i) = 0 \, , \quad i = 1, \ldots, n \, , \tag{10.37}$$

by means of the Lagrangian multipliers α_i, and obtain the following equation for the minimizing function f:

$$\boldsymbol{\nabla}(\nabla^2 f + \gamma^2 f) = \sum_{j=1}^{n} \alpha_j \boldsymbol{\nabla} f_j \, , \tag{10.38}$$

subject to the boundary conditions

$$\text{E mode:} \quad \nabla^2 f = 0 \, , \tag{10.39a}$$

$$\text{H mode:} \quad \frac{\partial}{\partial n} f = 0 \, . \tag{10.39b}$$

The Lagrangian multipliers may be shown to vanish as before. We multiply (10.38) by $\boldsymbol{\nabla} f_i$ and make use of the assumed orthogonality of the first n eigenfunctions to obtain

$$\begin{aligned}
\alpha_i \int_\sigma d\sigma \, (\boldsymbol{\nabla} f_i)^2 &= \int_\sigma d\sigma \, \boldsymbol{\nabla} f_i \cdot \boldsymbol{\nabla}(\nabla^2 + \gamma^2) f \\
&= \int_\sigma d\sigma \, \boldsymbol{\nabla}(\nabla^2 + \gamma^2) f_i \cdot \boldsymbol{\nabla} f \\
&\quad + \oint_C ds \left(\nabla^2 f \frac{\partial}{\partial n} f_i - \nabla^2 f_i \frac{\partial}{\partial n} f \right) = 0 \, , \tag{10.40}
\end{aligned}$$

in consequence of the boundary conditions shared by the minimizing function and the first n eigenfunctions, and of the orthogonality condition (10.35) combined with the wave equation satisfied by the eigenfunctions f_i. Thus the minimizing function f_{n+1} is a solution of the wave equation and the boundary condition. The argument which demonstrates the monotonic nature of the successive eigenvalues is equally applicable to this method of construction.

We have thus far ignored the possibility that the first eigenfunction yielded by Rayleigh's principle may correspond to $\gamma = 0$ and therefore be inadmissible as a waveguide mode function. It is apparent that the absolute minimum of (10.18a) is indeed $\gamma = 0$, which is attained with f equal to a constant. This possibility is negated by the E-mode boundary condition, $f = 0$ on C; the mode of zero cutoff wavenumber is a spurious H mode. Hence, the first physically significant H mode is the second mode yielded by Rayleigh's principle. To exclude the mode with $\gamma = 0$, we must require that the function f be orthogonal to a constant, or

$$\int d\sigma \, f = 0 \, . \tag{10.41}$$

The eigenfunction found by minimizing (10.18a) subject to this condition and to (10.39b) is the dominant H mode of the guide and will henceforth be considered the first mode. The second H mode is found among those functions that obey (10.41) and are orthogonal to the dominant mode function, and the process continues as before. A similar difficulty confronts Rayleigh's principle, applied to (10.18b). The absolute minimum of (10.18b) is zero, attained by a function f_0 that satisfies

$$\nabla^2 f_0 = 0 \, . \tag{10.42}$$

When dealing with H modes, the function f_0 is constrained by the boundary condition

$$\frac{\partial}{\partial n} f_0 = 0 \text{ on } C \, , \tag{10.43}$$

and this implies that ∇f_0 vanishes identically:

$$\int_\sigma d\sigma \, (\nabla f_0)^2 = \oint_C ds \, f_0 \frac{\partial}{\partial n} f_0 - \int_\sigma d\sigma \, f_0 \nabla^2 f_0 = 0 \, . \tag{10.44}$$

Therefore, only a spurious E mode exists. To effectively remove this mode, only such functions f are admitted that are orthogonal to f_0 in the sense of (10.36):

$$\int_\sigma d\sigma \, \nabla f \cdot \nabla f_0 = \oint_C ds \, f \frac{\partial}{\partial n} f_0 = 0 \, . \tag{10.45}$$

Now, since f_0 is restricted only by (10.42), the normal derivative of f_0 can be arbitrarily assigned on the boundary and the general validity of (10.45) can only be assured if f vanishes, or is constant [see (6.28)], on the boundary. Hence, despite the fact that (10.18b) is stationary with respect to arbitrary variations about an E-mode function, the function f must satisfy the E-mode boundary condition if the absolute minimum of γ is to be the cutoff wavenumber for the dominant E mode.

To summarize, for the validity of Rayleigh's principle applied to the expression (10.18b), it is necessary that f satisfies E- or H-mode boundary conditions. When applied to the expression (10.18a), the E-mode boundary condition must not be violated, but no restriction is placed on H-mode functions other than (10.41).[3]

---

[3] We are excluding here the consideration of T modes, which can occur for coaxial lines, for example. See Chap. 6.

10.3 Proof of Completeness

It will now be shown that the set of eigenfunctions obtained with the aid of Rayleigh's principle is closed and complete. A function set $\{f_n\}$ is called closed if there exists no function orthogonal to every member of the set, that is, if

$$\int d\sigma\, f f_n = 0\,, \quad n = 1, 2, \ldots \tag{10.46}$$

implies that $f = 0$. A set of eigenfunctions is considered complete if any function, subject to certain continuity and boundary conditions, can be approximated in the mean with arbitrary precision by a linear combination of eigenfunctions. This statement is to be understood in the sense that constants c_n can be associated with any function f such that

$$\lim_{N \to \infty} \int d\sigma \left[f - \sum_{n=1}^{N} c_n f_n \right]^2 = 0\,, \tag{10.47}$$

which may be written

$$\mathrm{l.i.m}_{N \to \infty} \sum_{n=1}^{N} c_n f_n(\mathbf{r}) = f(\mathbf{r})\,, \tag{10.48}$$

where l.i.m denotes the limit in the mean. We first note that the best approximation in the mean with a finite sum of eigenfunctions is obtained by minimizing

$$\int d\sigma \left[f - \sum_{n=1}^{N} c_n f_n \right]^2 = \int d\sigma\, f^2 - 2 \sum_{n=1}^{N} c_n \int d\sigma\, f_n f + \sum_{n=1}^{N} c_n^2 \int d\sigma\, f_n^2\,, \tag{10.49}$$

considered as a function of the constants c_n. This yields

$$c_n = \frac{\int d\sigma\, f_n f}{\int d\sigma\, f_n^2}\,, \tag{10.50}$$

or, more simply,

$$c_n = \int d\sigma\, f_n f\,, \tag{10.51a}$$

if the eigenfunctions are normalized by

$$\int d\sigma\, f_n^2 = 1\,. \tag{10.51b}$$

[Note we are using a different normalization than in (6.57) and (6.59).] Hence, the coefficients are chosen in accordance with Fourier's rule and are independent of N, the number of terms in the approximating function. For these values of c_n,

$$\int d\sigma \left[f - \sum_{n=1}^{N} c_n f_n \right]^2 = \int d\sigma \, f^2 - \sum_{n=1}^{N} c_n^2 \geq 0 , \tag{10.52}$$

from which we deduce that

$$\sum_{n=1}^{N} c_n^2 \leq \int d\sigma \, f^2 , \tag{10.53}$$

a relation that is known as the Bessel inequality. Since this result is valid for any N, we infer the convergence of the sum of the squares of the Fourier coefficients associated with a function that is quadratically integrable:

$$\sum_{n=1}^{\infty} c_n^2 \leq \int d\sigma \, f^2 . \tag{10.54}$$

In order to demonstrate the completeness property of the eigenfunctions, we must show that the latter relation is, in fact, an equality. For this purpose we consider in more detail the difference between an arbitrary function f and the approximating function constructed from the first N eigenfunctions:

$$g_N = f - \sum_{n=1}^{N} c_n f_n , \tag{10.55}$$

which has the property

$$\int d\sigma \, f_n g_N = 0 , \quad n = 1, \ldots, N , \tag{10.56}$$

in consequence of (10.51a). Since g_N is orthogonal to the first N eigenfunctions, the value of γ^2 computed from (10.18a) necessarily exceeds or equals γ_{N+1}, in accordance with Rayleigh's principle:

$$\frac{\int d\sigma \, (\nabla g_N)^2}{\int d\sigma \, g_N^2} \geq \gamma_{N+1}^2 . \tag{10.57}$$

In order that the conditions of Rayleigh's principle be met, the function f must vanish on the boundary if E-mode functions are employed; no boundary restriction on f is necessary if the eigenfunctions are H-mode functions. Now

$$\int d\sigma \, (\nabla g_N)^2 = \int d\sigma \, (\nabla f)^2 - 2 \sum_{n=1}^{N} c_n \int d\sigma \, \nabla f_n \cdot \nabla f + \sum_{n=1}^{N} \gamma_n^2 c_n^2 , \tag{10.58}$$

for

$$\int d\sigma \, \nabla f_n \cdot \nabla f_m = \gamma_n^2 \int d\sigma \, f_n f_m = \gamma_n^2 \delta_{nm} . \tag{10.59}$$

Furthermore,

$$\int_\sigma \mathrm{d}\sigma\, \boldsymbol{\nabla} f_n \cdot \boldsymbol{\nabla} f = \oint_C \mathrm{d}s\, f \frac{\partial}{\partial n} f_n + \gamma_n^2 \int_\sigma \mathrm{d}\sigma\, f_n f = \gamma_n^2 c_n \,, \qquad (10.60)$$

since the line integral vanishes both for E modes, where $f = 0$ on C, and for H modes, where $\frac{\partial}{\partial n} f_n = 0$ on C. Hence from (10.58)

$$\int \mathrm{d}\sigma\, (\boldsymbol{\nabla} g_N)^2 = \int \mathrm{d}\sigma\, (\boldsymbol{\nabla} f)^2 - \sum_{n=1}^N \gamma_n^2 c_n^2 \geq 0 \,, \qquad (10.61)$$

which implies that

$$\sum_{n=1}^N \gamma_n^2 c_n^2 \leq \int \mathrm{d}\sigma\, (\boldsymbol{\nabla} f)^2 \,. \qquad (10.62)$$

Therefore, if the gradient of f is quadratically integrable, the limit of the summation occurring in (10.62) is convergent:

$$\sum_{n=1}^\infty \gamma_n^2 c_n^2 \leq \int \mathrm{d}\sigma\, (\boldsymbol{\nabla} f)^2 \,. \qquad (10.63)$$

Equation (10.57), deduced from Rayleigh's principle, now informs us that

$$\int \mathrm{d}\sigma\, g_N^2 \leq \frac{\int \mathrm{d}\sigma\, (\boldsymbol{\nabla} g_N)^2}{\gamma_{N+1}^2} < \frac{\int \mathrm{d}\sigma\, (\boldsymbol{\nabla} f)^2}{\gamma_{N+1}^2} \,. \qquad (10.64)$$

An essential remark required for the completion of the proof is that the successive eigenvalues have no upper bound, but increase monotonically without limit:

$$\lim_{n \to \infty} \gamma_n = \infty \,. \qquad (10.65)$$

No detailed demonstration of this statement will be given;[4] the evidence offered by the particular problems treated in Chaps. 7 and 8 is a sufficient assurance of its general validity. Granted this property of the eigenvalues, the proof is complete, for according to (10.64), $\int \mathrm{d}\sigma\, g_N^2$ has an upper bound that approaches zero as $N \to \infty$:

$$\lim_{N \to \infty} \int \mathrm{d}\sigma\, g_N^2 = \lim_{N \to \infty} \int \mathrm{d}\sigma\, \left[f - \sum_{n=1}^N c_n f_n \right]^2 = 0 \,. \qquad (10.66)$$

Hence, from (10.52),

$$\sum_{n=1}^\infty c_n^2 = \int \mathrm{d}\sigma\, f^2 \,, \qquad (10.67)$$

a result known as Parseval's theorem, or the completeness relation. A valuable symbolic form of the completeness relation can be constructed by rewriting (10.67) as

---

[4] However, see the end of this section.

$$\int d\sigma \, f^2(\mathbf{r}) = \sum_{n=1}^{\infty} \left[\int d\sigma \, f_n(\mathbf{r}) f(\mathbf{r}) \right]^2 ,$$

$$= \int d\sigma \, d\sigma' \, f(\mathbf{r}) \left[\sum_{n=1}^{\infty} f_n(\mathbf{r}) f_n(\mathbf{r}') \right] f(\mathbf{r}') , \qquad (10.68)$$

whence

$$\sum_{n=1}^{\infty} f_n(\mathbf{r}) f_n(\mathbf{r}') = \delta(\mathbf{r} - \mathbf{r}') , \qquad (10.69)$$

since (10.68) is valid for an arbitrary function, $f(\mathbf{r})$. The closure property (10.46) of the eigenfunction set $\{f_n\}$ follows immediately from the completeness relation. If a function f is orthogonal to every eigenfunction, all Fourier coefficients vanish and $\int d\sigma \, f^2 = 0$, whence $f = 0$.

Further results can be obtained by using (10.18b) in conjunction with Rayleigh's principle. If f satisfies E- or H-mode boundary conditions, in view of (10.59) and (10.60),

$$\frac{\int d\sigma \, (\nabla^2 g_N)^2}{\int d\sigma \, (\boldsymbol{\nabla} g_N)^2} \geq \gamma_{N+1}^2 . \qquad (10.70)$$

The numerator of the left-hand member may be rewritten by the following sequence of operations

$$\nabla^2 g_N = \nabla^2 f + \sum_{n=1}^{N} \gamma_n^2 c_n f_n , \qquad (10.71a)$$

$$\int d\sigma \, (\nabla^2 g_N)^2 = \int d\sigma \, (\nabla^2 f)^2 + 2 \sum_{n=1}^{N} \gamma_n^2 c_n \int d\sigma \, f_n \nabla^2 f + \sum_{n=1}^{N} \gamma_n^4 c_n^2$$

$$= \int d\sigma \, (\nabla^2 f)^2 - \sum_{n=1}^{N} \gamma_n^4 c_n^2 \geq 0 , \qquad (10.71b)$$

since

$$\int d\sigma \, f_n \nabla^2 f = \int d\sigma \, f \nabla^2 f_n = -\gamma_n^2 c_n . \qquad (10.72)$$

Therefore,

$$\sum_{n=1}^{N} \gamma_n^4 c_n^2 \leq \int d\sigma \, (\nabla^2 f)^2 , \qquad (10.73)$$

which implies the convergence of the corresponding infinite series if f is such that $\nabla^2 f$ is quadratically integrable:

$$\sum_{n=1}^{\infty} \gamma_n^4 c_n^2 \leq \int d\sigma \, (\nabla^2 f)^2 . \qquad (10.74)$$

We may now write from (10.70) that

$$\int d\sigma \, (\boldsymbol{\nabla} g_N)^2 \le \frac{\int d\sigma \, (\nabla^2 g_N)^2}{\gamma_{N+1}^2} < \frac{\int d\sigma \, (\nabla^2 f)^2}{\gamma_{N+1}^2} \to 0 \,, \quad N \to \infty \,, \quad (10.75)$$

which demonstrates from (10.61) that the inequality (10.63) is truly an equality:

$$\sum_{n=1}^{\infty} \gamma_n^2 c_n^2 = \int d\sigma \, (\boldsymbol{\nabla} f)^2 \,. \tag{10.76}$$

We have shown that the set of functions supplied by Rayleigh's principle is closed and complete, and is therefore identical with the totality of solutions of the wave equation and the associated boundary condition. Conversely, if the completeness of the eigenfunctions be assumed, Rayleigh's principle can be deduced in a simple manner. In accordance with (10.76) and (10.67), the expression of the completeness property of the eigenfunction system reads

$$\gamma^2 = \frac{\int d\sigma \, (\boldsymbol{\nabla} f)^2}{\int d\sigma \, f^2} = \frac{\sum_{n=1}^{\infty} \gamma_n^2 c_n^2}{\sum_{n=1}^{\infty} c_n^2} \,. \tag{10.77}$$

If we seek the function f that renders γ^2 stationary by regarding the Fourier coefficients of f as arbitrary constants subject to independent variation, we obtain

$$(\gamma^2 - \gamma_n^2) c_n = 0 \,, \quad n = 1, 2, \ldots \,. \tag{10.78}$$

Hence the possible values of γ^2 are just the set of eigenvalues, and for each such value the constants c_n associated with other eigenvalues vanish; the functions that possess the stationary property are the eigenfunctions and the stationary values of γ^2 are the associated eigenvalues. To prove the second part of Rayleigh's principle we write

$$\gamma^2 = \gamma_1^2 + \frac{\sum_{n=2}^{\infty} (\gamma_n^2 - \gamma_1^2) c_n^2}{\sum_{n=1}^{\infty} c_n^2} \ge \gamma_1^2 \,, \tag{10.79}$$

that is, the value of γ^2 computed from an arbitrary function can never be less than the minimum eigenvalue. The equality sign holds only if $c_2 = c_3 = \ldots = 0$; the function that makes (10.79) an absolute minimum is the eigenfunction associated with the smallest eigenvalue. Furthermore, if f is restricted by

$$c_n = \int d\sigma \, f_n f = 0 \,, \quad n = 1, \ldots, N-1 \,, \tag{10.80}$$

it follows that

$$\gamma^2 = \frac{\sum_{n=N}^{\infty} \gamma_n^2 c_n^2}{\sum_{n=N}^{\infty} c_n^2} = \gamma_N^2 + \frac{\sum_{n=N+1}^{\infty} (\gamma_n^2 - \gamma_N^2) c_n^2}{\sum_{n=N}^{\infty} c_n^2} \ge \gamma_N^2 \,. \tag{10.81}$$

The minimum, $\gamma = \gamma_N$, is attained with $f = f_N$.

This formulation of Rayleigh's principle enables one to offer a proof of the statement that there exist no largest eigenvalue, (10.65). For if such an upper limit existed:

$$\gamma_n^2 < \gamma_\infty^2 < \infty , \quad n = 1, 2, \ldots , \tag{10.82}$$

it would follow from

$$\gamma^2 = \gamma_\infty^2 - \frac{\sum_{n=1}^\infty (\gamma_\infty^2 - \gamma_n^2) c_n^2}{\sum_{n=1}^\infty c_n^2} \leq \gamma_\infty^2 \tag{10.83}$$

that the value of γ^2 computed from (10.77) with an arbitrary function f could not exceed the finite upper limit γ_∞^2. Since it is possible, however, to exhibit an admissible function that makes γ^2 arbitrarily large, the assumption of a finite upper bound is disproved. To exhibit such a function, we locate the origin of coordinates in the interior of the region formed by the guide boundary and choose a distance a which is less than the minimum distance from the origin to the boundary. An admissible E- or H-mode function is defined by

$$f = \begin{cases} a^2 - r^2 , & r \leq a , \\ 0 , & r \geq a , \end{cases} \tag{10.84}$$

since it is continuous, has sectionally continuous derivatives, and satisfies the boundary conditions. The value of γ^2 obtained,

$$\gamma^2 = \frac{4 \int_0^a 2\pi r \, dr \, r^2}{\int_0^a 2\pi r \, dr \, (a^2 - r^2)^2} = \frac{6}{a^2} , \tag{10.85}$$

which tends to infinity as a diminishes to zero.

10.4 Variation–Iteration Method

We shall now discuss a method of successive approximations, employed in conjunction with Rayleigh's principle, which yields steadily improving approximations to the dominant mode cutoff wavenumber and mode function, the approximations ultimately converging to the true values.[5] It will be convenient in what follows to use a more compact symbol for the eigenvalue γ^2. We shall denote it by λ; no confusion with the wavelength is possible. In order to obtain an approximation to the dominant mode eigenvalue λ_1, we choose a function $F_1^{(0)}$ that satisfies the boundary condition in the E-mode case, and is subject only to (10.41) if H modes are under discussion. In accordance with Rayleigh's principle:

$$\lambda_1^{(0)} = \frac{\int d\sigma \left(\boldsymbol{\nabla} F_1^{(0)} \right)^2}{\int d\sigma \left(F_1^{(0)} \right)^2} > \lambda_1 , \tag{10.86}$$

---

[5] A discussion of this method in electrostatics is given in [9].

and $\lambda_1^{(0)}$ forms the zeroth approximation to the eigenvalue λ_1, derived from $F_1^{(0)}$, the zeroth approximation to the dominant mode function f_1. The equality sign has been omitted from (10.86) on the assumption that the probability of picking the correct dominant mode function by inspection is negligible. A first approximation to f_1 is now defined by

$$\nabla^2 F_1^{(1)} + F_1^{(0)} = 0 \ . \tag{10.87}$$

It will be supposed that this equation can be solved for $F_1^{(1)}$ subject to either E- or H-mode boundary conditions with the additional restriction (10.41) for the H modes. A first approximation to λ_1 is computed from $F_1^{(1)}$ by Rayleigh's principle:

$$\lambda_1^{(1)} = \frac{\int d\sigma \left(\boldsymbol{\nabla} F_1^{(1)} \right)^2}{\int d\sigma \left(F_1^{(1)} \right)^2} > \lambda_1 \ . \tag{10.88}$$

This process continues in the obvious manner; the nth approximation function $F_1^{(n)}$ is derived from $F_1^{(n-1)}$ by

$$\nabla^2 F_1^{(n)} + F_1^{(n-1)} = 0 \ , \tag{10.89}$$

subject to the various conditions already mentioned, and the nth approximation to λ_1 is calculated from

$$\lambda_1^{(n)} = \frac{\int d\sigma \left(\boldsymbol{\nabla} F_1^{(n)} \right)^2}{\int d\sigma \left(F_1^{(n)} \right)^2} > \lambda_1 \ . \tag{10.90}$$

We may note that

$$\int_\sigma d\sigma \left(\boldsymbol{\nabla} F_1^{(n)} \right)^2 = \oint_C ds \, F_1^{(n)} \frac{\partial}{\partial n} F_1^{(n)} - \int_\sigma d\sigma \, F_1^{(n)} \nabla^2 F_1^{(n)}$$
$$= \int_\sigma d\sigma \, F_1^{(n)} F_1^{(n-1)} \ , \quad n \ge 1 \ , \tag{10.91}$$

since the functions that enter the line integral are constructed to satisfy E- or H-mode boundary conditions. Hence

$$\lambda_1^{(n)} = \frac{\int d\sigma \, F_1^{(n)} F_1^{(n-1)}}{\int d\sigma \left(F_1^{(n)} \right)^2} > \lambda_1 \ , \quad n \ge 1 \ . \tag{10.92}$$

It is easily shown that the successive approximations to λ_1 decrease steadily. If we place $f = F_1^{(n+1)}$ in the relation (10.27), we obtain

$$\frac{\int d\sigma \left(\boldsymbol{\nabla} F_1^{(n)} \right)^2}{\int d\sigma \left(F_1^{(n)} \right)^2} \ge \frac{\int d\sigma \left(F_1^{(n)} \right)^2}{\int d\sigma \left(\boldsymbol{\nabla} F_1^{(n+1)} \right)^2} \ge \frac{\int d\sigma \left(\boldsymbol{\nabla} F_1^{(n+1)} \right)^2}{\int d\sigma \left(F_1^{(n+1)} \right)^2} \ , \tag{10.93}$$

or

$$\lambda_1^{(n)} \geq \lambda_1^{(n+1/2)} \geq \lambda_1^{(n+1)} , \tag{10.94a}$$

where

$$\lambda_1^{(n+1/2)} = \frac{\int d\sigma \left(F_1^{(n)}\right)^2}{\int d\sigma \left(\boldsymbol{\nabla} F_1^{(n+1)}\right)^2} = \frac{\int d\sigma \left(F_1^{(n)}\right)^2}{\int d\sigma \, F_1^{(n+1)} F_1^{(n)}} , \quad n \geq 0 , \tag{10.94b}$$

is naturally called the $n+\frac{1}{2}$th approximation to λ_1. Now, since the successive approximations to λ_1 decrease monotonically but can never be less than λ_1, the sequence $\{\lambda_1^{(n)}\}$ must approach a limit. We shall prove that this limit is no other than λ_1, unless $F_1^{(0)}$, the generating function of the sequence of approximating functions, has unhappily been chosen orthogonal to the dominant mode function f_1. In this event, the sequence of eigenvalue approximations will approach the smallest eigenvalue associated with those eigenfunctions to which $F_1^{(0)}$ is not orthogonal.

To demonstrate this assertion, we construct the function

$$F_1(\mathbf{r}) = \sum_{n=0}^{\infty} \alpha^n F_1^{(n+1)}(\mathbf{r}) , \tag{10.95}$$

where α is an arbitrary parameter. In order to investigate the convergence domain of (10.95) as a function of α, we require some information as to the magnitude of $F_1^{(n+1)}(\mathbf{r})$. It will be shown in Problem 10.8 that

$$\left[F_1^{(n+1)}(\mathbf{r})\right]^2 < C(\mathbf{r}) \int d\sigma \left(F_1^{(n)}\right)^2 , \tag{10.96}$$

where $C(\mathbf{r})$ is independent of n. Hence the series is absolutely and uniformly convergent whenever

$$\sum_{n=0}^{\infty} \alpha^n \left[\int d\sigma \left(F_1^{(n)}\right)^2\right]^{1/2} \tag{10.97}$$

converges. The restriction thereby imposed on α is

$$\alpha \lim_{n \to \infty} \left[\frac{\int d\sigma \left(F_1^{(n+1)}\right)^2}{\int d\sigma \left(F_1^{(n)}\right)^2}\right]^{1/2} < 1 . \tag{10.98}$$

However,

$$\frac{\int d\sigma \left(F_1^{(n)}\right)^2}{\int d\sigma \left(F_1^{(n+1)}\right)^2} = \lambda_1^{(n+1/2)} \lambda_1^{(n+1)} , \tag{10.99}$$

whence

$$\lim_{n\to\infty} \frac{\int d\sigma \left(F_1^{(n)}\right)^2}{\int d\sigma \left(F_1^{(n+1)}\right)^2} = \mu^2 \,, \tag{10.100}$$

where μ is the limit approached by the sequence of eigenvalue approximations. Therefore, (10.95) actually defines a function $F_1(\mathbf{r})$ for $\alpha < \mu$. The function $F_1(\mathbf{r})$ is a solution of the differential equation

$$\nabla^2 F_1 = -\sum_{n=0}^{\infty} \alpha^n F_1^{(n)} = -\alpha F_1 - F_1^{(0)} \,. \tag{10.101}$$

On multiplying this equation by the eigenfunction f_1 and integrating over the guide cross section, we obtain

$$\int d\sigma \, F_1^{(0)} f_1 = -\int d\sigma \, F_1(\nabla^2 + \alpha)f_1 = -(\alpha - \lambda_1)\int d\sigma \, F_1 f_1 \,. \tag{10.102}$$

It can now be shown that μ actually equals λ_1, provided the generating function $F_1^{(0)}$ is not orthogonal to the eigenfunction f_1, a requirement that is easily met in practice. If it be assumed that the sequence $\{\lambda_1^{(n)}\}$ approaches a limit that exceeds λ_1, $\mu > \lambda_1$, then the series (10.95) converges for $\alpha = \lambda_1$ and (10.102) supplies the information that

$$\int d\sigma \, F_1^{(0)} f_1 = 0 \,, \tag{10.103}$$

which contradicts the hypothesis that $F_1^{(0)}$ is not orthogonal to f_1. Therefore, $\mu = \lambda_1$. If $F_1^{(0)}$ is orthogonal to f_1, but not to f_2, the eigenvalue sequence will converge to λ_2, and more generally, it will converge to the smallest eigenvalue not excluded by the orthogonality properties of $F_1^{(0)}$.

We have shown that the method of successive approximations produces a sequence of eigenvalue approximations which eventually converges to λ_1. The error at every stage is positive, that is, each member of the set $\{\lambda_1^{(n)}\}$ necessarily exceeds λ_1, and therefore provides a steadily decreasing upper limit to λ_1. For the practical utilization of the method, it is desirable to supply an estimate of the rate of convergence as well as of the maximum error at each stage of the process. We shall show that the answer to both of these problems involves the value of λ_2, the second eigenvalue. To this end, we seek the function f that minimizes

$$\int d\sigma \, (\nabla^2 f + \lambda_1 f)(\nabla^2 f + \lambda_2 f) \,, \tag{10.104}$$

under the hypothesis that f satisfies either E- or H-mode boundary conditions, and that, as usual, f is restricted by (10.41) in the event that H modes are under discussion. On performing the variation, we easily find that the minimizing function satisfies

$$(\nabla^2 + \lambda_1)(\nabla^2 + \lambda_2)f = 0 , \tag{10.105}$$

and an additional boundary condition, which requires that $\nabla^2 f$ satisfies the same boundary condition as f. The minimizing function is obviously a linear combination of the first two eigenfunctions f_1 and f_2:

$$f = \alpha f_1 + \beta f_2 . \tag{10.106}$$

The value computed for (10.104) with this minimizing function is zero, which may be confirmed by direct calculation or by invoking a theorem stated in connection with (10.16a) and (10.16b): a homogeneous expression in f assumes the value zero for those functions that render it stationary. Hence we may assert that

$$\int d\sigma \, (\nabla^2 f + \lambda_1 f)(\nabla^2 f + \lambda_2 f) \geq 0 , \tag{10.107}$$

and the equality sign holds only if f is of the form (10.106). An alternative proof employs the completeness relation applied to the expansion of f and $\nabla^2 f$ in terms of the eigenfunctions f_n. We proceed formally by substituting the expansions

$$f = \sum_{n=1}^{\infty} c_n f_n , \qquad \nabla^2 f = -\sum_{n=1}^{\infty} c_n \lambda_n f_n \tag{10.108}$$

in (10.104), which yields

$$\int d\sigma \, (\nabla^2 f + \lambda_1 f)(\nabla^2 f + \lambda_2 f) = \sum_{n=3}^{\infty} (\lambda_n - \lambda_1)(\lambda_n - \lambda_2) c_n^2 \geq 0 . \tag{10.109}$$

The equality sign obtains only if $c_3 = c_4 = \ldots = 0$, which implies, as before, that the minimizing function is a linear combination of f_1 and f_2. A similar theorem states that

$$\int d\sigma \, \boldsymbol{\nabla}(\nabla^2 f + \lambda_1 f) \cdot \boldsymbol{\nabla}(\nabla^2 f + \lambda_2 f) \geq 0 , \tag{10.110}$$

provided that, for H modes, $\frac{\partial}{\partial n} f = 0$ on C, and for E modes, $f = \nabla^2 f = 0$ on C. The minimizing function satisfies

$$\boldsymbol{\nabla}(\nabla^2 + \lambda_1)(\nabla^2 + \lambda_2)f = 0 , \tag{10.111}$$

and the additional boundary conditions: for H modes, $\frac{\partial}{\partial n} \nabla^2 f = 0$ on C, and for E modes, $\nabla^2 \nabla^2 f = 0$ on C. As in the previous theorem, the minimum is reached when f is a linear combination of f_1 and f_2, save for a trivial additive constant (Problem 10.9).

In consequence of the boundary conditions imposed on f in the two inequalities (10.107) and (10.110), they may be rewritten as

$$\int d\sigma\,(\nabla^2 f)^2 - (\lambda_1 + \lambda_2)\int d\sigma\,(\nabla f)^2 + \lambda_1\lambda_2\int d\sigma\, f^2 \geq 0\,,$$

$$(10.112\text{a})$$

$$\int d\sigma\,(\nabla\nabla^2 f)^2 - (\lambda_1 + \lambda_2)\int d\sigma\,(\nabla^2 f)^2 + \lambda_1\lambda_2\int d\sigma\,(\nabla f)^2 \geq 0\,,$$

$$(10.112\text{b})$$

or

$$\left[\int d\sigma\,(\nabla^2 f)^2 - \lambda_1\int d\sigma\,(\nabla f)^2\right] - \lambda_2\left[\int d\sigma\,(\nabla f)^2 - \lambda_1\int d\sigma\, f^2\right]$$
$$\geq 0\,,$$

$$(10.113\text{a})$$

$$\left[\int d\sigma\,(\nabla\nabla^2 f)^2 - \lambda_1\int d\sigma\,(\nabla^2 f)^2\right] - \lambda_2\left[\int d\sigma\,(\nabla^2 f)^2 - \lambda_1\int d\sigma\,(\nabla f)^2\right]$$
$$\geq 0\,.$$

$$(10.113\text{b})$$

It is a consequence of (10.27) and Rayleigh's principle that the four quantities enclosed in brackets are never negative. Hence the two inequalities imply that

$$\lambda_2 \leq \frac{\int d\sigma\,(\nabla^2 f)^2 - \lambda_1\int d\sigma\,(\nabla f)^2}{\int d\sigma\,(\nabla f)^2 - \lambda_1\int d\sigma\, f^2} = \frac{\int d\sigma\,(\nabla f)^2}{\int d\sigma\, f^2}\,\frac{\dfrac{\int d\sigma\,(\nabla^2 f)^2}{\int d\sigma\,(\nabla f)^2} - \lambda_1}{\dfrac{\int d\sigma\,(\nabla f)^2}{\int d\sigma\, f^2} - \lambda_1}\,,$$

$$(10.114\text{a})$$

$$\lambda_2 \leq \frac{\int d\sigma\,(\nabla\nabla^2 f)^2 - \lambda_1\int d\sigma\,(\nabla^2 f)^2}{\int d\sigma\,(\nabla^2 f)^2 - \lambda_1\int d\sigma\,(\nabla f)^2}$$
$$= \frac{\int d\sigma\,(\nabla^2 f)^2}{\int d\sigma\,(\nabla f)^2}\,\frac{\dfrac{\int d\sigma\,(\nabla\nabla^2 f)^2}{\int d\sigma\,(\nabla^2 f)^2} - \lambda_1}{\dfrac{\int d\sigma\,(\nabla^2 f)^2}{\int d\sigma\,(\nabla f)^2} - \lambda_1}\,.$$

$$(10.114\text{b})$$

Thus we have obtained two alternative forms of Rayleigh's principle applied to the second eigenvalue, which have the distinctive feature that no restrictions, other than boundary conditions, are imposed on the function f. These relations supply the convergence rate information that we need. By virtue of the postulated method of construction, the functions $f = F_1^{(n+1)}$, $n \geq 0$, satisfy the boundary conditions necessary for the applicability of (10.114a) and (10.114b). This statement is valid for both types of modes. A glance at the definitions of the eigenvalue approximations $\lambda_1^{(n)}$ (10.90) and $\lambda_1^{(n+1/2)}$ (10.94b), shows that with this choice for f, the inequalities (10.114a) and (10.114b) become

$$\lambda_2 \leq \lambda_1^{(n+1)}\frac{\lambda_1^{(n+1/2)} - \lambda_1}{\lambda_1^{(n+1)} - \lambda_1}\,,$$

$$(10.115\text{a})$$

$$\lambda_2 \leq \lambda_1^{(n+1/2)} \frac{\lambda_1^{(n)} - \lambda_1}{\lambda_1^{(n+1/2)} - \lambda_1} . \qquad (10.115b)$$

It will be noticed that (10.115a) can be formally obtained from (10.115b) on replacing n by $n + 1/2$. The inequalities can be combined by multiplication into

$$\frac{\lambda_1^{(n)} - \lambda_1}{\lambda_1^{(n+1)} - \lambda_1} \geq \frac{\lambda_2^2}{\lambda_1^{(n+1/2)} \lambda_1^{(n+1)}} , \qquad (10.116)$$

which states that the error of the nth approximation to λ_1, divided by the error of the $(n + 1)$th approximation, exceeds a number which approaches $(\lambda_2/\lambda_1)^2 = (\gamma_2/\gamma_1)^4$ as the approximations proceed. Thus the rapidity of convergence of the successive approximation method is essentially determined by the magnitude of the second eigenvalue relative to the first eigenvalue, the convergence being the more rapid, the larger this ratio. Indeed, the number $(\gamma_2/\gamma_1)^2$ is usually rather large; it has, for example, the values 27.8 and 70.3 for the E and H modes of a circular guide, respectively.

It is important to realize in this connection that the second eigenvalue referred to in the convergence criterion may exceed the true second eigenvalue of the guide. This situation will arise whenever the guide possesses special symmetry properties that permit the decomposition of the eigenfunctions into various symmetry classes. If the generating function $F_1^{(0)}$ possesses the proper symmetry characteristics of the eigenfunction f_1, so also will the successive approximations $F_1^{(n)}$. Every member of this sequence will be automatically orthogonal to the eigenfunctions of other symmetry classes, and the relevant second mode is that possessing the same symmetry as f_1. Furthermore, in consequence of the automatic orthogonality between members of different symmetry classes, our methods are applicable independently to the dominant mode of each symmetry type. To illustrate these remarks, we may consider a guide with circular symmetry. Each mode function has the angle dependence $\genfrac{}{}{0pt}{}{\sin}{\cos} m\phi$, and mode functions associated with different values of m are automatically orthogonal, irrespective of the r dependence of the mode functions. Thus the modes associated with each value of m form a symmetry class, and Rayleigh's principle is applicable to each class individually. Hence in applying Rayleigh's principle to construct the dominant E mode of a circular guide, E_{01}, the second mode that determines the rapidity of convergence is not the second E mode of the guide, E_{11}, but E_{02}, the second mode with the same symmetry as E_{01}. Similarly, in the construction of the dominant H mode, H_{11}, the relevant second H mode is H_{12}, not H_{21}. It is with this in mind that the numbers mentioned in the preceding paragraph have been obtained.

10.4.1 Error Estimates

The inequalities (10.115a) and (10.115b) not only furnish a criterion for the convergence rate of the successive approximation method, but also supply an estimate of error for each stage of the process, provided the second eigenvalue, or a reasonably accurate lower limit to it, is known. To demonstrate this, we remark that the inequalities can be rewritten in the form

$$\left(\lambda_2 - \lambda_1^{(n+1/2)}\right)\lambda_1^{(n+1)} \leq \lambda_1'\left(\lambda_2 - \lambda_1^{(n+1)}\right), \qquad (10.117a)$$

$$\left(\lambda_2 - \lambda_1^{(n)}\right)\lambda_1^{(n+1/2)} \leq \lambda_1\left(\lambda_2 - \lambda_1^{(n+1/2)}\right). \qquad (10.117b)$$

Now at some stage of the process, the eigenvalue approximations become less than λ_2, and all the quantities in parentheses are positive. This will ordinarily be true even for the 0th approximation. To test the validity of this assumption an estimate of λ_2, in the form of a lower limit, must be available. Assuming that this circumstance is ensured, we deduce that

$$\lambda_1 \geq \lambda_1^{(n+1)}\frac{\lambda_2 - \lambda_1^{(n+1/2)}}{\lambda_2 - \lambda_1^{(n+1)}} = \lambda_1^{(n+1)} - \frac{\lambda_1^{(n+1/2)} - \lambda_1^{(n+1)}}{\frac{\lambda_2}{\lambda_1^{(n+1)}} - 1}, \qquad (10.118a)$$

$$\lambda_1 \geq \lambda_1^{(n+1/2)}\frac{\lambda_2 - \lambda_1^{(n)}}{\lambda_2 - \lambda_1^{(n+1/2)}} = \lambda_1^{(n+1/2)} - \frac{\lambda_1^{(n)} - \lambda_1^{(n+1/2)}}{\frac{\lambda_2}{\lambda_1^{(n+1/2)}} - 1}, \qquad (10.118b)$$

which constitute lower limits for λ_1. We may therefore state that, at the $n+1$st stage of the approximation process, the eigenvalue λ_1 has been located within the limits:

$$\lambda_1^{(n+1)} \geq \lambda_1 \geq \lambda_1^{(n+1)} - \frac{\lambda_1^{(n+1/2)} - \lambda_1^{(n+1)}}{\frac{\lambda_2}{\lambda_1^{(n+1)}} - 1}, \qquad (10.119)$$

which tend to coincidence as n increases. The similar approximant to the $n+\frac{1}{2}$ stage is obtained on replacing $n + 1$ with $n + \frac{1}{2}$. The sense of the inequality (10.118b) is not affected if λ_2 is replaced by a smaller quantity, provided the lower bound to λ_2 exceeds $\lambda_1^{(n+1/2)}$.

A closely related theorem, which is somewhat wider in scope, results from a consideration of the quantity

$$\mu^2 = \frac{\int d\sigma\,(\nabla^2 f + \lambda f)^2}{\int d\sigma\, f^2}, \qquad (10.120)$$

where λ is an arbitrary parameter. The function that minimizes μ^2 among the class of functions that satisfy E- or H-mode boundary conditions is such that

$$(\nabla^2 + \lambda)^2 f = \mu^2 f, \qquad (10.121)$$

where $\nabla^2 f$ satisfies the same boundary conditions as f. On rewriting (10.121) as

$$(\nabla^2 + \lambda + \mu)(\nabla^2 + \lambda - \mu)f = 0, \qquad (10.122)$$

it is apparent that the minimizing function is to be found among the eigenfunctions $\{f_n\}$. Since the stationary values of μ^2 are then

$$\mu^2 = (\lambda_n - \lambda)^2 \, , \tag{10.123}$$

(10.120) is minimized by that eigenfunction f_n, whose eigenvalue is nearest to λ, in the sense that $|\lambda_n - \lambda|$ is smallest. With λ_n understood to have this significance, we may assert that

$$\frac{\int d\sigma \, (\nabla^2 f + \lambda f)^2}{\int d\sigma \, f^2} \geq (\lambda_n - \lambda)^2 \, , \tag{10.124}$$

or

$$\lambda_n \leq \lambda + \left[\frac{\int d\sigma \, (\nabla^2 f + \lambda f)^2}{\int d\sigma \, f^2} \right]^{1/2} , \quad \lambda_n \geq \lambda \, , \tag{10.125a}$$

$$\lambda_n \geq \lambda - \left[\frac{\int d\sigma \, (\nabla^2 f + \lambda f)^2}{\int d\sigma \, f^2} \right]^{1/2} , \quad \lambda_n \leq \lambda \, . \tag{10.125b}$$

Therefore, if an arbitrary choice of λ is less than the nearest eigenvalue, (10.125a) provides an upper limit for that eigenvalue; on the other hand, if λ has been chosen greater than the nearest eigenvalue, (10.125b) constitutes a lower limit for that eigenvalue. The difficulty in applying this theorem lies, of course, in the task of establishing which of these situations is realized, in the absence of any knowledge of the eigenvalues. No difficulty exists, however, in establishing an upper limit to the first eigenvalue, for if λ is not positive the condition $\lambda < \lambda_1$ is assured, since all eigenvalues are positive. Hence, for $\lambda \leq 0$,

$$\lambda_1 < \lambda + \left[\lambda^2 - 2\lambda \frac{\int d\sigma \, (\nabla f)^2}{\int d\sigma \, f^2} + \frac{\int d\sigma \, (\nabla^2 f)^2}{\int d\sigma \, f^2} \right]^{1/2}$$

$$= \lambda + \left\{ \left[\frac{\int d\sigma \, (\nabla f)^2}{\int d\sigma \, f^2} - \lambda \right]^2 \right.$$

$$\left. + \frac{\int d\sigma \, (\nabla f)^2}{\int d\sigma \, f^2} \left[\frac{\int d\sigma \, (\nabla^2 f)^2}{\int d\sigma \, (\nabla f)^2} - \frac{\int d\sigma \, (\nabla f)^2}{\int d\sigma \, f^2} \right] \right\}^{1/2} . \tag{10.126}$$

Now the right-hand member of (10.126) is a monotonically increasing function of λ, since its derivative with respect to λ,

$$1 - \left(\frac{\int d\sigma \, (\nabla f)^2}{\int d\sigma \, f^2} - \lambda \right) \left\{ \left[\frac{\int d\sigma \, (\nabla f)^2}{\int d\sigma \, f^2} - \lambda \right]^2 \right.$$

$$\left. + \frac{\int d\sigma \, (\nabla f)^2}{\int d\sigma \, f^2} \left[\frac{\int d\sigma \, (\nabla^2 f)^2}{\int d\sigma \, (\nabla f)^2} - \frac{\int d\sigma \, (\nabla f)^2}{\int d\sigma \, f^2} \right] \right\}^{-1/2} , \tag{10.127}$$

is essentially positive, in consequence of (10.27). Hence the upper limit steadily decreases as λ approaches $-\infty$, and the most stringent upper limit deduced from (10.126) is the familiar expression of Rayleigh's principle,

$$\lambda_1 < \frac{\int d\sigma \, (\nabla f)^2}{\int d\sigma \, f^2} \, . \tag{10.128}$$

If a value of λ can be selected in the range

$$\lambda_1 \leq \lambda \leq \frac{\lambda_1 + \lambda_2}{2} \, , \tag{10.129}$$

(10.125b) offers a lower limit to the first eigenvalue:

$$\lambda_1 > \lambda - \left[\lambda^2 - 2\lambda \frac{\int d\sigma \, (\nabla f)^2}{\int d\sigma \, f^2} + \frac{\int d\sigma \, (\nabla^2 f)^2}{\int d\sigma \, f^2} \right]^{1/2} \, . \tag{10.130}$$

It may be established, as in the preceding paragraph, that the right-hand member of (10.130) is a monotonically increasing function of λ. Thus, the most incisive lower limit is obtained from the largest value of λ that is compatible with (10.129). That is, we must place

$$\lambda = \frac{1}{2}(\underline{\lambda}_1 + \underline{\lambda}_2) \, , \tag{10.131}$$

where $\underline{\lambda}_1$ and $\underline{\lambda}_2$ are known to be lower limits to the first and second eigenvalues, respectively. Now (10.130) itself provides the best available value of $\underline{\lambda}_1$, and by combining the latter equation with (10.131), one obtains

$$\underline{\lambda}_1 = \frac{\int d\sigma \, (\nabla f)^2}{\int d\sigma \, f^2} \frac{\underline{\lambda}_2 - \dfrac{\int d\sigma \, (\nabla^2 f)^2}{\int d\sigma \, (\nabla f)^2}}{\underline{\lambda}_2 - \dfrac{\int d\sigma \, (\nabla f)^2}{\int d\sigma \, f^2}} \, . \tag{10.132}$$

On placing $f = F_2^{(n+1)}$, this result becomes identical with (10.118a), save that λ_2 is properly replaced by the lower limit $\underline{\lambda}_2$:

$$\underline{\lambda}_1 = \lambda_1^{(n+1)} \frac{\underline{\lambda}_2 - \lambda_1^{(n+1/2)}}{\underline{\lambda}_2 - \lambda_1^{(n+1)}} = \lambda_1^{(n+1)} - \frac{\lambda_1^{(n+1/2)} - \lambda_1^{(n+1)}}{\dfrac{\underline{\lambda}_2}{\lambda_1^{(n+1)}} - 1} \, . \tag{10.133}$$

In order that this result be valid, it is necessary that $\frac{1}{2}(\underline{\lambda}_1 + \underline{\lambda}_2) > \lambda_1$, which will be guaranteed if $\underline{\lambda}_1 + \underline{\lambda}_2 > 2\lambda_1^{(n+1)}$. This restriction, taken in conjunction with (10.133), determines the smallest lower limit to the second eigenvalue that will be admissible in the latter equation:

$$\underline{\lambda}_2 \geq \lambda_1^{(n+1)} + \left[\lambda_1^{(n+1)} \left(\lambda_1^{(n+1/2)} - \lambda_1^{(n+1)} \right) \right]^{1/2} \, . \tag{10.134}$$

That is, if a lower limit to the second eigenvalue is known, the formula (10.133) cannot be safely applied unless (10.134) is satisfied. In the absence of any information concerning the second eigenvalue, we may notice that the right-hand member of (10.134) eventually converges to λ_1 and must therefore becomes less than λ_2 at some stage in the approximation procedure. Hence (10.134) supplies a possible lower limit to λ_2, which on substitution in (10.133) yields

$$\underline{\lambda}_1 = \lambda_1^{(n+1)} - \left[\lambda_1^{(n+1)}\left(\lambda_1^{(n+1/2)} - \lambda_1^{(n+1)}\right)\right]^{1/2} . \tag{10.135}$$

Note that this result is obtained immediately from (10.130) on placing $\lambda = \lambda_1^{(n+1)}$ and $f = F_1^{(n+1)}$, using (10.94b). This lower limit eventually converges to λ_1. However, it is a much poorer approximation than would be obtained from (10.133) with a fixed value of $\underline{\lambda}_2$, since, in the latter event, the difference between the upper limit $\overline{\lambda}_1 = \lambda_1^{(n+1)}$ and the lower limit (10.133) approaches zero, as n increases, in the following manner:

$$\overline{\lambda}_1 - \underline{\lambda}_1 \rightarrow \frac{\lambda_1^{(n+1/2)} - \lambda_1^{(n+1)}}{\frac{\lambda_2}{\lambda_1} - 1} , \tag{10.136}$$

which is much more rapid than that described by (10.135):

$$\overline{\lambda}_1 - \underline{\lambda}_1 \rightarrow \left[\lambda_1\left(\lambda_1^{(n+1/2)} - \lambda_1^{(n+1)}\right)\right]^{1/2} . \tag{10.137}$$

Similar results can be found from a consideration of the quantity

$$\mu^2 = \frac{\int d\sigma\, [\boldsymbol{\nabla}(\nabla^2 f + \lambda^2 f)]^2}{\int d\sigma\, (\boldsymbol{\nabla} f)^2} . \tag{10.138}$$

We obtain in this way an equation analogous to (10.126) which leads to the second form of Rayleigh's principle for the lowest eigenvalue:

$$\lambda_1 < \frac{\int d\sigma\, (\nabla^2 f)^2}{\int d\sigma\, (\boldsymbol{\nabla} f)^2} , \tag{10.139}$$

and a lower limit to λ_1 that is identical with (10.133).

10.5 Problems for Chap. 10

1. A conducting body embedded in the vacuum is characterized by the constant potential ϕ_0 on its surface S, which carries charge $Q = \oint_S dS\,\sigma$. The condition that the potential in the region V around the body,

$$\phi(\mathbf{r}) = \oint_S \frac{dS'}{4\pi} \frac{\sigma(\mathbf{r}')}{|\mathbf{r} - \mathbf{r}'|} , \tag{10.140}$$

becomes the specified constant ϕ_0 on S is an integral equation that determines σ. Prove that a solution to this surface integral equation is *the* solution by

(a) comparing appropriate upper and lower limits to the energy, and
(b) examining $\int_V (d\mathbf{r})(\boldsymbol{\nabla}\varphi)^2$, where φ is the difference of two solutions.

2. The two-dimensional distance between two points in polar coordinates is

$$P = \sqrt{\rho^2 + \rho'^2 - 2\rho\rho' \cos \varPhi} \,, \tag{10.141}$$

where $\varPhi$ is the angle between the two directions. Use Bessel function properties to show that

$$\int_0^{2\pi} \frac{d\varPhi}{2\pi} \frac{1}{P} = \int_0^{2\pi} \int_0^{2\pi} \frac{d\phi\, d\phi'}{2\pi\ 2\pi} \pi\delta(\rho\cos\phi + \rho\cos\phi') \,. \tag{10.142}$$

3. Let x' is the x-coordinate of a variable point on the surface of a sphere of radius a. Verify the following solid angle integral over the surface of the sphere:

$$\int d\Omega'\, \delta(x' + x) = \frac{2\pi}{a} \,, \tag{10.143}$$

where x is a fixed quantity, such that $|x| < a$. Hint: change the coordinate system.

4. We will now use the preceding two problems to solve the integral equation that determines the charge density on a conducting disk. Let the disk have radius a. Write the integral equation of Problem 10.1 in polar coordinates ρ, ϕ. Cylindrical symmetry indicates that $\sigma(\mathbf{r}) = \sigma(\rho)$. Introduce new variables: $\rho = a \sin \theta$, $\rho' = a \sin \theta'$ where θ and θ' go from 0 to $\pi/2$. Now combine the results from Problems 10.2 and 10.3. Recognize a simple solution of the integral equation; according to Problem 10.1, it is *the* solution. Compute the total charge Q and identify the capacitance $C = Q/\phi_0$.

5. Show that the E-mode cutoff wavenumbers are given by

$$\gamma'^2 = \frac{\int_\sigma d\sigma\, (\boldsymbol{\nabla}_\perp\varphi)^2 - 2\oint_C ds \frac{\partial\varphi}{\partial n}\varphi}{\int_\sigma d\sigma\, \varphi^2} \,, \tag{10.144}$$

which is stationary for variations about

$$(\nabla_\perp^2 + \gamma'^2)\varphi = 0 \,, \quad \varphi = 0 \quad \text{on} \quad C \,, \tag{10.145}$$

without requiring that φ obey this boundary condition (i.e., $\delta\varphi$ need not vanish on C). Conclude that any outward (inward) displacement of the boundary lowers (raises) the value of γ'^2.

6. Illustrate the above theorem by considering, for E modes, a circular guide of radius a bounded by inscribed and circumscribed square guides. Show that the lowest value of $\gamma'a$ is indeed intermediate between those of the two squares.

7. Test the inequality of (10.23) for a one-dimensional problem, parallel plate conductors, with trial function for the E mode

$$\varphi_0(x) = x(1 - x/a) , \quad 0 < x < a .$$ (10.146)

Then, how does it work out with the function produced by iteration, the solution of

$$-\frac{d^2}{dx^2}\varphi(x) = \text{const. } x(1 - x/a) ?$$ (10.147)

8. Solve the iteration equation (10.89) for the function $F^{(n+1)}$ in terms of $F^{(n)}$ and the Green's function for the Laplacian operator,

$$F^{(n+1)}(\mathbf{r}) = \int d\sigma \, G(\mathbf{r}, \mathbf{r}')F^{(n)}(\mathbf{r}') .$$ (10.148)

Then use the Cauchy–Schwarz inequality (10.22) to derive (10.96).

9. Prove that the minimum of (10.110) is achieved when f is a linear combination of the two lowest mode functions.

10. Prove (10.139) and the remark following concerning the lower limit for λ_1.

Examples of Variational Calculations for Circular Guide

We will now illustrate the method proposed in the previous chapter in a simple realistic context.

11.1 E Modes

We consider a circular cylinder of radius a and discuss the mode of lowest eigenvalue with $m = 0$:

$$\varphi_{01}(\rho, \phi) = \frac{1}{(2\pi)^{1/2}} P_{01}(\rho) , \quad P_{01}(a) = 0 . \tag{11.1}$$

The equation defining the successive iterations (10.89) is (henceforth the subscripts are omitted)

$$-\frac{1}{\rho} \frac{\mathrm{d}}{\mathrm{d}\rho} \rho \frac{\mathrm{d}}{\mathrm{d}\rho} P^{(n+1)}(\rho) = P^{(n)}(\rho) , \tag{11.2}$$

and the required integrals are

$$[m + n] = \int_0^a \mathrm{d}\rho \, \rho \, P^{(m)}(\rho) \, P^{(n)}(\rho) . \tag{11.3}$$

Note that we may integrate by parts to show this is truly a function only of $m + n$.

Inasmuch as a is the natural unit of length, and the various operations of differentiation and integration are even in ρ, it is expedient to introduce the variable

$$t = (\rho/a)^2 , \tag{11.4}$$

and define the functions

$$T^{(n)}(t) = \left(\frac{4}{a^2} \right)^n P^{(n)}(\rho) , \tag{11.5}$$

which obey the boundary condition

$$T^{(n)}(1) = 0 \,, \tag{11.6}$$

and the differential equation

$$-\frac{\mathrm{d}}{\mathrm{d}t}t\frac{\mathrm{d}}{\mathrm{d}t}T^{(n+1)}(t) = T^{(n)}(t) \,. \tag{11.7}$$

The integrals (11.3) now appear as

$$[m+n] = 2\left(\frac{a^2}{4}\right)^{m+n+1}\langle m+n\rangle \,, \tag{11.8}$$

with

$$\langle m+n\rangle = \int_0^1 \mathrm{d}t \, T^{(m)}(t)\, T^{(n)}(t) \,. \tag{11.9}$$

In this notation, the nth approximation to the desired eigenvalue is given by

$$\lambda^{(n)} = \frac{[2n-1]}{[2n]} = \frac{4}{a^2}\frac{\langle 2n-1\rangle}{\langle 2n\rangle} \,, \tag{11.10}$$

which is valid both for integer and integer $+\frac{1}{2}$ values of n.

We shall find it desirable to introduce a preiteration function, $T^{(-1)}(t)$, according to

$$-\frac{\mathrm{d}}{\mathrm{d}t}t\frac{\mathrm{d}}{\mathrm{d}t}T^{(0)}(t) = T^{(-1)}(t) \,, \tag{11.11}$$

which function does not satisfy the boundary condition. Indeed, we use

$$T^{(-1)}(t) = 1 \,, \tag{11.12}$$

for that makes it the beginning of the line:

$$-\frac{\mathrm{d}}{\mathrm{d}t}t\frac{\mathrm{d}}{\mathrm{d}t}T^{(-1)}(t) = T^{(-2)}(t) = 0 \,. \tag{11.13}$$

To see what benefits emerge thereby, consider

$$\begin{aligned}
\langle k\rangle &= \int_0^1 \mathrm{d}t\, T^{(k)}(t)T^{(0)}(t) \\
&= \int_0^1 \mathrm{d}t\left[-\frac{\mathrm{d}}{\mathrm{d}t}t\frac{\mathrm{d}}{\mathrm{d}t}T^{(k+1)}(t)\right]T^{(0)}(t) \\
&= \int_0^1 \mathrm{d}t\, T^{(k+1)}(t)T^{(-1)}(t) \,,
\end{aligned} \tag{11.14}$$

for both $T^{(k+1)}$ and $T^{(0)}$ obey the boundary condition. Then, with the choice (11.12), we reach the simple evaluation

$$\langle k \rangle = \int_0^1 dt \left[-\frac{d}{dt} t \frac{d}{dt} T^{(k+2)}(t) \right] = -\frac{d}{dt} T^{(k+2)}(1) \,. \tag{11.15}$$

In carrying out the integrations required to produce the successive $T^{(n)}(t)$, it suffices to note that

$$-\frac{d}{dt} t \frac{d}{dt} \left[-\frac{t^{n+1}}{(n+1)^2} \right] = t^n \,. \tag{11.16}$$

Thus, the solution of (11.11) that obeys the boundary condition (11.6) is (here $n = 0$)

$$T^{(0)}(t) = 1 - t = T^{(-1)}(t) - t \,. \tag{11.17}$$

Then, we get in succession,

$$T^{(1)}(t) = T^{(0)}(t) + \frac{1}{2^2} \left(t^2 - T^{(-1)}(t) \right), \tag{11.18a}$$

$$T^{(2)}(t) = T^{(1)}(t) - \frac{1}{2^2} T^{(0)}(t) - \frac{1}{2^2} \frac{1}{3^2} \left(t^3 - T^{(-1)}(t) \right), \tag{11.18b}$$

$$T^{(3)}(t) = T^{(2)}(t) - \frac{1}{(2!)^2} T^{(1)}(t) + \frac{1}{(3!)^2} T^{(0)}(t) + \frac{1}{(4!)^2} \left(t^4 - T^{(-1)}(t) \right), \tag{11.18c}$$

and, in general,

$$T^{(n)}(t) = \sum_{l=1}^{n+1} \frac{(-1)^{l-1}}{(l!)^2} T^{(n-l)}(t) + (-1)^{n-1} \frac{t^{n+1}}{[(n+1)!]^2} \,. \tag{11.19}$$

From this we deduce recurrence relations for the integrals as evaluated in (11.15),

$$\langle k \rangle = \sum_{l=1}^{k+3} \frac{(-1)^{l-1}}{(l!)^2} \langle k - l \rangle + (-1)^k \frac{1}{(k+2)!\,(k+3)!} \,. \tag{11.20}$$

The quantities appearing here with negative numbers, $\langle -3 \rangle$, $\langle -2 \rangle$, $\langle -1 \rangle$, are to be understood in the sense of the last version of (11.14), and therefore are integrals of products with one factor of $T^{(-1)}(t)$,

$$\langle -3 \rangle = \int_0^1 dt\, T^{(-2)}(t) T^{(-1)}(t) = 0 \,, \tag{11.21a}$$

$$\langle -2 \rangle = \int_0^1 dt\, T^{(-1)}(t) T^{(-1)}(t) = 1 \,, \tag{11.21b}$$

$$\langle -1 \rangle = \int_0^1 dt\, T^{(0)}(t) T^{(-1)}(t) = \frac{1}{2} \,, \tag{11.21c}$$

and then we get

$$\langle 0 \rangle = \frac{1}{2} - \frac{1}{(2!)^2} + \frac{1}{2!\,3!} = \frac{1}{3} \,, \tag{11.22a}$$

$$\langle 1 \rangle = \frac{1}{3} - \frac{1}{(2!)^2}\frac{1}{2} + \frac{1}{(3!)^2} - \frac{1}{3!\,4!} = \frac{11}{48} \,, \tag{11.22b}$$

$$\langle 2 \rangle = \frac{11}{48} - \frac{1}{(2!)^2}\frac{1}{3} + \frac{1}{(3!)^2}\frac{1}{2} - \frac{1}{(4!)^2} + \frac{1}{4!\,5!} = \frac{19}{120} \,, \tag{11.22c}$$

$$\langle 3 \rangle = \frac{19}{120} - \frac{1}{(2!)^2}\frac{11}{48} + \frac{1}{(3!)^2}\frac{1}{3} - \frac{1}{(4!)^2}\frac{1}{2} + \frac{1}{(5!)^2} - \frac{1}{5!\,6!} = \frac{473}{4320} \,, \tag{11.22d}$$

$$\langle 4 \rangle = \frac{473}{4320} - \frac{1}{(2!)^2}\frac{19}{120} + \frac{1}{(3!)^2}\frac{11}{48} - \frac{1}{(4!)^2}\frac{1}{3} + \frac{1}{(5!)^2}\frac{1}{2} - \frac{1}{(6!)^2} + \frac{1}{6!\,7!}$$
$$= \frac{229}{3024} \,, \tag{11.22e}$$

$$\langle 5 \rangle = \frac{229}{3024} - \frac{1}{(2!)^2}\frac{473}{4320} + \frac{1}{(3!)^2}\frac{19}{120} - \frac{1}{(4!)^2}\frac{11}{48} + \frac{1}{(5!)^2}\frac{1}{3}$$
$$- \frac{1}{(6!)^2}\frac{1}{2} + \frac{1}{(7!)^2} - \frac{1}{7!\,8!} = \frac{101369}{1935360} \,, \tag{11.22f}$$

$$\langle 6 \rangle = \frac{946523}{26127360} \,, \tag{11.22g}$$

$$\langle 7 \rangle = \frac{64567219}{261273600} \,; \tag{11.22h}$$

these numbers are sufficient to produce four iterations with integer n, and four with integer n plus $1/2$.

These successive values of $\lambda^{(n)}a^2$ are

$$n = 0 : \quad 4\frac{\langle -1 \rangle}{\langle 0 \rangle} = 6 = 6.0 \,, \tag{11.23a}$$

$$n = \frac{1}{2} : \quad 4\frac{\langle 0 \rangle}{\langle 1 \rangle} = \frac{64}{11} = 5.8181818 \,, \tag{11.23b}$$

$$n = 1 : \quad 4\frac{\langle 1 \rangle}{\langle 2 \rangle} = \frac{110}{19} = 5.7894737 \,, \tag{11.23c}$$

$$n = \frac{3}{2} : \quad 4\frac{\langle 2 \rangle}{\langle 3 \rangle} = \frac{2736}{473} = 5.7843552 \,, \tag{11.23d}$$

$$n = 2 : \quad 4\frac{\langle 3 \rangle}{\langle 4 \rangle} = \frac{6622}{1145} = 5.7834061 \,, \tag{11.23e}$$

$$n = \frac{5}{2} : \quad 4\frac{\langle 4 \rangle}{\langle 5 \rangle} = \frac{586240}{101369} = 5.7832276 \,, \tag{11.23f}$$

$$n = 3 : \quad 4\frac{\langle 5 \rangle}{\langle 6 \rangle} = \frac{5473926}{946523} = 5.7831939 \,, \tag{11.23g}$$

$$n = \frac{7}{2} : \quad 4\frac{\langle 6 \rangle}{\langle 7 \rangle} = \frac{378609200}{65467219} = 5.7831875 \,. \tag{11.23h}$$

11.1.1 Bounds on Second Eigenvalue

In order to produce corresponding lower limits, we turn our attention to λ_2. The best result for $\lambda_2 a^2$ so far is obtained by comparison with a square, see Problem 10.6. To do better, we apply the technique developed in Chap. 23 of [9]. We first state, and then derive the following equation referring to the $m = 0$ modes of the circle of radius a:

$$\sum_{n=1}^{\infty} \frac{P_{0n}(\rho)P_{0n}(\rho')}{\gamma_{0n}^2} = \log \frac{a}{\rho_>} , \tag{11.24}$$

where

$$P_{mn}(\rho) = \frac{\sqrt{2}}{a} \frac{J_m(\gamma_{mn}\rho)}{J'_m(\gamma_{mn}a)} . \tag{11.25}$$

We recognize here, for $m = 0$, the $k \to 0$ limit of the relation expressing the equality of the eigenfunction expansion and the closed-form expressions for the reduced Green's function for the Coulomb problem in cylindrical coordinates (see Problem 11.1):

$$\sum_{n=1}^{\infty} \frac{P_{mn}(\rho)P_{mn}(\rho')}{k^2 + \gamma_{mn}^2} = I_m(k\rho_<)\left[K_m(k\rho_>) - I_m(k\rho_>)\frac{K_m(ka)}{I_m(ka)}\right] , \tag{11.26}$$

for $\rho, \rho' < a$, subject to Dirichlet boundary conditions at $\rho = a$. Now according to Problem 8.8, we have

$$t \ll 1: \quad K_0(t) \sim \log \frac{1}{t} + \text{const.} , \quad I_0(t) \sim 1 , \tag{11.27}$$

from which (11.24) follows.

This verification can, of course, be performed more directly. The left-hand side of (11.24) – call it $g_0(\rho, \rho')$ – obeys

$$-\frac{1}{\rho}\frac{\partial}{\partial\rho}\rho\frac{\partial}{\partial\rho}g_0(\rho, \rho') = \frac{1}{\rho}\delta(\rho - \rho') , \tag{11.28a}$$

$$g_0(a, \rho') = 0 . \tag{11.28b}$$

Then we see that $\log(a/\rho_>)$ vanishes for $\rho_> = a$, and that

$$-\rho\frac{\partial}{\partial\rho}\log\frac{a}{\rho_>} = \begin{cases} 1, \rho > \rho' \\ 0, \rho < \rho' \end{cases} ; \tag{11.29}$$

the derivative of this discontinuous function produces the required delta function.

Now we derive, by integrating (11.24) with $\rho = \rho'$

$$\sum_{n=1}^{\infty} \frac{1}{\gamma_{0n}^2} = \int_0^a d\rho\,\rho\,\log\frac{a}{\rho} = \frac{a^2}{4}\int_0^1 dt\,\log\frac{1}{t} = \frac{a^2}{4} , \tag{11.30}$$

in which we have introduced the variable (11.4). Such a relation provides a lower limit to λ_2 in terms of an upper limit to λ_1:

$$\frac{1}{\overline{\lambda}_1 a^2} + \frac{1}{\underline{\lambda}_2 a^2} = \frac{1}{4} \, . \tag{11.31}$$

If we use the best upper limit in (11.23h), that is, for $n = 7/2$, we get $\underline{\lambda}_2 a^2 = 12.97$, which is a very poor result. Accordingly, we try the next stage, which involves integrating the square of (11.24):

$$\sum_{n=1}^{\infty} \frac{1}{\gamma_{0n}^4} = 2 \int_0^a \mathrm{d}\rho\, \rho \int_0^\rho \mathrm{d}\rho'\, \rho' \left(\log \frac{a}{\rho} \right)^2$$

$$= \frac{a^4}{8} \int_0^1 \mathrm{d}t \int_0^t \mathrm{d}t' \left(\log \frac{1}{t} \right)^2$$

$$= \frac{a^4}{8} \int_0^1 \mathrm{d}t\, t \left(\log \frac{1}{t} \right)^2 = \frac{a^4}{32} \, , \tag{11.32}$$

the latter integral, like that of (11.30), being performed by partial integration, or by a change of variable: $t = \exp(-x)$. Now we get

$$\frac{1}{(\overline{\lambda}_1 a^2)^2} + \frac{1}{(\underline{\lambda}_2 a^2)^2} = \frac{1}{32} \, , \tag{11.33}$$

from which we get

$$\underline{\lambda}_2 a^2 = 27.213 \, , \tag{11.34}$$

considerably better than the 19.74 produced by comparison with a square.

But before we examine how well (11.34) performs, let us see what we can learn from (11.26) by putting $\rho = \rho'$ and integrating, namely:

$$\sum_{n=1}^{\infty} \frac{1}{k^2 + \gamma_{mn}^2} = \int_0^a \mathrm{d}\rho\, \rho\, I_m(k\rho) \overline{K}_m(k\rho) \, , \tag{11.35}$$

where

$$\overline{K}_m(t) = K_m(t) - I_m(t) \frac{K_m(ka)}{I_m(ka)} \, . \tag{11.36}$$

The two functions that enter the integral of (11.35) obey the same differential equation, for $\rho > 0$,

$$\left[\frac{1}{\rho} \frac{\mathrm{d}}{\mathrm{d}\rho} \left(\rho \frac{\mathrm{d}}{\mathrm{d}\rho} \right) - k^2 - \frac{m^2}{\rho^2} \right] \left\{ \begin{array}{c} I_m(k\rho) \\ K_m(k\rho) \end{array} \right\} = 0 \, . \tag{11.37}$$

We also need the differential equations obeyed by

$$\frac{\partial}{\partial k} I_m(k\rho) = \rho I'_m(k\rho) \, ; \tag{11.38}$$

it is

$$\left[\frac{1}{\rho} \frac{\mathrm{d}}{\mathrm{d}\rho} \left(\rho \frac{\mathrm{d}}{\mathrm{d}\rho} \right) - k^2 - \frac{m^2}{\rho^2} \right] \rho I'_m(k\rho) = 2k I_m(k\rho) \; . \tag{11.39}$$

Then cross multiplication between the latter equation and that for $\overline{K}_m$ yields

$$\frac{\mathrm{d}}{\mathrm{d}\rho} \left[\overline{K}_m(k\rho) \rho \frac{\mathrm{d}}{\mathrm{d}\rho} \rho I'_m(k\rho) - \rho I'_m(k\rho) \rho \frac{\mathrm{d}}{\mathrm{d}\rho} \overline{K}_m(k\rho) \right] = 2k \rho I_m(k\rho) \overline{K}_m(k\rho) \; , \tag{11.40}$$

so that the integrand of (11.35) is a total differential.

At the upper limit of the integral, $\rho = a$,

$$\overline{K}_m(ka) = 0 \; , \tag{11.41}$$

and

$$\rho \frac{\mathrm{d}}{\mathrm{d}\rho} \overline{K}_m(k\rho) \bigg|_{\rho=a} = ka \left[K'_m - I'_m \frac{K_m}{I_m} \right](ka) = -\frac{1}{I_m(ka)} \; , \tag{11.42}$$

according to the Wronskian (Problem 8.8). To handle the lower limit, $\rho = 0$, we recognize that the structure being differentiated in (11.40) can, apart from a factor of $1/k$, be presented as

$$K_m(t) \left(t \frac{\mathrm{d}}{\mathrm{d}t} \right)^2 I_m(t) - t \frac{\mathrm{d}}{\mathrm{d}t} I_m(t) \, t \frac{\mathrm{d}}{\mathrm{d}t} K_m(t) \; , \tag{11.43}$$

for it is only through the singularity of $K_m(t)$ at $t = 0$ that a finite contribution can emerge. Now, for small values of t, $I_m(t) \sim t^m$, and

$$t \ll 1 : \quad t \frac{\mathrm{d}}{\mathrm{d}t} I_m(t) \sim m I_m(t) \; , \tag{11.44}$$

so that (11.43) becomes

$$mt \left[K_m(t) I'_m(t) - I_m(t) K'_m(t) \right] = m \; , \tag{11.45}$$

which again employs the Wronskian. Thus the integral in (11.35) equals

$$\frac{1}{2k} \left[a \frac{I'_m(ka)}{I_m(ka)} - \frac{m}{k} \right] \; . \tag{11.46}$$

The result, presented as

$$\frac{I'_m(t)}{I_m(t)} = \frac{m}{t} + 2t \sum_{n=1}^{\infty} \frac{1}{t^2 + (\gamma_{mn} a)^2} \; , \tag{11.47}$$

is not unfamiliar; we have derived the pole expansion of the logarithmic derivative of $I_m(t)$ in terms of the behavior at $t = 0$ and the imaginary roots at $\pm i\gamma_{mn}$. Then, the initial terms of the power series expansion

$$m \geq 0: \quad I_m(t) = \frac{\left(\frac{1}{2}t\right)^m}{m!} \left[1 + \frac{\left(\frac{1}{2}t\right)^2}{m+1} + \frac{1}{2}\frac{\left(\frac{1}{2}t\right)^4}{(m+1)(m+2)} + \cdots \right] , \quad (11.48)$$

with its logarithmic consequence

$$\frac{I_m'(t)}{I_m(t)} = \frac{m}{t} + \frac{1}{2}\frac{t}{m+1} - \frac{1}{8}\frac{t^3}{(m+1)^2(m+2)} + \cdots , \quad (11.49)$$

give the first two of an infinite set of summations as

$$\sum_{n=1}^{\infty} \frac{1}{\lambda_{mn}a^2} = \frac{1}{4}\frac{1}{m+1} , \quad (11.50a)$$

$$\sum_{n=1}^{\infty} \frac{1}{(\lambda_{mn}a^2)^2} = \frac{1}{16}\frac{1}{(m+1)^2(m+2)} . \quad (11.50b)$$

The results already obtained in (11.30) and (11.32) for $m = 0$ are repeated, and, with the substitution $m \to l + \frac{1}{2}$, we regain the spherical Bessel function summations given in [9], (23.40) and (23.44).

We record, for convenience, the consecutive differences, $\lambda^{(n)}a^2 - \lambda^{(n+1/2)}a^2$, with $n = 0, 1/2, \ldots, 2$, and the ratios of adjacent differences,

$$\frac{\lambda^{(n)}a^2 - \lambda^{(n+1/2)}a^2}{\lambda^{(n+1/2)}a^2 - \lambda^{(n+1)}a^2} , \quad (11.51)$$

as found in (11.23a)–(11.23f):

$$\lambda^{(0)}a^2 - \lambda^{(1/2)}a^2 = 0.1818182$$

$$: \quad 6.333 , \quad (11.52a)$$

$$\lambda^{(1/2)}a^2 - \lambda^{(1)}a^2 = 0.0287081$$

$$: \quad 5.609 , \quad (11.52b)$$

$$\lambda^{(1)}a^2 - \lambda^{(3/2)}a^2 = 0.0051185$$

$$: \quad 5.393 , \quad (11.52c)$$

$$\lambda^{(3/2)}a^2 - \lambda^{(2)}a^2 = 0.0009491$$

$$: \quad 5.317 \quad (11.52d)$$

$$\lambda^{(2)}a^2 - \lambda^{(5/2)}a^2 = 0.0001785$$

$$: \quad 5.288 \quad (11.52e)$$

$$\lambda^{(5/2)}a^2 - \lambda^{(3)}a^2 = 0.0000338$$

$$: \quad 5.277 . \quad (11.52f)$$

$$\lambda^{(3)}a^2 - \lambda^{(7/2)}a^2 = 0.0000064$$

Now let us adopt, provisionally, the lower limit of λ_2 given in (11.34). The following exhibits the lower bounds [(10.133)] thereby produced for $n = \frac{1}{2}, \ldots, \frac{7}{2}$, along with the corresponding upper bounds:

$$n = \frac{1}{2}: \quad 5.8181818 > \lambda_1 a^2 > 5.7687375\,, \tag{11.53a}$$

$$n = 1: \quad 5.7894737 > \lambda_1 a^2 > 5.7817156\,, \tag{11.53b}$$

$$n = \frac{3}{2}: \quad 5.7843552 > \lambda_1 a^2 > 5.7829735\,, \tag{11.53c}$$

$$n = 2: \quad 5.7834061 > \lambda_1 a^2 > 5.7831500\,, \tag{11.53d}$$

$$n = \frac{5}{2}: \quad 5.7832276 > \lambda_1 a^2 > 5.7831794\,, \tag{11.53e}$$

$$n = 3: \quad 5.7831939 > \lambda_1 a^2 > 5.7831848\,, \tag{11.53f}$$

$$n = \frac{7}{2}: \quad 5.7831875 > \lambda_1 a^2 > 5.7831858\,. \tag{11.53g}$$

We also give, analogously to (11.52a)–(11.52f), the consecutive differences and their ratios:

$$\underline{\lambda}^{(1)} a^2 - \underline{\lambda}^{(1/2)} a^2 = 0.0129781$$

$$: \quad 10.3\,, \tag{11.54a}$$

$$\underline{\lambda}^{(3/2)} a^2 - \underline{\lambda}^{(1)} a^2 = 0.0012579$$

$$: \quad 7.11\,, \tag{11.54b}$$

$$\underline{\lambda}^{(2)} a^2 - \underline{\lambda}^{(3/2)} a^2 = 0.0001769$$

$$: \quad 6.02\,, \tag{11.54c}$$

$$\underline{\lambda}^{(5/2)} a^2 - \underline{\lambda}^{(2)} a^2 = 0.0000294$$

$$: \quad 5.44\,, \tag{11.54d}$$

$$\underline{\lambda}^{(3)} a^2 - \underline{\lambda}^{(5/2)} a^2 = 0.0000054$$

$$: \quad 5.4\,. \tag{11.54e}$$

$$\underline{\lambda}^{(7/2)} a^2 - \underline{\lambda}^{(3)} a^2 = 0.0000010$$

All is as expected: With each additional iteration the upper bound decreases and the lower bound increases. Notice also in (11.53a)–(11.53g) that the net increase of the lower bound, 0.01445, is less than half of the net decrease in the upper bound, 0.03499. This means that we have made not too bad a choice of $\underline{\lambda}_2$. A related and more striking observation is the contrast between the smooth convergence of the upper limit ratios in (11.52a)–(11.52f), and the initially more rapid descent of the lower limit ratios in (11.54a)–(11.54e); this shows that the second mode is more suppressed in the lower limit, becoming dominant only after several iterations have been performed.

And now the stage is set for an internal determination of λ_2. According to Problem 11.2, the asymptotic value for the ratios displayed in (11.52a)–(11.52f) is λ_2/λ_1. The evident convergence makes it plain that $\lambda_2/\lambda_1 \leq 5.277$, or, using the six significant figures already established for λ_1 in (11.53g), that $\lambda_2 a^2 \leq 30.52$.

First we test whether, as claimed in connection with (25.111) in [9], the use of a $\underline{\lambda}_2$ value greater that λ_2 will be made apparent by a qualitative change in the iteration process. Displayed below are the results of lower limit computations employing $\lambda_2 a^2 = 30.52$, along with the consecutive differences and their ratios:

$$n = \frac{1}{2}: \quad 5.7753570$$

$$: \quad 0.0073961$$

$$n = 1: \quad 5.7827530 \qquad\qquad : \quad 18.3\,,$$

$$: \quad 0.0004052$$

$$n = \frac{3}{2}: \quad 5.7831582 \qquad\qquad : \quad 15.6\,,$$

$$: \quad 0.0000260$$

$$n = 2: \quad 5.7831842 \qquad\qquad : \quad 16\,, \qquad\qquad (11.55)$$

$$: \quad 0.00000017$$

$$n = \frac{5}{2}: \quad 5.78318588 \qquad\qquad : \quad 20\,,$$

$$: \quad 0.000000084$$

$$n = 3: \quad 5.78318597 \qquad\qquad : \quad -70\,.$$

$$: \quad -0.0000000012$$

$$n = \frac{7}{2}: \quad 5.783185965$$

The contrast with the ratios in (11.54a)–(11.54e) is eloquent. Clearly the anticipated has happened: After an initial iterative increase in the "lower limits," that rise has ceased and the convergence to λ_1 from above has begun. As a result we learn that $\lambda_2 a^2 < 30.52$, and that

$$\lambda_1 a^2 < 5.783185965\,, \qquad\qquad\qquad (11.56)$$

A computer search program would be most effective in the last step, the quest for the transition from the qualitative behavior of (11.55) to that of (11.54a)–(11.54e), which identifies λ_2. Instead, we present just one example, where the number 30.52 is reduced about 0.2% to 30.45:

$$n = \frac{1}{2}: \quad 5.7752353$$

$$: \quad 0.0074987$$

$$n = 1: \quad 5.7827340 \qquad\qquad : \quad 17.8\,,$$

$$: \quad 0.0004209$$

$$n = \frac{3}{2}: \quad 5.7831548 \qquad\qquad : \quad 14.6\,,$$

$$: \quad 0.0000288$$

$$n = 2: \quad 5.7831836 \qquad\qquad : \quad 13.2\,, \qquad\qquad (11.57)$$

$$: \quad 0.00000217$$

$$n = \frac{5}{2}: \quad 5.7831858 \qquad\qquad : \quad 12.1\,,$$

$$: \quad 0.00000018$$

$$n = 3 : \quad 5.78318594 \qquad\qquad : \quad 10.6 \; .$$

$$: \quad 0.000000017$$

$$n = \frac{7}{2} : \quad 5.783185961$$

The situation has become normal, and so we know that $\lambda_2 a^2 > 30.45$, and that

$$\lambda_1 a^2 > 5.783185961 \; . \tag{11.58}$$

What have we accomplished? The best determination of $\lambda_1 a^2$ in (11.53g) can be presented as

$$\lambda_1 a^2 = 5.7831866 \pm 0.0000009 \; , \tag{11.59}$$

an accuracy of about one part in a ten million. Now, without any additional input, $\lambda_1 a^2$ has, according to (11.56) and (11.58), been located at

$$\lambda_1 a^2 = 5.783185963 \pm 0.000000002 \; , \tag{11.60}$$

an accuracy of three parts in ten billion.

We have refrained from explicit use of the true values of λ_1 and λ_2, which are the squares of γ_{01} and γ_{02}, respectively:

$$\lambda_1 a^2 = 5.78318596297 \; , \tag{11.61a}$$

$$\lambda_2 a^2 = 30.4712623438 \; . \tag{11.61b}$$

That our choice of $\underline{\lambda}_2$ in (11.57) is very close to λ_2 is quite apparent in the significantly increased rate of convergence there, as compared with (11.54a)–(11.54c). What happens if we use the actual λ_2 value? The result of $\underline{\lambda}^{(7/2)}$ increases almost indiscernibly, the successive ratios somewhat more, in particular, the final ratio, 10.6, is raised to 13.2. And that is reasonable, for then these ratios are converging toward (Problem 11.3)

$$\frac{\lambda_3}{\lambda_1} = 12.949092 \; . \tag{11.62}$$

11.2 H Modes

The restriction (10.41) on H modes is automatically satisfied for all but the H_{0n} modes of a circular guide. In the construction of the dominant mode of the latter type, H_{01}, which is actually the third highest H mode, after H_{11} and H_{21}, each approximant function must obey the requirement

$$\int_0^a R_1^{(n)}(r) \, r \, dr = 0 \; , \quad n = 0, 1, 2, \ldots . \tag{11.63}$$

The successive approximant functions obtained from the defining equation

$$\left(\frac{d^2}{dr^2} + \frac{1}{r}\frac{d}{dr}\right) R_1^{(n)}(r) = \frac{1}{r}\frac{d}{dr} r \frac{d}{dr} R_1^{(n)}(r) = -R_1^{(n-1)}(r) \qquad (11.64)$$

automatically satisfies the boundary condition (10.43) in consequence of (11.63). This is proved by multiplying (11.64) by r and integrating with respect to r from 0 to a, whence

$$a\frac{d}{dr} R_1^{(n)}(a) = -\int_0^a R^{(n-1)}(r)\, r\, dr = 0 , \quad n = 1, 2, \dots . \qquad (11.65)$$

A simple generating function, consistent with (11.63), is

$$R_1^{(0)}(r) = a^2 - 2r^2 , \qquad (11.66)$$

and the first three approximation functions deduced from (11.64) subject to (11.63), are

$$R_1^{(1)}(r) = \frac{1}{8}\left(\frac{2}{3}a^4 - 2a^2r^2 + r^4\right) , \qquad (11.67a)$$

$$R_1^{(2)}(r) = \frac{1}{288}\left(\frac{7}{4}a^6 - 6a^4r^2 + \frac{9}{2}a^2r^4 - r^6\right) , \qquad (11.67b)$$

$$R_1^{(3)}(r) = \frac{1}{18432}\left(\frac{39}{5}a^8 - 28a^6r^2 + 24a^4r^4 - 8a^2r^6 + r^8\right) . \qquad (11.67c)$$

The successive approximations to $\gamma_1 a$ as calculated from (10.90) using these functions are

$$\gamma_1^{(0)} a = \sqrt{24} = 4.898980 , \qquad (11.68a)$$

$$\gamma_1^{(1/2)} a = \sqrt{16} = 4.0 , \qquad (11.68b)$$

$$\gamma_1^{(1)} a = \sqrt{15} = 3.872983 , \qquad (11.68c)$$

$$\gamma_1^{(3/2)} a = \sqrt{\frac{192}{13}} = 3.843076 , \qquad (11.68d)$$

$$\gamma_1^{(2)} a = \sqrt{\frac{1456}{99}} = 3.834980 , \qquad (11.68e)$$

$$\gamma_1^{(5/2)} a = \sqrt{\frac{9504}{647}} = 3.832667 , \qquad (11.68f)$$

$$\gamma_1^{(3)} a = \sqrt{\frac{116460}{7931}} = 3.831990 . \qquad (11.68g)$$

The third approximant exceeds the correct value, $\gamma_1 a = 3.8317060$, by 0.000284, an error of roughly 7 in 10^5. The 3/2 approximation is of sufficient accuracy for most purposes, being in error by 0.025%. The relatively slow convergence, in contrast to that found for the E_{01} mode, is attributable to the smaller value of $(\gamma_2/\gamma_1)^4$, 11.24. Successive upper limits to $\gamma_2 a$, deduced from (10.115a) and (10.115b) with the correct value of $\gamma_1 a$ are: 10.6355, 7.8845, 7.3368, 7.1505, 7.0756, 7.0433. The last result exceeds the true value of $\gamma_2 a$, 7.015587, by 0.4%. A lower limit to the second eigenvalue is provided by the result, to be established in Problem 11.4, that

$$\sum_{n=1}^{\infty} \frac{1}{\lambda_n^2} = \frac{a^4}{192} \,, \tag{11.69}$$

whence

$$\frac{1}{\lambda_1^2} + \frac{1}{\lambda_2^2} < \frac{a^4}{192} \,. \tag{11.70}$$

If λ_1 is replaced by the upper bound supplied by the third approximation, we obtain $\gamma_2 a > 6.4701$. The lower limits to $\gamma_1 a$ deduced from (10.133) are: 3.324252, 3.800209, 3.826674, 3.830588, 3.831416, 3.831625. On combining the upper and lower limits, we find, as successive estimates to $\gamma_1 a$:

$$3.662126 \pm 0.337874 \,, \tag{11.71a}$$
$$3.836596 \pm 0.036387 \,, \tag{11.71b}$$
$$3.834875 \pm 0.008201 \,, \tag{11.71c}$$
$$3.832784 \pm 0.002196 \,, \tag{11.71d}$$
$$3.832042 \pm 0.000626 \,, \tag{11.71e}$$
$$3.831807 \pm 0.000183 \,. \tag{11.71f}$$

Thus the third approximation determines $\gamma_1 a$ with an uncertainty of ± 5 in 10^5.

11.3 Problems for Chap. 11

1. Work out the Coulomb Green's function in cylindrical coordinates, by finding the reduced Green's function (i.e., the part depending on radial coordinates) either directly or through an eigenfunction expansion. In this way establish the equality (11.26).
2. By developing a perturbative argument, show that for large n, the successive ratios of upper bounds,

$$\frac{\lambda^{(n-1/2)} - \lambda^{(n)}}{\lambda^{(n)} - \lambda^{(n+1/2)}} \tag{11.72}$$

approach λ_2/λ_1 from above.
3. In general, show that the successive ratios of lower bounds approach the same limit, but more slowly. However, if the lower limit on λ_2 is replaced by the exact value of λ_2, the convergence is much more rapid, and the limiting value is λ_3/λ_1.
4. Establish the sum rule for the H-mode eigenvalues, (11.69).

Steady Currents and Dissipation

12.1 Variational Principles for Current

Although the focus of this book is on radiation, the interaction of electro-magnetic radiation with matter is described in terms of electric currents and dissipation, so it is useful to first remind ourselves of the steady-state context. Suppose we have a medium with a conductivity $\sigma(\mathbf{r})$, which is defined in terms of the linear relation between the electric field and the electric current:

$$\mathbf{J} = \sigma \mathbf{E} \,. \tag{12.1}$$

In statics, the electric field is derived from a scalar potential,

$$\mathbf{E} = -\boldsymbol{\nabla}\phi \,, \tag{12.2}$$

and in addition the current density is divergenceless,

$$\boldsymbol{\nabla} \cdot \mathbf{J} = 0 \,, \tag{12.3}$$

which is the static version of the local conservation of charge, (1.14). This implies that the potential obeys the equation

$$\boldsymbol{\nabla} \cdot \sigma(-\boldsymbol{\nabla}\phi) = 0 \,. \tag{12.4}$$

The rate at which the field does work on the current, or the power of Joule heating, is given by

$$P = \int (\mathrm{d}\mathbf{r})\mathbf{J} \cdot \mathbf{E} = \int (\mathrm{d}\mathbf{r})\,\sigma E^2 = \int (\mathrm{d}\mathbf{r})\frac{J^2}{\sigma} \,. \tag{12.5}$$

The following construction of the power,

$$P[\mathbf{J}, \phi] = \int (\mathrm{d}\mathbf{r}) \left[2\mathbf{J} \cdot (-\boldsymbol{\nabla}\phi) - \frac{J^2}{\sigma} \right] \,, \tag{12.6}$$

is stationary with respect to small variations in the current density and the potential, leading to the following equations holding in the volume

$$\delta \mathbf{J} : \quad \mathbf{J} = \sigma(-\boldsymbol{\nabla}\phi) = \sigma \mathbf{E} , \tag{12.7a}$$

$$\delta \phi : \quad \boldsymbol{\nabla} \cdot \mathbf{J} = 0 . \tag{12.7b}$$

Omitted here was a surface contribution

$$\delta P = -2 \int (\mathrm{d}\mathbf{r}) \boldsymbol{\nabla} \cdot (\mathbf{J}\,\delta\phi) = -2 \oint_S \mathrm{d}S \mathbf{n} \cdot \mathbf{J}\,\delta\phi , \tag{12.8}$$

where S is the surface surrounding the region of interest, and $\mathbf{n}$ is the outward normal. Let us suppose that we insert, as part of that boundary, conducting surfaces S_i, electrodes, on which the potential is specified:

$$\phi = \phi_i = \text{const on } S_i , \tag{12.9}$$

that is, $\delta\phi$ is zero on S_i. Let us denote the rest of the boundary as S':

$$S = S' + \sum_i S_i , \tag{12.10}$$

so that (12.8) is

$$\delta P = -2 \int_{S'} \mathrm{d}S \, \mathbf{n} \cdot \mathbf{J}\,\delta\phi , \tag{12.11}$$

where $\delta\phi$ is arbitrary. This will be zero, $\delta P = 0$, if

$$\mathbf{n} \cdot \mathbf{J} = 0 \text{ on } S' , \tag{12.12}$$

which is to say, no current flows out of or into the region of interest, except through the electrodes.

With prescribed potentials ϕ_i on the electrode surfaces S_i, $P[\mathbf{J}, \phi]$ is stationary for variations about $\mathbf{J} = \sigma\mathbf{E}$, $\boldsymbol{\nabla} \cdot \mathbf{J} = 0$, $\mathbf{n} \cdot \mathbf{J} = 0$ on S'. Suppose we now accept (12.1) as given. Then the two terms in (12.6) combine to give

$$P[\phi] = \int (\mathrm{d}\mathbf{r})\sigma(-\boldsymbol{\nabla}\phi)^2 , \tag{12.13}$$

which when varied with respect to ϕ yields the current conditions (12.3) and (12.12). Because $P[\phi]$ is stationary for infinitesimal variations, for finite variations about the true potential ϕ_0,

$$P[\phi_0 + \delta\phi] = P + \int (\mathrm{d}\mathbf{r})\sigma(-\boldsymbol{\nabla}\delta\phi)^2 > P ; \tag{12.14}$$

the correct potential minimizes the power loss, with given potentials ϕ_i on the electrodes. On the other hand, if we accept (12.3) and (12.12), we have

$$P[\mathbf{J}] = -2\sum_i \phi_i \int_{S_i} dS\,\mathbf{n} \cdot \mathbf{J} - \int (d\mathbf{r})\frac{J^2}{\sigma}\,, \qquad (12.15)$$

which is stationary for infinitesimal variations in $\mathbf{J}$ (Problem 12.1), while for finite variations,

$$P[\mathbf{J}_0 + \delta\mathbf{J}] = P - \int (d\mathbf{r})\frac{(\delta\mathbf{J})^2}{\sigma} < P\,, \qquad (12.16)$$

so the correct current makes P a maximum, for given ϕ_i.

From the latter form (12.15) by rescaling the current $\mathbf{J} \to \lambda\mathbf{J}$ and applying the stationary principle at $\lambda = 1$ we learn that

$$0 = -2\sum_i \phi_i \int_{S_i} dS\,\mathbf{n} \cdot \mathbf{J} - 2\int (d\mathbf{r})\frac{J^2}{\sigma}\,, \qquad (12.17)$$

or

$$P = \int (d\mathbf{r})\frac{J^2}{\sigma} = \sum_i \phi_i I_i\,, \qquad (12.18)$$

where

$$I_i = -\int_{S_i} dS\,\mathbf{n} \cdot \mathbf{J} \qquad (12.19)$$

is the current input at the ith electrode. Of course,

$$\sum_i I_i = -\int_S dS\,\mathbf{n} \cdot \mathbf{J} = -\int (d\mathbf{r})\boldsymbol{\nabla} \cdot \mathbf{J} = 0\,. \qquad (12.20)$$

And therefore only potential differences matter in (12.18), that is,

$$\sum_i (\phi_i + \text{const.})I_i = \sum_i \phi_i I_i\,. \qquad (12.21)$$

Another stationary principle begins with

$$P = 2\sum_i \phi_i I_i - \int (d\mathbf{r})\left[2\mathbf{J} \cdot (-\boldsymbol{\nabla}\phi) - \frac{J^2}{\sigma}\right]$$

$$= -2\int_S dS\,\mathbf{n} \cdot \mathbf{J}\phi + \int (d\mathbf{r})\left[2\boldsymbol{\nabla} \cdot (\mathbf{J}\phi) - 2\boldsymbol{\nabla} \cdot \mathbf{J}\phi + \frac{J^2}{\phi}\right] \qquad (12.22)$$

which leads us to

$$P[\mathbf{J}, \phi] = \int (d\mathbf{r})\left[\frac{J^2}{\sigma} - 2\phi\boldsymbol{\nabla} \cdot \mathbf{J}\right]\,. \qquad (12.23)$$

Here

$$I_i = -\int_{S_i} dS\,\mathbf{n} \cdot \mathbf{J} \text{ is specified, and } \mathbf{n} \cdot \mathbf{J} = 0 \text{ on } S'\,. \qquad (12.24)$$

If (12.23) undergoes infinitesimal variations, we recover the appropriate equations

$$\delta\phi : \quad \boldsymbol{\nabla} \cdot \mathbf{J} = 0 , \quad \delta\mathbf{J} : \quad \mathbf{J} = \sigma(-\boldsymbol{\nabla}\phi) . \tag{12.25}$$

The latter follows from the volume variation. What is left is a surface integral:

$$\delta P = -2 \int (\mathrm{d}\mathbf{r})\boldsymbol{\nabla} \cdot (\phi\, \delta\mathbf{J}) = -2 \int \mathrm{d}S\, \phi\, \mathbf{n} \cdot \delta\mathbf{J} , \tag{12.26}$$

subject to

$$\int_{S_i} \mathrm{d}S\, \mathbf{n} \cdot \delta\mathbf{J} = 0 , \quad \mathbf{n} \cdot \delta\mathbf{J} = 0 \text{ on } S' . \tag{12.27}$$

Therefore

$$\delta P = -2 \sum_i \int_{S_i} \mathrm{d}S\, \mathbf{n} \cdot \delta\mathbf{J}\, \phi = 0 , \tag{12.28}$$

subject to the first constraint in (12.27), from which we conclude that on S_i, $\phi = \phi_i$, a constant.

Next, impose

$$\mathbf{J} = \sigma(-\boldsymbol{\nabla}\phi) \tag{12.29}$$

in (12.23). Then we have

$$\begin{aligned}
P[\phi] &= \int (\mathrm{d}\mathbf{r}) \left[\sigma(-\boldsymbol{\nabla}\phi)^2 - 2\phi\boldsymbol{\nabla} \cdot (\sigma(-\boldsymbol{\nabla}\phi)) \right] \\
&= -2 \int_S \mathrm{d}S\, \phi\, \mathbf{n} \cdot \sigma(-\boldsymbol{\nabla}\phi) - \int (\mathrm{d}\mathbf{r})\, \sigma(-\boldsymbol{\nabla}\phi)^2 \\
&= -2 \sum_i \int_{S_i} \mathrm{d}S\, \phi\, \mathbf{n} \cdot \sigma(-\boldsymbol{\nabla}\phi) - \int (\mathrm{d}\mathbf{r})\sigma(-\boldsymbol{\nabla}\phi)^2 .
\end{aligned} \tag{12.30}$$

Here we have used on S'

$$0 = \mathbf{n} \cdot \mathbf{J} = \mathbf{n} \cdot \sigma(-\boldsymbol{\nabla}\phi) . \tag{12.31}$$

Now under a finite variation around the true potential ϕ_0

$$P[\phi_0 + \delta\phi] = P - 2 \sum_i \int_{S_i} \mathrm{d}S\, \delta\phi\, \mathbf{n} \cdot \sigma(-\boldsymbol{\nabla}\delta\phi) - \int (\mathrm{d}\mathbf{r})\, \sigma(-\boldsymbol{\nabla}\delta\phi)^2 . \tag{12.32}$$

Now if we only demand that $\delta\phi$ be constant on S_i, that is, ϕ is constant but not necessarily the correct potential, the first integral here vanishes:

$$-\int_{S_i} \mathrm{d}S\, \delta\phi\, \mathbf{n} \cdot \sigma(-\boldsymbol{\nabla}\delta\phi) = -\delta\phi \int_{S_i} \mathrm{d}S\, \mathbf{n} \cdot \sigma(-\boldsymbol{\nabla}\delta\phi) = \delta\phi\, \delta I_i = 0 , \tag{12.33}$$

so $P[\phi] < P$; for a given I_i the correct ϕ makes P maximum. On the other hand, when we set $\boldsymbol{\nabla} \cdot \mathbf{J} = 0$ in (12.23), then

$$P[\mathbf{J}] = \int (d\mathbf{r}) \frac{J^2}{\sigma} \, , \tag{12.34}$$

and under a finite variation

$$P[\mathbf{J} + \delta\mathbf{J}] = P + \int (d\mathbf{r}) \frac{(\delta\mathbf{J})^2}{\sigma} > P \, , \tag{12.35}$$

so the correct $\mathbf{J}$ makes P a minimum, again subject to given I_i. These results are opposite to those obtaining in (12.14) and (12.16), in which the potentials on the electrodes, not the currents supplied by them, are specified.

12.2 Green's Functions

Suppose we now introduce a Green's function for the potential, so that the steady current condition

$$\boldsymbol{\nabla} \cdot \mathbf{J} = \boldsymbol{\nabla} \cdot \sigma(-\boldsymbol{\nabla}\phi) = 0 \, , \tag{12.36}$$

corresponds to the following Green's function equation

$$-\boldsymbol{\nabla} \cdot [\sigma \boldsymbol{\nabla} G(\mathbf{r}, \mathbf{r}')] = \delta(\mathbf{r} - \mathbf{r}') \, . \tag{12.37}$$

The boundary conditions on the Green's function are

$$\mathbf{n} \cdot \boldsymbol{\nabla} G = 0 \text{ on } S' \, , \quad G = 0 \text{ on } S_i \, , \tag{12.38}$$

Dirichlet boundary conditions on S_i and Neumann ones on S'. Multiply the Green's function equation (12.37) by $\phi(\mathbf{r})$ and the potential equation (12.36) by $G(\mathbf{r}, \mathbf{r}')$ and subtract:

$$-\phi(\mathbf{r})\boldsymbol{\nabla} \cdot [\sigma \boldsymbol{\nabla} G(\mathbf{r}, \mathbf{r}')] + G(\mathbf{r}, \mathbf{r}')\boldsymbol{\nabla} \cdot [\sigma \boldsymbol{\nabla}\phi(\mathbf{r})] = \delta(\mathbf{r} - \mathbf{r}')\phi(\mathbf{r}') \, . \tag{12.39}$$

The left-hand side of this equation is a total divergence:

$$\boldsymbol{\nabla} \cdot [G(\mathbf{r}, \mathbf{r}')\sigma \boldsymbol{\nabla}\phi(\mathbf{r}) - \phi(\mathbf{r})\sigma \boldsymbol{\nabla} G(\mathbf{r}, \mathbf{r}')] \, , \tag{12.40}$$

so when we integrate over the volume we obtain

$$\phi(\mathbf{r}') = \int_{S'+\sum_i S_i} dS \, \mathbf{n} \cdot [G(\mathbf{r}, \mathbf{r}')\sigma \boldsymbol{\nabla}\phi(\mathbf{r}) - \phi(\mathbf{r})\sigma \boldsymbol{\nabla} G(\mathbf{r}, \mathbf{r}')]$$
$$= -\sum_i \phi_i \int_{S_i} dS \, \sigma(\mathbf{r})\mathbf{n} \cdot \boldsymbol{\nabla} G(\mathbf{r}, \mathbf{r}') \, , \tag{12.41}$$

because on S'

$$\mathbf{n} \cdot \boldsymbol{\nabla}\phi = 0 \, , \quad \mathbf{n} \cdot \boldsymbol{\nabla} G(\mathbf{r}, \mathbf{r}') = 0 \, , \tag{12.42}$$

while on S_i

$$\phi = \phi_i , \qquad G = 0 . \tag{12.43}$$

Changing variables, and using the reciprocity relation (Problem 12.2)

$$G(\mathbf{r}, \mathbf{r}') = G(\mathbf{r}', \mathbf{r}) , \tag{12.44}$$

we obtain

$$\phi(\mathbf{r}) = \sum_j \phi_j \int_{S_j} \mathrm{d}S' \, \sigma(\mathbf{r}')(-\mathbf{n}' \cdot \boldsymbol{\nabla}')G(\mathbf{r}, \mathbf{r}') . \tag{12.45}$$

Now compute the current input by the ith electrode,

$$I_i = -\int_{S_i} \mathrm{d}S \, \mathbf{n} \cdot \mathbf{J} = -\int_{S_i} \mathrm{d}S \, \sigma \, \mathbf{n} \cdot (-\boldsymbol{\nabla}\phi) . \tag{12.46}$$

Inserting (12.45) into this we obtain a linear relation between the current input at the ith electrode and the potential on the jth:

$$I_i = \sum_j G_{ij}\phi_j \tag{12.47}$$

where the "coefficients of conductance" are given by

$$G_{ij} = -\int_{S_i} \mathrm{d}S \int_{S_j} \mathrm{d}S' \, \sigma(\mathbf{r})\mathbf{n} \cdot \boldsymbol{\nabla}\sigma(\mathbf{r}')\mathbf{n}' \cdot \boldsymbol{\nabla}'G(\mathbf{r}, \mathbf{r}') = G_{ji} . \tag{12.48}$$

Then we can write the power as

$$P = \sum_i \phi_i I_i = \sum_{ij} \phi_i G_{ij}\phi_j . \tag{12.49}$$

From this we see

$$\frac{\partial P}{\partial \phi_i} = 2I_i , \tag{12.50}$$

which is consistent with (12.11). We already know from (12.20) that $\sum_i I_i = 0$, for arbitrary ϕ_i. Therefore,

$$\sum_i G_{ij} = 0 . \tag{12.51}$$

A direct proof of this is as follows:

$$\begin{aligned}
\sum_i G_{ij} &= \int_{S_j} \mathrm{d}S' \, \sigma(\mathbf{r}')\mathbf{n}' \cdot \boldsymbol{\nabla}' \int_S \mathrm{d}S \, \mathbf{n} \cdot \sigma(\mathbf{r})(-\boldsymbol{\nabla})G(\mathbf{r}, \mathbf{r}') \\
&= \int_{S_j} \mathrm{d}S' \, \sigma(\mathbf{r}')\mathbf{n}' \cdot \boldsymbol{\nabla}' \int (\mathrm{d}\mathbf{r})\boldsymbol{\nabla} \cdot [\sigma(\mathbf{r})(-\boldsymbol{\nabla})G(\mathbf{r}, \mathbf{r}')] \\
&= 0 , \tag{12.52}
\end{aligned}$$

since the last volume integral is 1.

Consider the case of two electrodes. Then

$$G_{11} + G_{21} = 0 , \qquad G_{12} + G_{22} = 0 . \tag{12.53}$$

Thus there is a single conductance, which we can take to be

$$G = -G_{12} = -G_{21} = G_{11} = G_{22} . \tag{12.54}$$

The current in the two electrodes is

$$I_1 = -I_2 = G_{11}\phi_1 + G_{12}\phi_2 = G(\phi_1 - \phi_2) = GV = I . \tag{12.55}$$

The resistance is defined as the inverse of the conductance, $G = 1/R$. Then the power is

$$P = \sum_i I_i\phi_i = IV = GV^2 = RI^2 . \tag{12.56}$$

As an example consider a cylindrical conductor, of arbitrary cross section, with σ varying arbitrarily across the section. The Green's function equation (12.37) is

$$-\frac{\partial}{\partial z}\sigma\frac{\partial}{\partial z}G - \boldsymbol{\nabla}_\perp \cdot (\sigma\boldsymbol{\nabla}_\perp G) = \delta(z - z')\delta(\mathbf{r}_\perp - \mathbf{r}'_\perp) . \tag{12.57}$$

Now when we integrate this over the cross section of the cylinder we encounter

$$\int (d\mathbf{r}_\perp)\boldsymbol{\nabla}_\perp \cdot (\sigma\boldsymbol{\nabla}_\perp G) = \oint_C ds\,\sigma\,\mathbf{n} \cdot \boldsymbol{\nabla}_\perp G = 0 , \tag{12.58}$$

where the line integral is extended over the circumference C of the cylinder, on which (12.42) applies, and $\mathbf{n}$ is perpendicular to the walls of the cylinder. We are left with, under the assumption that σ does not depend on z,

$$-\frac{\partial^2}{\partial z^2}\int (d\mathbf{r}_\perp)\,\sigma G = \delta(z - z') . \tag{12.59}$$

The solution of this equation is (Problem 12.3)

$$\int (d\mathbf{r}_\perp)\,\sigma(\mathbf{r}_\perp)G(\mathbf{r}, \mathbf{r}') = \frac{z_<(L - z_>)}{L} , \tag{12.60}$$

where L is the length of the cylinder. We have imposed the boundary condition (12.43), $G = 0$ at $z = 0, L$. Then the conductance is obtained from (12.48)

$$G = \frac{1}{R} = -\int_{z,z'=0} (d\mathbf{r}'_\perp)\,\sigma(\mathbf{r}'_\perp)\frac{\partial}{\partial z}\frac{\partial}{\partial z'}\left[-\frac{zz'}{L}\right] = \frac{1}{L}\int (d\mathbf{r}_\perp)\sigma(\mathbf{r}_\perp) , \tag{12.61}$$

which is reminiscent of the corresponding formula for the capacitance,

$$C = \frac{1}{a}\int (d\mathbf{r}_\perp)\,\varepsilon(\mathbf{r}_\perp) , \tag{12.62}$$

a being the separation between the parallel plates of a capacitor, filled with a dielectric of permittivity ε.

12.3 Problems for Chap. 12

1. Use Lagrange multipliers to show that $P[J]$, (12.15), is stationary for infinitesimal variations.
2. Prove the reciprocity relation (12.44) directly from the differential equation (12.37) and the boundary conditions (12.38).
3. Directly solve (12.59) for (12.60).

13

The Impedance Concept in Waveguides

13.1 Waveguides and Équivalent Transmission Lines

In preceding chapters we have discussed the nature of the fields in a waveguide, oriented along the z-axis,

$$\mathbf{E}_\perp(x,y,z) = \boldsymbol{\mathcal{E}}(x,y)\left(\alpha e^{i\kappa z} + \beta e^{-i\kappa z}\right) , \tag{13.1}$$

$$\mathbf{H}_\perp(x,y,z) = \boldsymbol{\mathcal{H}}(x,y)\left(\alpha e^{i\kappa z} - \beta e^{-i\kappa z}\right) . \tag{13.2}$$

In the mode description given in Chap. 6, for an H mode, in (6.33a) and (6.33b),

$$\boldsymbol{\mathcal{H}}_\perp = -\boldsymbol{\nabla}\psi , \qquad \boldsymbol{\mathcal{E}}_\perp = \mathbf{e} \times \boldsymbol{\nabla}\psi , \tag{13.3a}$$

while for an E mode, (6.32a) and (6.32b) give

$$\boldsymbol{\mathcal{E}}_\perp = -\boldsymbol{\nabla}\varphi , \qquad \boldsymbol{\mathcal{H}}_\perp = -\mathbf{e} \times \boldsymbol{\nabla}\varphi . \tag{13.3b}$$

We discussed the analogy between the z-variation of the transverse electric and magnetic field with that of voltage and current in a transmission line. Factors in that identification are essentially arbitrary, and for example, the characteristic impedance of the equivalent transmission line is dictated by convenience only, although absolute factors are naturally fixed by the complex power identification, (6.76) and (6.77),

$$\frac{1}{2}\int d\boldsymbol{\sigma} \cdot \mathbf{E} \times \mathbf{H}^* = \frac{1}{2}VI^* . \tag{13.4}$$

The analogy is applicable to more than one propagating mode and also to attenuated modes, where the characteristic impedance and propagation constant κ become imaginary. The analogy is also immediately applicable to pure electrical discontinuities, that is, where the dielectric constant has discontinuities in z. (See Chap. 9, for example.) There the continuity of the tangential components of $\mathbf{E}$ and $\mathbf{H}$ become the continuity of the current and voltage if the characteristic impedance is $\propto k/\kappa$ (for H modes), the arbitrary factor being the same on both sides of the discontinuity. Simplification of this result lies in the possibility of satisfying the boundary conditions with only one mode. The equivalent transmission line problem is, rigorously, given by specifying the impedance as $Z = 1$ for $z < 0$, and $Z = \kappa/\kappa'$ for $z > 0$, the discontinuity lying at $z = 0$. This describes the actual fields everywhere. This concept is less trivial, and its utility more apparent, in connection with geometrical discontinuities, to which we now turn.

13.2 Geometrical Discontinuities and Equivalent Circuits

We now consider geometrical discontinuities, such as those provided by an abrupt change in the radius of the guide, or a partial barrier, such as an iris. What is the nature of the fields in the vicinity of such a geometrical discontinuity? Higher mode fields must necessarily be present in order to satisfy the boundary conditions. At a sufficient distance from the discontinuity the field consists only of a propagating mode or modes. The problem can be formulated as that of finding the amplitude of the fields moving away from the discontinuity in terms of those falling on it, that is, to find all the reflection and transmission coefficients. This formulation has two disadvantages:

- Restrictive conditions imposed by conservation of energy have a complicated expression.
- The treatment of more than one discontinuity (which is the whole point of the subject) is complicated, involving multiple reflections, etc.

Instead of dealing with incident and reflected lowest mode fields, let us introduce total electric and magnetic fields of the lowest modes in the various guides, that is, the voltages and currents. The relation between incident and reflected fields becomes a relation between currents and voltages which have the form of a circuit equation. Conservation of energy finds its usual simple expression for a dissipationless network (or a dissipative one, for that matter), and the combinatorial problems are reduced to the conventional rules of impedance combinations and transformations.

13.2.1 S-Matrix

In more detail, consider a general situation with n channels, all mutually connected, with an arbitrary geometrical configuration within the mutual region, as illustrated in Fig. 13.1. The far fields in the ith guide are

$$\mathbf{E}_\perp(x,y,z) = \boldsymbol{\mathcal{E}}_i(x_i,y_i)\left(\alpha_i e^{i\kappa_i z_i} + \beta_i e^{-i\kappa_i z_i}\right), \tag{13.5a}$$

$$\mathbf{H}_\perp(x,y,z) = \boldsymbol{\mathcal{H}}_i(x_i,y_i)\left(\alpha_i e^{i\kappa_i z_i} - \beta_i e^{-i\kappa_i z_i}\right), \tag{13.5b}$$

z_i being a coordinate measured along the ith guide and with arbitrary origin in or near the discontinuity region. The α_is are arbitrary amplitudes of incident

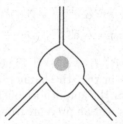

Fig. 13.1. Generic sketch of junction between waveguides

fields, and the β_is are the amplitudes of reflected fields. In view of the linearity of the problem,

$$\beta_i = \sum_j S_{ij}\alpha_j \ . \tag{13.6}$$

The scattering matrix S_{ij} contains the ensemble of reflection and transmission coefficients. In matrix form, with

$$\begin{pmatrix} \alpha_1 \\ \alpha_2 \\ . \\ . \\ . \end{pmatrix} = \psi \ , \qquad \begin{pmatrix} \beta_1 \\ \beta_2 \\ . \\ . \\ . \end{pmatrix} = \phi \ , \quad \phi = \mathsf{S}\psi \ . \tag{13.7}$$

A general restriction on S is provided by the reciprocity theorem. If in each guide, the fields are normalized such that

$$\int d\boldsymbol{\sigma}_i \cdot \boldsymbol{\mathcal{E}}_i \times \boldsymbol{\mathcal{H}}_i^* = 1 \tag{13.8}$$

[this is the same normalization given in (6.56) and (6.59)], or at least the integral is the same for all guides, then S is symmetric

$$S_{ij} = S_{ji} \ , \qquad \mathsf{S}^T = \mathsf{S} \ , \tag{13.9}$$

where T denotes transposition. If the system is dissipationless, it is required, with this normalization, that

$$\sum_i |\beta_i|^2 = (\phi, \phi) = \sum_i |\alpha_i|^2 = (\psi, \psi) \ , \tag{13.10}$$

that is, the matrix S is unitary,

$$\mathsf{S}^\dagger \mathsf{S} = \mathsf{S}^* \mathsf{S} = 1 \ . \tag{13.11}$$

(The proof of this assertion appears in the following paragraph.) Introduce now the current and voltage,

$$V_i(z_i) = \alpha_i e^{i\kappa_i z_i} + \beta_i e^{-i\kappa_i z_i} , \tag{13.12a}$$
$$I_i(z_i) = \alpha_i e^{i\kappa_i z_i} - \beta_i e^{-i\kappa_i z_i} , \tag{13.12b}$$

where at $z_i = 0$, $V_i = \alpha_i + \beta_i$, $I_i = \alpha_i - \beta_i$. These are essentially the actual electric and magnetic fields in the ith channel, a wavelength back from the origin. They do not represent the amplitudes of the actual fields at the origin, since here the higher modes will be present. Now

$$V = \begin{pmatrix} V_1 \\ V_2 \\ . \\ . \\ . \end{pmatrix} = \psi + \phi , \quad I = \begin{pmatrix} I_1 \\ I_2 \\ . \\ . \\ . \end{pmatrix} = \psi - \phi , \tag{13.13}$$

or

$$V = (1 + S)\psi , \qquad I = (1 - S)\psi , \tag{13.14}$$

that is,

$$V = \frac{1 + S}{1 - S} I = ZI . \tag{13.15}$$

Thus we obtain the relation of circuit form between the voltage and current, in terms of the impedance matrix $Z = (1+S)/(1-S)$. Now it is apparent that

$$Z^T = Z , \qquad Z^* = -Z , \tag{13.16}$$

which are restatements of the reciprocity theorem (13.9) and of energy conservation (13.10), respectively. Thus energy conservation here states simply that all the elements of the impedance matrix are imaginary. Hence the reflectivity property of the junction can be described in terms of a purely reactive network as defined by Z. The reflections induced by the network on the transmission lines will be identical with those induced in the actual guide by the geometrical discontinuity.

Notice that we have arbitrarily chosen the characteristic impedance of each guide as 1. This is neither necessary nor natural. Introduce a transformed expression by [cf. (6.104a) and (6.104b)]

$$V_i' = \sqrt{Z_i} V_i , \qquad I_i' = \frac{1}{\sqrt{Z_i}} I_i . \tag{13.17}$$

Then the new characteristic impedance in the ith guide is Z_i. In matrix form, introduce the diagonal matrix

$$z^{1/2} = \begin{pmatrix} Z_i^{1/2} & 0 & 0 & . & . \\ 0 & Z_2^{1/2} & 0 & . & . \\ 0 & 0 & Z_3^{1/2} & . & . \\ . & . & . & . & . \\ . & . & . & . & . \end{pmatrix} , \tag{13.18}$$

so that

$$V' = z^{1/2}V \ , \qquad I' = z^{-1/2}I \ . \tag{13.19}$$

Hence, $V' = z^{1/2}Zz^{1/2}I'$, or the new impedance matrix is

$$Z' = z^{1/2}Zz^{1/2} \ , \qquad Z'_{ij} = Z_{ij}\sqrt{Z_i Z_j} \ , \tag{13.20}$$

which is still symmetric and purely imaginary, and represents an equally valid description. It would, in fact, be better to introduce an arbitrary impedance, from the beginning. Thus for the distant fields in the ith guide, we write, instead of (13.5a) and (13.5b),

$$\mathbf{E}_\perp = \boldsymbol{\mathcal{E}}_i Z_i^{1/2}\left(\alpha_i e^{i\kappa_i z_i} + \beta_i e^{-i\kappa_i z_i}\right) \ , \tag{13.21a}$$

$$\mathbf{H}_\perp = \boldsymbol{\mathcal{H}}_i Z_i^{-1/2}\left(\alpha_i e^{i\kappa_i z_i} - \beta_i e^{-i\kappa_i z_i}\right) \ , \tag{13.21b}$$

with the normalization (13.8), with the normal toward the junction. Define

$$V_i(z_i) = Z_i^{1/2}\left(\alpha_i e^{i\kappa_i z_i} + \beta_i e^{-i\kappa_i z_i}\right) \ , \tag{13.22a}$$

$$I_i(z_i) = Z_i^{-1/2}\left(\alpha_i e^{i\kappa_i z_i} - \beta_i e^{-i\kappa_i z_i}\right) \ , \tag{13.22b}$$

with the $z_i = 0$ values

$$V_i = Z_i^{1/2}(\alpha_i + \beta_i) \ , \tag{13.23a}$$

$$I_i = Y_i^{1/2}(\alpha_i - \beta_i) \ , \qquad Y_i = 1/Z_i \ . \tag{13.23b}$$

The net power flowing into the junction is

$$\frac{1}{2}\mathrm{Re}\int d\boldsymbol{\sigma} \cdot \mathbf{E} \times \mathbf{H}^* = \frac{1}{2}\sum_i \left(|\alpha_i|^2 - |\beta_i|^2\right) = \frac{1}{2}\left[(\psi, \psi) - (\phi, \phi)\right]$$

$$= \frac{1}{2}\mathrm{Re}\sum_i V_i I_i^* \ . \tag{13.24}$$

As before, if the system is dissipationless, this must vanish, so $\mathsf{S}^*\mathsf{S} = 1$, or $\mathrm{Re}\, Z_{ij} = 0$. Now

$$z^{-1/2}V = (1 + \mathsf{S})\psi \ , \qquad z^{1/2}I = (1 - \mathsf{S})\psi \ , \tag{13.25}$$

which implies

$$V = z^{1/2}\frac{1 + \mathsf{S}}{1 - \mathsf{S}}z^{1/2}I \ , \tag{13.26}$$

that is, the impedance and admittance matrices are given by

$$Z = z^{1/2}\frac{1 + \mathsf{S}}{1 - \mathsf{S}}z^{1/2} \ , \qquad Y = y^{1/2}\frac{1 - \mathsf{S}}{1 + \mathsf{S}}y^{1/2} \ , \tag{13.27}$$

which may be solved for the scattering matrix:

$$S = z^{-1/2}\frac{Z-z}{Z+z}z^{1/2} = y^{-1/2}\frac{y-Y}{y+Y}y^{1/2} \,, \qquad (13.28)$$

a matrix generalization of the usual simple relation between reflection coefficients and impedance or admittance. Notice that in addition to the arbitrariness in the assigned value of Z_i, one also has the possibility of reversing the positive direction of current and voltage in any channel. This does not change the characteristic impedance and is equivalent to changing the sign of $Z_i^{1/2}$, which of course leaves the impedance matrix symmetric and imaginary. Hence in the matrix $z^{1/2}$, the signs of $Z_i^{1/2}$ can be chosen arbitrarily. A further degree of arbitrariness lies in the choice of reference point in each of the arms. One is at liberty (and it may be convenient) to shift the choice in reference point, thereby altering the definition of current and voltage and transforming the impedance matrix, but of course without altering in the slightest its general properties or the information contained therein.

13.3 Normal Modes

This transformation and other algebraic manipulations with the impedance matrix are facilitated by introducing the concept of normal modes for the system. By a normal mode we shall understand an assignment of incident fields in each guide such that the amplitude of the reflected field in each guide equals the amplitude of the incident field times a constant which is the same for all guides, that is, the entire reflected field is a multiple of the incident field, $\beta_i = C\alpha_i$. If energy conservation is to hold, there can only be a phase shift, that is, $C = e^{2i\vartheta}$. Thus if we choose a vector ψ such that $\phi = C\psi$, this means from (13.7) that we have chosen a characteristic vector (or eigenvector) of the scattering matrix. One knows that for an n-dimensional unitary matrix there are n such and they can be chosen normalized and orthogonal. Hence, let the characteristic vectors of the scattering matrix be $\psi^{(\lambda)}$, with

$$(\psi^{(\lambda')}, \psi^{(\lambda)}) = \delta_{\lambda\lambda'} \,, \qquad (13.29)$$

and let the characteristic values be

$$S^{(\lambda)} = e^{2i\vartheta_\lambda} \,. \qquad (13.30)$$

It now follows that a normal mode is also such that the input impedance in each branch is the same, relative to the characteristic impedance of that branch. For, since from (13.25)

$$V = z^{1/2}(1+S)\psi \,, \qquad I = z^{-1/2}(1-S)\psi \,, \qquad (13.31)$$

in the λ normal mode,

$$V = z^{1/2}(1+S^{(\lambda)})\psi^{(\lambda)}\,, \quad I = z^{-1/2}(1-S^{(\lambda)})\psi^{(\lambda)}\,, \quad \text{and} \quad V = z\frac{1+S^{(\lambda)}}{1-S^{(\lambda)}}I\,.$$
$$(13.32)$$

Thus, the impedance matrix is diagonal, with the λth element

$$\left(\frac{Z}{z}\right)^{(\lambda)} = \frac{1+S^{(\lambda)}}{1-S^{(\lambda)}}\,, \quad \left(\frac{Y}{y}\right)^{(\lambda)} = \frac{1-S^{(\lambda)}}{1+S^{(\lambda)}}\,. \tag{13.33}$$

The utility of this concept comes from the fact that a matrix may be constructed directly if its characteristic values and characteristic vectors are given. Further, any algebraic function of that matrix can be constructed with equal ease. Thus, we assert that

$$S_{ij} = \sum_\lambda S^{(\lambda)}\psi_i^{(\lambda)}\psi_j^{(\lambda)*}\,. \tag{13.34}$$

We must show that the matrix thus defined also has the correct characteristic vectors and values. Now

$$(S\psi^{(\lambda)})_i = \sum_j S_{ij}\psi_j^{(\lambda)} = \sum_{\lambda',j} S^{(\lambda')}\psi_i^{(\lambda')}\psi_j^{(\lambda')*}\psi_j^{(\lambda)} = S^{(\lambda)}\psi_i^{(\lambda)}\,, \tag{13.35}$$

in virtue of the orthogonality and normalization of the $\psi^{(\lambda)}$s, (13.29). Any algebraic function of S, $f(S)$ has the same characteristic vectors, and the characteristic values are $f(S^{(\lambda)})$, that is,

$$[f(S)]_{ij} = \sum_\lambda f(S^{(\lambda)})\psi_i^{(\lambda)}\psi_j^{(\lambda)*}\,. \tag{13.36}$$

In particular

$$z^{-1/2}Zz^{-1/2} = \frac{1+S}{1-S} \quad \text{and} \quad y^{-1/2}Yy^{-1/2} = \frac{1-S}{1+S} \tag{13.37}$$

are algebraic functions of S and therefore

$$\left[z^{-1/2}Zz^{-1/2}\right]_{ij} = \frac{Z_{ij}}{\sqrt{Z_iZ_j}} = \sum_\lambda \frac{1+S^{(\lambda)}}{1-S^{(\lambda)}}\psi_i^{(\lambda)}\psi_j^{(\lambda)*}$$
$$= i\sum_\lambda \cot\vartheta_\lambda\psi_i^{(\lambda)}\psi_j^{(\lambda)*}\,, \tag{13.38a}$$

$$\left[y^{-1/2}Yy^{-1/2}\right]_{ij} = \frac{Y_{ij}}{\sqrt{Y_iY_j}} = \sum_\lambda \frac{1-S^{(\lambda)}}{1+S^{(\lambda)}}\psi_i^{(\lambda)}\psi_j^{(\lambda)*}$$
$$= -i\sum_\lambda \tan\vartheta_\lambda\psi_i^{(\lambda)}\psi_j^{(\lambda)*}\,. \tag{13.38b}$$

This method is particularly useful in situations of high symmetry where the characteristic vectors can be inferred by inspection.

13.3.1 Shift of Reference Point

Consider now the change in the form of the impedance matrix produced by a shift in the reference point. If the reference point in the ith guide is shifted back a distance l_i, that is, $z_i = z_i' - l_i$, the voltage (13.22a) changes by

$$V_i(z) \rightarrow Z_i^{1/2} \left(\alpha_i' e^{i\kappa_i z_i'} + \beta_i' e^{-i\kappa_i z_i'} \right) , \tag{13.39}$$

where

$$\alpha_i' = \alpha_i e^{-i\kappa_i l_i} , \qquad \beta_i' = \beta_i e^{i\kappa_i l_i} . \tag{13.40}$$

In terms of the unitary matrix defined by $(\mathsf{U})_{ij} = e^{i\kappa_i l_i} \delta_{ij}$, $\psi' = \mathsf{U}^{-1}\psi$, $\phi' = \mathsf{U}\phi$, and therefore the scattering matrix is transformed into $\mathsf{S}' = \mathsf{U}\mathsf{S}\mathsf{U}$, which is *not* a unitary transformation. In general, this leads to a very complicated transformation of the impedance matrix. Consider first, however, the simple situation in which all the $\kappa_i l_i = \varphi$ are the same. Then $\mathsf{U} = e^{i\varphi}\mathbf{1}$ and $\mathsf{S}' = e^{2i\varphi}\mathsf{S}$. Since the scattering matrix is changed only by a factor, the characteristic vectors are unaltered and the characteristic values are changed to

$$S^{(\lambda)'} = e^{2i\varphi} S^{(\lambda)} = e^{2i(\vartheta_\lambda + \varphi)} , \tag{13.41}$$

and therefore the new impedance matrix is given simply by

$$\frac{Z_{ij}'}{\sqrt{Z_i Z_j}} = i \sum_\lambda \cot(\vartheta_\lambda + \varphi) \psi_i^{(\lambda)} \psi_j^{(\lambda)*} . \tag{13.42}$$

This is a matrix generalization of the usual impedance transformation from one point to another. For just one guide, with $Z_i \rightarrow 3$, this becomes

$$Z' = i3 \cot(\vartheta + \varphi) = i3\frac{\cot\vartheta \cot\varphi - 1}{\cot\vartheta + \cot\varphi} = i3\frac{\frac{Z}{i3} - \tan\kappa l}{1 + \frac{Z}{i3}\tan\kappa l} , \tag{13.43}$$

or

$$Z' = \frac{Z - i3 \tan\kappa l}{1 - i\frac{Z}{3}\tan\kappa l} , \tag{13.44}$$

which is the standard transformation.

Consider next the situation in which only one reference point is shifted, that is,

$$\mathsf{U} = \begin{pmatrix} e^{i\varphi} & 0 & 0 & . & . \\ 0 & 1 & 0 & . & . \\ 0 & 0 & 1 & . & . \\ . & . & . & . & . \end{pmatrix} . \tag{13.45}$$

Now the scattering matrix is essentially altered and the characteristic vectors are altered. We shall proceed by trying to find the new characteristic vectors. Let a new characteristic vector ψ' be expressed in terms of the old $\psi^{(\lambda)}$s by

$$\psi' = \sum_\lambda c_\lambda U^{-1} \psi^{(\lambda)} \, . \tag{13.46}$$

The equation to determine the new characteristic values of S is

$$S'\psi' = S'\psi' = USU\psi' = US \sum_\lambda c_\lambda \psi^{(\lambda)} = U \sum_\lambda S^{(\lambda)} c_\lambda \psi^{(\lambda)} \, , \tag{13.47}$$

or

$$S' \sum_\lambda c_\lambda \psi^{(\lambda)} = U^2 \sum_\lambda S^{(\lambda)} c_\lambda \psi^{(\lambda)} \, . \tag{13.48}$$

Hence

$$S'c_\lambda = \sum_{\lambda'} S^{(\lambda')} c_{\lambda'} (\psi^{(\lambda)}, U^2 \psi^{(\lambda')}) \, , \tag{13.49}$$

or

$$\left[S' - S^{(\lambda)} \right] c_\lambda = \sum_{\lambda'} S^{(\lambda')} c_{\lambda'} \left(e^{2i\varphi} - 1 \right) \psi_1^{(\lambda)*} \psi_1^{(\lambda')} \, . \tag{13.50}$$

Hence, we may put

$$\left[S' - S^{(\lambda)} \right] c_\lambda = \psi_1^{(\lambda)*} A \, , \tag{13.51}$$

and the characteristic value equation to determine the new S's is

$$1 = \left[\sum_\lambda S^{(\lambda)} \frac{|\psi_1^{(\lambda)}|^2}{S' - S^{(\lambda)}} \right] \left(e^{2i\varphi} - 1 \right) \, . \tag{13.52}$$

Having found these, the characteristic vectors are given from (13.46) by

$$\psi' = A \sum_\lambda \frac{\psi_1^{(\lambda)*}}{S' - S^{(\lambda)}} U^{-1} \psi^{(\lambda)} \, , \tag{13.53}$$

with A determined by $(\psi', \psi') = 1$, or

$$
1 = |A|^2 \sum_\lambda \frac{|\psi_1^{(\lambda)}|^2}{|S' - S^{(\lambda)}|^2} = -|A|^2 S' \sum_\lambda \frac{S^{(\lambda)} |\psi_1^{(\lambda)}|^2}{(S' - S^{(\lambda)})^2}
$$
$$
= |A|^2 S' \frac{d}{dS'} \frac{1}{e^{2i\phi} - 1} = -|A|^2 S' \frac{2i e^{2i\varphi}}{(e^{2i\varphi} - 1)^2} \frac{d\varphi}{dS'} \, , \tag{13.54}
$$

using (13.52) and the fact that $|S'|^2 = |S^{(\lambda)}|^2 = 1$. The normalization factor is therefore given by

$$|A|^2 = -2i \sin^2 \varphi \frac{1}{S'} \frac{dS'}{d\varphi} \, , \tag{13.55}$$

or writing $S' = e^{2i\vartheta'}$,

$$|A|^2 = 4 \sin^2 \varphi \frac{d\vartheta'}{d\varphi} \, . \tag{13.56}$$

13.3.2 Lumped Network Description

This process combined with the previous ones enables one to shift each reference point. However, the above processes are somewhat intricate and it may be advantageous to make use of the fact that shifting a reference point back a distance l is equivalent to adding just this length of transmission line to the system being represented by a lumped network. Now a length of transmission line can be represented by a network which can then be added to the original network and visually reduced back to the original form. Consider then the simple illustrative problem of an equivalent network for a simple waveguide of length l. For convenience choose the origin at the center. The voltage and current will be of standard form

$$V(z) = 3^{1/2} \left(\alpha e^{i\kappa z} + \beta e^{-i\kappa z} \right) , \qquad (13.57a)$$

$$I(z) = 3^{-1/2} \left(\alpha e^{i\kappa z} - \beta e^{-i\kappa z} \right) . \qquad (13.57b)$$

At either end of the waveguide one sees incident and reflected waves, of respective amplitudes

$$\alpha_1 = \alpha e^{-i\kappa l/2} , \qquad \beta_1 = \beta e^{i\kappa l/2} , \qquad (13.58a)$$

$$\alpha_2 = \beta e^{-i\kappa l/2} , \qquad \beta_2 = \alpha e^{i\kappa l/2} . \qquad (13.58b)$$

Thus

$$\beta_1 = e^{i\kappa l} \alpha_2 , \qquad \beta_2 = e^{i\kappa l} \alpha_1 , \qquad (13.59)$$

and the scattering matrix therefore has the form

$$S = \begin{pmatrix} 0 & e^{i\kappa l} \\ e^{i\kappa l} & 0 \end{pmatrix} . \qquad (13.60)$$

It is evident from symmetry that the characteristic vectors of this matrix are

$$\psi^{(1)} = \frac{1}{\sqrt{2}} \begin{pmatrix} 1 \\ 1 \end{pmatrix} , \qquad \psi^{(2)} = \frac{1}{\sqrt{2}} \begin{pmatrix} 1 \\ -1 \end{pmatrix} , \qquad (13.61)$$

and the characteristic values are

$$S^{(1)} = e^{i\kappa l} , \qquad S^{(2)} = -e^{i\kappa l} . \qquad (13.62)$$

Thus the characteristic values of the relative impedance matrix are from (13.33)

$$\left(\frac{Z}{3} \right)^{(1)} = i \cot \frac{\kappa l}{2} , \qquad \left(\frac{Z}{3} \right)^{(2)} = -i \tan \frac{\kappa l}{2} . \qquad (13.63)$$

The elements of the relative impedance matrix are from (13.38a)

$$\frac{Z_{11}}{3} = \frac{Z_{22}}{3} = \frac{1}{2} \left[\left(\frac{Z}{3} \right)^{(1)} + \left(\frac{Z}{3} \right)^{(2)} \right] = i \cot \kappa l , \qquad (13.64a)$$

$$\frac{Z_{12}}{3} = \frac{Z_{21}}{3} = \frac{1}{2} \left[\left(\frac{Z}{3} \right)^{(1)} - \left(\frac{Z}{3} \right)^{(2)} \right] = i \csc \kappa l . \qquad (13.64b)$$

Similarly, the elements of the relative admittance matrix are from (13.38b)

$$\frac{Y_{11}}{y} = \frac{Y_{22}}{y} = i \cot \kappa l , \qquad \frac{Y_{12}}{y} = \frac{Y_{21}}{y} = -i \csc \kappa l .\qquad (13.65)$$

In terms of the impedance matrix, the length of line can be represented by the T section, shown in Fig. 13.2. Employing the admittance matrix we obtain a

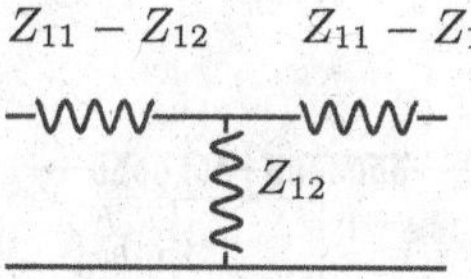

$$Z_{11} - Z_{12} \qquad Z_{11} - Z_{12}$$

Z_{12}

Fig. 13.2. T section. Here, $Z_{11} - Z_{12} = -i3 \tan \kappa l/2$ and $Z_{12} = i3 \csc \kappa l$

Π-section representation, shown in Fig. 13.3. (See Problem 13.2.) Either of

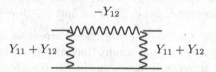

$$-Y_{12}$$

$$Y_{11} + Y_{12} \qquad\qquad Y_{11} + Y_{12}$$

Fig. 13.3. Π section. Here, $-Y_{12} = iy \csc \kappa l$ and $Y_{11} + Y_{12} = -iy \tan \kappa l/2$

these networks added to one of the terminals of an original network representing the system will produce a new network which corresponds to the reference point of that transmission line moved back a distance l. Of course, l can be negative (which is the same as shifting back a distance $\lambda_g/2 - l$). It is interesting to consider the form of the circuit for a length l such that $\kappa l \ll 1$. Then the T section becomes that shown in Fig. 13.2, with $Z_{11} - Z_{12} \to -i3\kappa l/2$ and $Z_{12} \to i3/\kappa l$. If the field quantities $\mathcal{E}$ and $\mathcal{H}$ are such that all the frequency dependence is absorbed into the characteristic impedance 3, then for an H mode (i.e., the lowest mode), from (6.36d),

$$3 = \frac{K}{c} \frac{\omega}{\kappa c} = \frac{K}{c} \frac{\omega}{\sqrt{\omega^2 - \omega_0^2}} , \qquad (13.66)$$

where ω_0 is the cutoff frequency and K is a pure number, an arbitrary function of the geometry. Hence the series impedance is

$$-i3\frac{\kappa l}{2} = -i\frac{Kl}{2c^2}\omega , \qquad (13.67)$$

which represents a pure inductance, see (5.5b). The shunt admittance is

$$-\mathrm{i}y\kappa l = -\mathrm{i}\frac{l}{K}\left(\omega - \frac{\omega_0^2}{\omega}\right) , \qquad (13.68)$$

a parallel combination of capacitance and inductance, see (5.6b). [Recall the discussion after (6.35b), where identical results are obtained if $K = 1/\varepsilon$.] Hence, the circuit for a short length of waveguide, sustaining an H mode, which describes its properties for all frequencies, subject to $\kappa l \ll 1$, is given by Fig. 13.4. Thus cutoff occurs at the parallel resonant frequency of the shunt

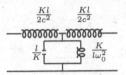

Fig. 13.4. Equivalent circuit for an H mode with $\kappa l \ll 1$

element. It is interesting that this circuit is a direct pictorial description of the property of an H mode. An H mode possesses longitudinal and transverse **H** fields and a transverse **E** field. Thus we find the direct correspondence:

- Longitudinal **H** $\equiv$ transverse conduction current $\Rightarrow$ shunt inductance,
- Transverse **H** $\equiv$ longitudinal conduction current $\Rightarrow$ series inductance,
- Transverse **E** $\equiv$ transverse displacement current $\Rightarrow$ shunt capacitance.

The cutoff is now seen to arise from the shorting action of the transverse conduction current. In the fundamental mode of a coax, there is no transverse conduction current and no cutoff. The Π-section representation of the above is shown in Fig. 13.5.

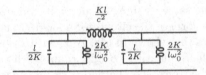

Fig. 13.5. Equivalent Π-section circuit for an H mode with $\kappa l \ll 1$

In a similar way for an E mode, the characteristic admittance is of the form, from (6.36b),

$$y = \frac{c}{K'}\frac{\omega}{\sqrt{\omega^2 - \omega_0^2}} , \qquad (13.69)$$

and the equivalent Π section is given in Fig. 13.6. Here the cutoff arises from the series resonance of the series element. The T-section representation is given in Fig. 13.7. [Identity with the results stated after (6.35b) is again achieved if $K' = 1/\varepsilon$.] We again find a correspondence between the properties of the mode and the circuit:

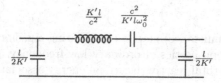

Fig. 13.6. Equivalent Π-section circuit for an E mode with $\kappa l \ll 1$

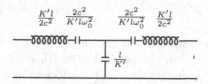

Fig. 13.7. Equivalent T-section circuit for an E mode with $\kappa l \ll 1$

- Longitudinal **E** $\equiv$ longitudinal displacement current $\Rightarrow$ series capacitance,
- Transverse **E** $\equiv$ transverse displacement current $\Rightarrow$ shunt capacitance,
- Transverse **H** $\equiv$ longitudinal conduction current $\Rightarrow$ series inductance.

We must show how to include the prescription of the incident field into the circuit. At $z = 0$, or λ_g back,

$$V = V^{\text{inc}} + V^{\text{refl}} , \qquad I = \mathrm{y}(V^{\text{inc}} - V^{\text{refl}}) , \qquad (13.70)$$

or

$$V + {}_3I = 2V^{\text{inc}} , \qquad I + \mathrm{y}V = 2I^{\text{inc}} . \qquad (13.71)$$

The fact that V^{inc} is independent of I, or I^{inc} independent of V can be represented by saying that the incident field is that produced by a constant voltage generated in series with the line with generator voltage $2V^{\text{inc}}$, or a constant current generator in shunt with the line, of generator current $2I^{\text{inc}}$, as shown in Fig. 13.8. In the first case, since the generator is of constant

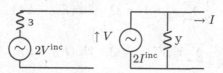

Fig. 13.8. Equivalent voltage and current generators

voltage it must be of zero impedance and

$$V = 2V^{\text{inc}} - {}_3I , \quad \text{or} \quad V + {}_3I = 2V^{\text{inc}} . \qquad (13.72)$$

In the second case, since the generator is of constant current it must be of zero admittance and

$$I = 2I^{\text{inc}} - \mathrm{y}V , \quad \text{or} \quad I + \mathrm{y}V = 2I^{\text{inc}} . \qquad (13.73)$$

13.3.3 Energy

The impedance (admittance) matrix representing a geometrical discontinuity has all the general properties of ordinary low-frequency impedances (admittances). Define the total reactive electrical energy associated with the discontinuity by

$$E_E = \frac{\varepsilon}{4} \int (\mathrm{dr}) |\mathbf{E}|^2 - \sum_i l_i \frac{k}{8c\kappa_i} \left(Y_i |V_i|^2 + Z_i |I_i|^2 \right) , \tag{13.74}$$

where the integral is extended over the entire region of the junction and out to a distance l_i away from the origin in the ith guide. We take l_i to be a half-integral number of guide wavelengths, sufficient in number that all higher modes are attenuated. Under this condition, E_E is independent of l_i. Similarly, the total reactive magnetic energy is

$$E_H = \frac{\mu}{4} \int (\mathrm{dr}) |\mathbf{H}|^2 - \sum_i l_i \frac{k}{8c\kappa_i} \left(Y_i |V_i|^2 + Z_i |I_i|^2 \right) . \tag{13.75}$$

Of course, the additional terms represent the total electric or magnetic energy associated with the propagating modes in the various guides, see (6.111). In addition, we shall define a quantity $\mathcal{Q}$ as the average amount of energy being dissipated per unit time in the region of the junction. Since we have neglected dissipation along the guides themselves (but see Sect. 13.6), it is not necessary to be precise about the region in question. Therefore, the dissipation (which is really the extra dissipation associated with the junction) is assumed to be localized. Now [see (6.115) and (6.106)]

$$\frac{1}{2} \sum_i V_i I_i^* = \mathcal{Q} - 2\mathrm{i}\omega(E_H - E_E) . \tag{13.76}$$

Hence,

$$\frac{1}{2} \sum_{ij} Z_{ij} I_i^* I_j = \mathcal{Q} - 2\mathrm{i}\omega(E_H - E_E) , \tag{13.77a}$$

$$\frac{1}{2} \sum_{ij} Y_{ij} V_i^* V_j = \mathcal{Q} + 2\mathrm{i}\omega(E_H - E_E) . \tag{13.77b}$$

Breaking the impedance into resistive and reactive parts (and the admittance into conductive and susceptive parts),

$$Z_{ij} = R_{ij} + \mathrm{i}X_{ij} , \qquad Y_{ij} = G_{ij} - \mathrm{i}B_{ij} , \tag{13.78}$$

we have

$$\frac{1}{2} \sum_{ij} R_{ij} I_i^* I_j = \frac{1}{2} \sum_{ij} G_{ij} V_i^* V_j = \mathcal{Q} , \tag{13.79a}$$

$$\frac{1}{2} \sum_{ij} X_{ij} I_i^* I_j = \frac{1}{2} \sum_{ij} B_{ij} V_i^* V_j = -2\omega(E_H - E_E) . \tag{13.79b}$$

Thus the resistive (conductive) matrix is positive definite ($Q > 0$). The reactance (susceptance) matrix is positive definite if the electric energy exceeds the magnetic. It is negative definite if the magnetic energy exceeds the electric. Furthermore, neglecting dissipation, from the energy theorem (see Problem 13.3)

$$\frac{1}{2}\sum_{ij}\frac{\partial X_{ij}}{\partial \omega}I_i^* I_j = -2(E_H + E_E)\,, \tag{13.80a}$$

$$\frac{1}{2}\sum_{ij}\frac{\partial B_{ij}}{\partial \omega}V_i^* V_j = 2(E_H + E_E)\,. \tag{13.80b}$$

Hence the frequency derivative of the reactance matrix is always negative definite, and the frequency derivative of the susceptance matrix is always positive definite. All these results are generalizations of well-known low-frequency theorems. (See also Sect. 6.6.) The essential difference is that at low frequencies there exist discontinuities (coils, condensers) with which are associated only magnetic or electric energy. This is not possible, except to a first approximation, when the dimensions are comparable with the wavelength. It is instructive to see how the assumption of pure magnetic or electric energy leads to the conventional results for a simple impedance. Thus if $E_H = 0$.

$$\frac{1}{2}B|V|^2 = 2\omega E_E\,, \quad \frac{1}{2}\frac{\partial B}{\partial \omega}|V|^2 = 2E_E\,, \quad \frac{\partial B}{\partial \omega} = \frac{B}{\omega}\,, \tag{13.81}$$

so $B = \omega C$ and we have a pure capacitive energy

$$E_E = \frac{1}{4}C|V|^2\,, \tag{13.82}$$

that is, $C > 0$. If $E_E = 0$,

$$\frac{1}{2}X|I|^2 = -2\omega E_H\,, \quad \frac{1}{2}\frac{\partial X}{\partial \omega}|I|^2 = -2E_H\,, \quad \frac{\partial X}{\partial \omega} = \frac{X}{\omega}\,, \tag{13.83}$$

so $X = -\omega L$ and we have a pure inductive energy

$$E_H = \frac{1}{4}L|I|^2\,, \tag{13.84}$$

that is, $L > 0$.

13.4 Variational Principle

The relation (13.79b) between reactance (susceptance) and the difference between magnetic and electric energy is one of the most fundamental equations of the theory for it forms the basis of the variational principle. Consider the change in the values of B_{ij} calculated from an electric field differing slightly from the true one, but with prescribed values of the voltages V_i, and satisfying correspondingly the boundary conditions on metallic surfaces. Thus

$$
\begin{aligned}
\frac{1}{2} \sum_{ij} \delta B_{ij} V_i^* V_j &= -\frac{\omega \varepsilon}{2} \delta \int (\mathrm{d}\mathbf{r}) \left[\frac{1}{k^2} |\mathbf{\nabla} \times \mathbf{E}|^2 - |\mathbf{E}|^2 \right] \\
&= -\frac{\omega \varepsilon}{2k^2} \int (\mathrm{d}\mathbf{r}) \left[\mathbf{\nabla} \times \delta \mathbf{E}^* \cdot \mathbf{\nabla} \times \mathbf{E} - k^2 \delta \mathbf{E}^* \cdot \mathbf{E} \right] + \text{c.c.} \\
&= -\frac{\omega \varepsilon}{2k^2} \Big\{ \oint \mathrm{d}\mathbf{S} \cdot \delta \mathbf{E}^* \times (\mathbf{\nabla} \times \mathbf{E}) \\
&\quad + \int (\mathrm{d}\mathbf{r}) \, \delta \mathbf{E}^* \cdot \left[\mathbf{\nabla} \times (\mathbf{\nabla} \times \mathbf{E}) - k^2 \mathbf{E} \right] + \text{c.c.} \Big\} .
\end{aligned}
$$

$$(13.85)$$

Since $\mathbf{E}$ is the correct field, satisfying the wave equation, the volume integral vanishes. The surface integral will vanish if $\mathbf{n} \times \mathbf{E}$ is zero on all metallic walls – the boundary condition must be satisfied – and if the asymptotic form of the fields is correct, for the value of $\mathbf{n} \times \mathbf{E}$ on the surface across a guide is fixed by the voltage and is thus prescribed. Under these conditions – which are that the deformed field still satisfies the boundary conditions on metallic walls and at large distances (i.e., that the asymptotic forms of the fields are correct), but need not satisfy the wave equation everywhere – the susceptance is stationary with respect to small variations for prescribed voltages. In a similar way, based on the magnetic field with the currents prescribed,

$$
\begin{aligned}
\frac{1}{2} \sum_{ij} \delta X_{ij} I_i^* I_j &= -\frac{\omega \mu}{2} \delta \int (\mathrm{d}\mathbf{r}) \left[|\mathbf{H}|^2 - \frac{1}{k^2} |\mathbf{\nabla} \times \mathbf{H}|^2 \right] \\
&= \frac{\omega \mu}{2k^2} \int (\mathrm{d}\mathbf{r}) \left[\mathbf{\nabla} \times \delta \mathbf{H}^* \cdot \mathbf{\nabla} \times \mathbf{H} - k^2 \delta \mathbf{H}^* \cdot \mathbf{H} \right] + \text{c.c.} \\
&= \frac{\omega \mu}{2k^2} \Big\{ \oint \mathrm{d}\mathbf{S} \cdot \delta \mathbf{H}^* \times (\mathbf{\nabla} \times \mathbf{H}) \\
&\quad + \int (\mathrm{d}\mathbf{r}) \, \delta \mathbf{H}^* \cdot \left[\mathbf{\nabla} \times (\mathbf{\nabla} \times \mathbf{H}) - k^2 \mathbf{H} \right] + \text{c.c.} \Big\} .
\end{aligned}
$$

$$(13.86)$$

Since $\mathbf{H}$ is the correct field, the volume integral vanishes, and $\mathbf{n} \times (\mathbf{\nabla} \times \mathbf{H})$ is zero on all metallic walls. In order that the integral across the section of a guide vanishes, it is necessary that the field have the correct asymptotic form for then $\mathbf{n} \times \delta \mathbf{H}$ is zero since the current is prescribed. Thus, under just the condition that the asymptotic form of the field be correct, but that neither the wave equation or the boundary condition be satisfied, the reactance is stationary with respect to small variations for prescribed currents.

13.5 Bifurcated Guide

Now consider a waveguide which bifurcates into two parallel guides at $z = 0$, as diagrammed in Fig. 13.9. The simplicity of this situation is two fold. First,

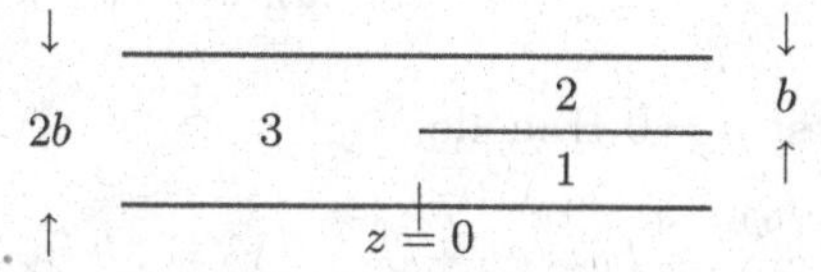

Fig. 13.9. Bifurcated guide. The width of the two guides on the right, labeled 1 and 2, is half that of the guide on the left, labeled 3. The extent of the guide in the x-direction, perpendicular to the page, is larger than the dimensions in the depicted y-direction

for y being the coordinate in the plane shown transverse to the guide axes,

$$\int_3 dy\, E_y = \int_1 dy\, E_y + \int_2 dy\, E_y , \tag{13.87}$$

where the integration is extended over the plane of the junction, where $z = 0$. Second, an H-mode wave moving to the right in region 3 is completely transmitted. In region 3, in terms of the Green's function vanishing on the boundaries of region 3, including the junction plane $z = 0$ (see Problem 13.4)

$$\psi = \alpha_3 \left(e^{i\kappa z} + e^{-i\kappa z} \right) - \frac{1}{2b} \sum_{n=-\infty}^{\infty} \int_3 dy' \frac{\cos\frac{n\pi y}{2b} \cos\frac{n\pi y'}{2b}}{i\sqrt{\kappa^2 - \left(\frac{n\pi}{2b}\right)^2}} e^{-i\sqrt{\kappa^2 - \left(\frac{n\pi}{2b}\right)^2}\, z} \frac{\partial\psi}{\partial z}(y') , \tag{13.88}$$

excluding the common x-dependence, so for the $n = 0$ mode

$$\beta_3 = \alpha_3 + \frac{i}{2\kappa b} \int_3 dy' \frac{\partial\psi}{\partial z} , \quad I_3 = \alpha_3 + \beta_3 , \quad V_3 = 2\kappa b(\alpha_3 - \beta_3) = -i \int_3 dy' \frac{\partial\psi}{\partial z} , \tag{13.89}$$

where we have made a convenient choice for the current and voltage, including the characteristic impedance. In region 1

$$\psi = \alpha_1 \left(e^{i\kappa z} + e^{-i\kappa z} \right) + \frac{1}{b} \sum_{n=-\infty}^{\infty} \int_1 dy' \frac{\cos\frac{n\pi y}{b} \cos\frac{n\pi y'}{b}}{i\sqrt{\kappa^2 - \left(\frac{n\pi}{b}\right)^2}} e^{i\sqrt{\kappa^2 - \left(\frac{n\pi}{b}\right)^2}\, z} \frac{\partial\psi}{\partial z}(y') , \tag{13.90}$$

so for the lowest H mode

$$\beta_1 = \alpha_1 - \frac{i}{\kappa b} \int_1 dy' \frac{\partial\psi}{\partial z} , \quad I_1 = \alpha_1 + \beta_1 , \quad V_1 = \kappa b(\alpha_1 - \beta_1) = i \int_1 dy' \frac{\partial\psi}{\partial z} . \tag{13.91}$$

In region 2

$$\psi = \alpha_2 \left(e^{i\kappa z} + e^{-i\kappa z}\right) + \frac{1}{b}\sum_{n=-\infty}^{\infty}\int_2 dy' \frac{\cos\frac{n\pi}{b}(y-b)\cos\frac{n\pi}{b}(y'-b)}{i\sqrt{\kappa^2 - \left(\frac{n\pi}{b}\right)^2}}$$
$$\times e^{i\sqrt{\kappa^2 - \left(\frac{n\pi}{b}\right)^2}z}\frac{\partial\psi}{\partial z}(y') , \qquad (13.92)$$

so again for the lowest $n = 0$ H mode

$$\beta_2 = \alpha_2 - \frac{i}{\kappa b}\int_2 dy'\frac{\partial\psi}{\partial z} , \quad I_2 = \alpha_2 + \beta_2 , \quad V_2 = \kappa b(\alpha_2 - \beta_2) = i\int_2 dy'\frac{\partial\psi}{\partial z} . \qquad (13.93)$$

Therefore, because $\partial\psi/\partial z$ is continuous at $z = 0$,

$$V_1 + V_2 + V_3 = 0 . \qquad (13.94)$$

The boundary condition that ψ be continuous at $z = 0$ reads

$$I_3 + \frac{1}{b}\sum_{n=1}^{\infty}\int_3 dy'\frac{\cos\frac{n\pi y}{2b}\cos\frac{n\pi y'}{2b}}{\sqrt{\left(\frac{n\pi}{2b}\right)^2 - \kappa^2}}\frac{\partial\psi}{\partial z} = I_{1,2} - \frac{2}{b}\sum_{n=1}^{\infty}\int_{1,2} dy'\frac{\cos\frac{n\pi y}{b}\cos\frac{n\pi y'}{b}}{\sqrt{\left(\frac{n\pi}{b}\right)^2 - \kappa^2}}\frac{\partial\psi}{\partial z} . \qquad (13.95)$$

Consider the even situation, $I_1 = I_2$, $V_1 = V_2$. The obvious solution is $\partial\psi/\partial z = $ constant. Then $I_3 = I_1$, which, together with $V_3 + 2V_1 = 0$ implies that $\beta_1 = \alpha_3$, $\beta_3 = \alpha_1$, that is, complete transmission. In the odd situation, $V_1 = -V_2$, $I_1 = -I_2$, $\partial\psi/\partial z(1) = -\partial\psi/\partial z(2)$ at corresponding points. Hence only odd modes are expected in the guide 3 and $V_3 = I_3 = 0$. The integral equation is

$$\frac{1}{b}\sum_{n\,\text{odd}}\int_3 dy'\frac{\cos\frac{n\pi y}{2b}\cos\frac{n\pi y'}{2b}}{\sqrt{\left(\frac{n\pi}{2b}\right)^2 - \kappa^2}}\frac{\partial\psi}{\partial z} = I_1 - \frac{2}{b}\sum_{n=1}^{\infty}\int_1 dy'\frac{\cos\frac{n\pi y}{b}\cos\frac{n\pi y'}{b}}{\sqrt{\left(\frac{n\pi}{b}\right)^2 - \kappa^2}}\frac{\partial\psi}{\partial z} ,$$
$$y < b , \qquad (13.96a)$$

$$= -I_1 - \frac{2}{b}\sum_{n=1}^{\infty}\int_2 dy'\frac{\cos\frac{n\pi y}{b}\cos\frac{n\pi y'}{b}}{\sqrt{\left(\frac{n\pi}{b}\right)^2 - \kappa^2}}\frac{\partial\psi}{\partial z} ,$$
$$y > b . \qquad (13.96b)$$

The admittance matrix has the form

$$Y = \begin{pmatrix} Y_{11} & Y_{12} & Y_{13} \\ Y_{12} & Y_{11} & Y_{13} \\ Y_{13} & Y_{13} & Y_{33} \end{pmatrix} . \qquad (13.97)$$

For the even case,

$$I_1 = (Y_{11} + Y_{12})V_1 + Y_{13}V_3 , \qquad (13.98a)$$
$$I_3 = 2Y_{13}V_1 + Y_{33}V_3 . \qquad (13.98b)$$

The solution of this must be $I_3 = I_1$, $V_3 = -2V_1$, so

$$(Y_{11} + Y_{12} - 2Y_{13})V_1 = (Y_{33} - Y_{13})V_3 = (2Y_{13} - 2Y_{33})V_1 , \tag{13.99}$$

so

$$\frac{Y_{11} + Y_{12}}{2} + Y_{33} - 2Y_{13} = 0 \tag{13.100}$$

is a necessary condition. Then

$$I_1 = -2(Y_{33} - Y_{13})V_1 + Y_{13}(V_3 + 2V_1) , \tag{13.101a}$$
$$I_3 = (Y_{33} - Y_{13})V_3 + Y_{13}(V_3 + 2V_1) , \tag{13.101b}$$

implying

$$I_1 - I_3 = (Y_{13} - Y_{33})(V_3 + 2V_1) . \tag{13.102}$$

Thus $V_3 + 2V_1 = 0$ implies $I_1 = I_3$, provided $Y_{13} - Y_{33}$ is finite, and if $Y_{13} - Y_{33}$ is finite, $|Y_{13}| = \infty$ requires $V_3 + 2V_1 = 0$. Thus necessary and sufficient conditions are that

$$|Y_{13}| = \infty , \quad \text{and} \quad \frac{Y_{11} + Y_{12}}{2} - Y_{13} = Y_{13} - Y_{33} \quad \text{is finite} . \tag{13.103}$$

In the odd case, of course,

$$V_1 = -V_2 , \quad I_1 = -I_2 , \quad I_3 = V_3 = 0 , \tag{13.104}$$

and

$$I_1 = (Y_{11} - Y_{12})V_1 . \tag{13.105}$$

Now describe the above even situation in terms of the general six-terminal circuit shown in Fig. 13.10. Here Y_{13} is infinite and the top arm can be replaced

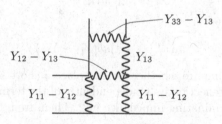

Fig. 13.10. General six-terminal circuit describing the bifurcated guide

by a single admittance of

$$Y_{12} + Y_{33} - 2Y_{13} = \left(\frac{Y_{11} + Y_{12}}{2} + Y_{33} - 2Y_{13} \right) - \frac{Y_{11} - Y_{12}}{2} = -\frac{Y_{11} - Y_{12}}{2} , \tag{13.106}$$

according to (13.100). Hence, introducing an impedance relative to guide 1,

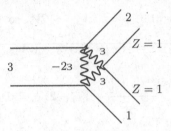

Fig. 13.11. Six-terminal circuit corresponding to Fig. 13.9

$$\frac{1}{3} = \kappa b(Y_{11} - Y_{12}) \,, \tag{13.107}$$

the circuit becomes that shown in Fig. 13.11. Let us first inquire whether it is possible to terminate 3 by a plunger corresponding to an impedance Z_t and match all the power from 1 to 2, with 2 matched of course. Hence from guide 1 we see the successive simplifications of impedances shown in Fig. 13.12, which implies the impedance

$$\begin{aligned}
Z &= \frac{3\left(\frac{3}{3+1} - \frac{23Z_t}{Z_t - 23}\right)}{3 + \frac{3}{3+1} - \frac{23Z_t}{Z_t - 23}} = 3\frac{\frac{1}{3+1} - \frac{2Z_t}{Z_t - 23}}{1 + \frac{1}{3+1} - \frac{2Z_t}{Z_t - 23}} \\
&= 3\frac{(Z_t - 23) - 2Z_t(3+1)}{(3+1)(Z_t - 23) + (Z_t - 23) - 2Z_t(3+1)} = \frac{Z_t(23 + 1) + 23}{Z_t + 2(3 + 2)} \,.
\end{aligned} \tag{13.108}$$

The condition that this be unity is $Z_t = 2/3$, which is pure imaginary! Hence, the plunger must be placed back a distance l such that

$$Z_t = -2\mathrm{i}\tan \kappa l = \frac{2}{3} \,, \tag{13.109}$$

or

$$\tan \kappa l = \frac{\mathrm{i}}{3} = \mathrm{i}\kappa b(Y_{11} - Y_{12}) \,. \tag{13.110}$$

Second, let us inquire whether we can place a short somewhere in 2 and produce complete reflection in 3 independently of the termination in 1. Let Z'_t be the required terminating impedance in 2. Then, from the point of view of

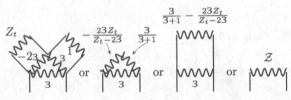

Fig. 13.12. Impedance of plunger, Z_t, such that all power from 1 is matched to 2

Fig. 13.13. Short Z'_t in 2 required to produce total reflection in 3

3, we have the situation shown in Fig. 13.13. Evidently, if 3 and Z'_t combine to give infinite impedance, no current flows in the second branch independently of the termination in 1, provided that one does not have "resonance." Thus

$$Z'_t = -3 = -\mathrm{i} \tan \kappa l' , \qquad (13.111)$$

or

$$\tan \kappa l' = -\mathrm{i}3 = \frac{3}{\mathrm{i}} = \cot \kappa l , \qquad (13.112)$$

so

$$l' = \frac{\pi}{2\kappa} - l = \frac{\lambda_g}{4} - l , \qquad (13.113)$$

or any equivalent length. The input impedance under these conditions is

$$-23 = -2\mathrm{i} \tan \kappa l' = -2\mathrm{i} \cot \kappa l . \qquad (13.114)$$

How far x back do we have to go in guide 3 to see zero impedance? For this we require

$$0 = \frac{-2\mathrm{i} \cot \kappa l - 2\mathrm{i} \tan \kappa x}{1 - \mathrm{i}(-\mathrm{i}) \cot \kappa l \tan \kappa x} , \quad \tan \kappa x = -\cot \kappa l , \quad x = \frac{\lambda_g}{4} + l . \quad (13.115)$$

This is illustrated in Fig. 13.14.

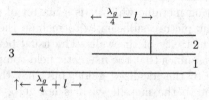

Fig. 13.14. The vertical arrow indicates the effective position of the short in guide 3, resulting from a short in guide 2

Let us combine these ideas into the waveguide shown in Fig. 13.15. Under these conditions all power is transferred from guide b to guide c, if the latter is matched, independent of the condition in a, because there is an effective short a distance l to the left of the b–c junction. Let us finally inquire what

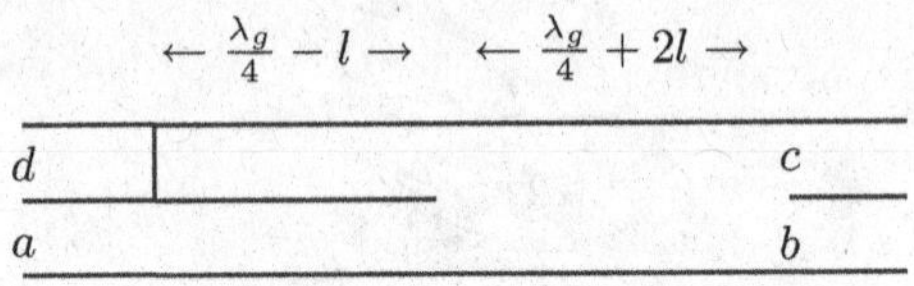

$$\leftarrow \frac{\lambda_g}{4} - l \rightarrow \qquad \leftarrow \frac{\lambda_g}{4} + 2l \rightarrow$$

$$
\begin{array}{ll}
d & c \\
a & b
\end{array}
$$

Fig. 13.15. Junction of two bifurcated waveguides with a short

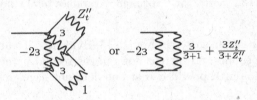

Fig. 13.16. Equivalent impedance for the waveguide configuration shown in Fig. 13.15 for reflection in a for b matched and an arbitrary terminating impedance in c

is the reflection in a for b matched and an arbitrary terminating impedance Z_t'' in c. At the b–c junction we see the impedance shown in Fig. 13.16. The equivalent impedance is

$$\mathcal{Z} = \frac{-23^2 \left(\frac{1}{3+1} + \frac{Z_t''}{3+Z_t''} \right)}{3 \left(-2 + \frac{1}{3+1} + \frac{Z_t''}{3+Z_t''} \right)} = 2 \frac{\frac{1}{3+1} + \frac{Z_t''}{3+Z_t''}}{\frac{1}{3+1} + \frac{1}{3+Z_t''}} \, . \tag{13.116}$$

We will let the reader draw the evident conclusions.

13.6 Imperfect Conducting Walls

The preceding discussion has dealt exclusively with idealized guides composed of perfectly conducting metallic walls. It is a matter of practical importance to extend the circuit picture sufficiently to include the attenuating results from the finite conductivity of the walls. The usual perturbative treatment of skin effect losses assumes that the magnetic field components do not differ sensibly from those of the idealized guide, but that the component of the electric field tangent to the metallic wall is not zero, being related to the tangential magnetic field component by (5.14), or

$$\mathbf{n} \times \mathbf{E} = \zeta \frac{k\delta}{2}(1 - \mathrm{i})\mathbf{H} \, , \tag{13.117}$$

where δ is the skin depth, related to the conductivity by

$$\delta = \sqrt{\frac{2}{\mu\omega\sigma}} \, , \tag{13.118}$$

and $\mathbf{n}$ is the normal to the wall. The attenuation is then obtained by computing the average flow of power into the walls, per unit length. We obtain the circuit representation of this consideration by introducing resistive elements into the distributive parameter circuit, but neglecting the changes in the reactive elements produced by the finite conductivity. This constitutes an excellent approximation in the usual circumstance that the ratio of the skin depth to the transverse guide dimension is an exceedingly small number.

The average power dissipated per unit length is obtained by evaluating the real part of the line integral, taken around the guide periphery, of the component of the complex Poynting vector normal to the walls, that is, (5.13), or

$$P_{\text{diss}} = \text{Re}\,\frac{1}{2}\oint_C ds\,(\mathbf{E}\times\mathbf{H}^*)\cdot\mathbf{n} = \frac{k\delta}{4}\zeta\oint_C ds\,|\mathbf{H}|^2\ . \qquad (13.119)$$

This integral, to be computed from the fields of the idealized guide, involves only the component of $\mathbf{H}$ tangential to the metal; the normal component is zero in virtue of the boundary condition on $\mathbf{E}$.

To illustrate these ideas, we consider the simpler situation of E modes, when only one component of $\mathbf{H}$ contributes to the dissipation, associated with the fact that there exists only a longitudinal conduction current. Hence from (6.32b), taking into account the normalization (6.57),

$$P_{\text{diss}} = \frac{k\delta}{4}\oint_C ds\,(\partial_n\varphi)^2\zeta|I|^2 = \frac{1}{4}k\delta\frac{\oint_C ds\,(\partial_n\varphi)^2}{\gamma'^2\int d\sigma\,\varphi^2}\zeta|I|^2 = \frac{1}{2}\mathcal{R}|I|^2 \qquad (13.120)$$

which defines a series resistance per unit length, generalizing (5.16),

$$\mathcal{R} = \frac{\zeta k\delta}{2}\frac{\oint_C ds\,(\partial_n\varphi)^2}{\gamma'^2\int d\sigma\,\varphi^2}\ . \qquad (13.121)$$

Accordingly, the distributed parameter circuit representing an E mode is now defined by

$$Z_s = \mathcal{R} - i\omega L + \frac{i}{\omega C'}\ , \qquad Y_\perp = -i\omega C\ , \qquad (13.122)$$

and the lumped constant network representing a short length Δz is shown in Fig. 13.17. The line constants κ and Z are no longer real, the imaginary part

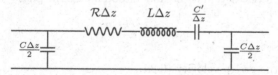

Fig. 13.17. Lumped waveguide section including dissipation in walls

of κ describing the attenuation associated with the skin-effect power loss. We shall write [cf. (5.20) and (5.22)]

$$-i\sqrt{(-Y_\perp)(-Z_s)} = \kappa + \frac{i\alpha}{2} \tag{13.123}$$

to retain the symbol κ for the real part of the propagation constant. Of course to the approximation considered here, it equals the propagation constant in the idealized guide. The current and voltage describing a wave propagating in the positive z-direction will now contain the factor

$$e^{i(\kappa+i\alpha/2)z} = e^{-\alpha z/2}e^{i\kappa z} . \tag{13.124}$$

Therefore, $\frac{1}{2}\alpha$ represents the attenuation constant for voltage or current, α is the power attenuation constant. The fact that the Z now possesses a small reactive component does not interest us, the only quantity of direct experimental value being the attenuation constant. We may regard the introduction of the quantity $\mathcal{R}$ into the series impedance per unit length as equivalent to changing the inductance by the small amount $i\mathcal{R}/\omega$. Hence

$$\alpha = \frac{2\mathcal{R}}{\omega}\frac{\partial\kappa}{\partial L} , \tag{13.125}$$

and remarking that

$$\kappa = \sqrt{LC\omega^2 - \frac{C}{C'}} , \tag{13.126}$$

we obtain using (6.36b), since $C = \varepsilon$,

$$\alpha = \frac{\omega}{\kappa}\mathcal{R}C = \frac{k}{\kappa}\frac{\mathcal{R}}{\zeta} = \frac{\mathcal{R}}{Z} , \tag{13.127}$$

which generalizes (5.22). The factor by which the power is reduced in a length of guide l of total resistance $R = \mathcal{R}l$, is $e^{-R/Z}$. Thus, the attenuation constant is the ratio of the series resistance per unit length to the characteristic impedance. We obtain another convenient representation by employing the time required to transverse a distance z, thus permitting the calculation of the conventional figure of merit, Q, as a measure of the power dissipated:

$$e^{-\alpha z} = e^{-\alpha vt} = e^{-\omega t/Q} . \tag{13.128}$$

Here v, the wave velocity, must be understood not as the phase velocity, (6.98),

$$u = \frac{\omega}{\kappa} = c\frac{k}{\kappa} > c , \tag{13.129}$$

but rather as the group velocity (6.99),

$$v = \frac{d\omega}{d\kappa} = c\frac{\kappa}{k} < c , \tag{13.130}$$

which is the measure of the velocity of energy transport. Hence

$$Q = \frac{\omega}{\alpha v} = \frac{\omega \zeta}{\mathcal{R}c} = \frac{\omega L}{\mathcal{R}} , \tag{13.131}$$

recalling that the series inductance of either an E or H mode is $L = \zeta/c = \mu$.

Now consider the change in the value of γ' on moving all points of the boundary inward a distance δn along the normal to the surface. Let φ be the original φ and $\overline{\varphi}$ be the new eigenfunction. Thus we have

$$\nabla \cdot (\overline{\varphi} \nabla \varphi - \varphi \nabla \overline{\varphi}) = (\overline{\gamma}'^2 - \gamma'^2) \varphi \overline{\varphi} , \tag{13.132}$$

which when integrated over the deformed region is

$$- \oint_C ds \, \varphi \partial_n \overline{\varphi} = (\overline{\gamma}'^2 - \gamma'^2) \int_\sigma d\sigma \, \varphi \overline{\varphi} , \tag{13.133}$$

because $\overline{\varphi}$ vanishes at the deformed boundary. At a point on the new boundary

$$\varphi = -\delta n \, \partial_n \varphi , \tag{13.134}$$

hence, as $\delta n \to 0$,

$$\delta n \oint_C ds \, (\partial_n \varphi)^2 = \delta \gamma'^2 \int_\sigma d\sigma \, \varphi^2 . \tag{13.135}$$

Therefore we have the variational statement

$$\frac{\oint_C ds (\partial_n \varphi)^2}{\int_\sigma d\sigma \, \varphi^2} = \frac{\delta \gamma'^2}{\delta n} , \tag{13.136}$$

which permits us to rewrite (13.121) as

$$\mathcal{R} = \frac{k \delta \zeta}{2} \frac{1}{\gamma'^2} \frac{\delta \gamma'^2}{\delta n} . \tag{13.137}$$

This permits us to write the following forms for the attenuation factor and the figure of merit:

$$\alpha = \frac{k}{\kappa} \frac{k \delta}{2} \frac{1}{\gamma'^2} \frac{\delta \gamma'^2}{\delta n} , \qquad \frac{1}{Q} = \frac{\delta}{2} \frac{1}{\gamma'^2} \frac{\delta \gamma'^2}{\delta n} . \tag{13.138}$$

Let us illustrate this for the case of a circular guide of radius a, with $\gamma' a = $ constant. Then

$$\frac{1}{\gamma'^2} \frac{\delta \gamma'^2}{\delta n} = -2 \frac{d}{da} \log \gamma' = \frac{2}{a} . \tag{13.139}$$

Then

$$Q = \frac{a}{\delta} , \qquad \alpha = \frac{k^2 \delta}{\kappa a} . \tag{13.140}$$

13.7 Conclusion

The reader will ask, how can we stop now, when we have barely begun! We certainly could go on to discuss more general junctions and discontinuities in waveguides. However, this is a subject of an already existing treatise [5], not to mention more engineering-oriented books, starting with [6]. To these we refer the reader, whose interest has hopefully been whetted, to learn more.

13.8 Problems for Chap. 13

1. Apply the ideas expressed in Sect. 13.1 to the purely electrical discontinuity of the form

$$\varepsilon(z) = \begin{cases} 1, & z < -a/2, \\ \epsilon, & -a/2 < z < a/2, \\ 1, & a/2 < z, \end{cases} \tag{13.141}$$

   in an otherwise uniform cylindrical waveguide.

2. Derive the T and Π sections shown in Figs. 13.2 and 13.3, by considering the flow of current through the network, and the corresponding low-frequency limits shown in Figs. 13.4–13.7.

3. Derive (13.80a) and (13.80b), specifying the frequency dependence of the reactance and the susceptance, from the energy theorem (6.90).

4. Derive the Green's function between two parallel planes, $y = 0$ and b, with normal derivative vanishing on those surfaces and also vanishing on the plane $z = 0$, and thereby derive (13.88).

5. Repeat the analysis of Sect. 13.6 for H modes. In this case, both the longitudinal and transverse components of $\mathbf{H}$ contribute to the dissipation, the corresponding contributions having the form

$$P_\perp = \frac{1}{2}\mathcal{R}|I|^2, \qquad P_s = \frac{1}{2}\mathcal{G}|V|^2. \tag{13.142}$$

   Give expressions for the resistance $\mathcal{R}$ and conductance $\mathcal{G}$ quantities in terms of line and area integrals of the mode function ψ. Consider an inward displacement of the boundary δn, and derive

$$\frac{k^2}{\gamma''^2}\frac{\mathcal{G}}{\eta} - \frac{\mathcal{R}}{\zeta} = \frac{k\delta}{2}\frac{1}{\gamma''^2}\frac{\delta\gamma''^2}{\delta n}. \tag{13.143}$$

   The quantities $\mathcal{R}$ and $\mathcal{G}$ thus are not determined independently in general, although they may be easily worked out for special circumstances. Thus, for a circular guide, in a mode with azimuthal mode number m, show that

$$\frac{\mathcal{G}}{\eta} = \frac{k\delta}{a}\frac{\gamma''^2}{k^2}\frac{1}{1 - (m/\gamma''a)^2}, \qquad \frac{\mathcal{R}}{\zeta} = \frac{k\delta}{a}\frac{(m/\gamma''a)^2}{1 - (m/\gamma''a)^2}. \tag{13.144}$$

Show that we now have a series impedance and shunt admittance,

$$Z_s = \mathcal{R} - i\omega L , \qquad Y_\perp = \mathcal{G} - i\omega C + \frac{i}{\omega L'} , \qquad (13.145)$$

so that the modified propagation constant is

$$i\kappa' = i\sqrt{\left(1 + \frac{i\mathcal{R}}{\omega L}\right)(\kappa^2 + i\omega L\mathcal{G})} \approx i\kappa - \frac{\alpha}{2} . \qquad (13.146)$$

Show that the attenuation constant may be written as

$$\alpha = \frac{k}{\kappa}\left(\frac{\mathcal{R}}{\zeta} + \frac{\gamma''^2}{k^2}\frac{k\delta}{2}\frac{1}{\gamma''^2}\frac{\delta\gamma''^2}{\delta n}\right) . \qquad (13.147)$$

In general

$$\frac{\mathcal{R}}{\zeta} = f\frac{k^2}{\gamma''^2}\frac{\mathcal{G}}{\eta} , \qquad (13.148)$$

where $f < 1$ is given as a ratio of integrals of the mode function ψ. Then derive the following expression for the attenuation constant and the Q factor,

$$\alpha = \frac{k}{\kappa}\frac{k\delta}{2}\frac{1}{\gamma''^2}\frac{\delta\gamma''^2}{\delta n}\left(\frac{\gamma''^2}{k^2} + \frac{f}{1-f}\right) , \qquad (13.149a)$$

$$\frac{1}{Q} = \frac{\delta}{2}\frac{1}{\gamma''^2}\frac{\delta\gamma''^2}{\delta n}\left(\frac{\gamma''^2}{k^2} + \frac{f}{1-f}\right) . \qquad (13.149b)$$

6. Construct an equivalent circuit for an E plane T junction and derive some properties of the junction from the circuit.

7. Prove that the two E plane obstacles of negligible thickness indicated by Figs. 13.18 and 13.19 have identical shunt admittances. (This is an example of Babinet's principle.)

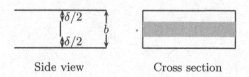

Side view   Cross section

Fig. 13.18. Waveguide with central obstacle

8. Formulate an integral equation and variational principle for an H plane obstacle of negligible thickness, as shown in Fig. 13.20.

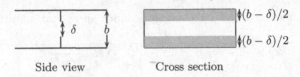

Fig. 13.19. Waveguide with complementary obstacles

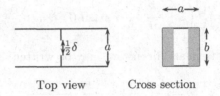

Fig. 13.20. Waveguide with H plane obstacles

Accelerators: Microtrons and Synchrotrons

14.1 The Microtron

The following introductory remarks are based on lectures given by Schwinger at the Los Alamos laboratory at the end of July 1945, about a week after the first nuclear test, the Trinity experiment at Alamogordo. There he pointed out ideas for a linear accelerator consisting of a succession of microwave cavities driven by a traveling electromagnetic wave of wavelength $\lambda \sim 10$ cm, where the length of each cavity was $\lambda/2$, so that the phases of the wave, and hence the polarities of the cavities, were reversed in successive cavities. Thus the electron always meets an accelerating field. The total voltage developed by the accelerator

$$\Delta V \sim \sqrt{P}\sqrt{n}\,, \tag{14.1}$$

where P is the power put in a single cavity, and n is the number of cavities. Schwinger estimated that to achieve 10^8 V, some 100 cavities were required. Of course, precisely this idea is utilized in present-day linear accelerators.

However, at the time, Schwinger was concerned with the number of high-power cavities, so he proposed an alternative. Rather than many cavities, why not use one, and recycle the electron as in the cyclotron, by using a magnetic field H. The schematic of the idea is sketched in Fig. 14.1. To bring the electron back in phase they need to return to the cavity after a multiple of the period T of the RF cavity. The time required for the electron of energy E to complete one circuit is $2\pi E/eHc^2$, which for an accelerating voltage per pass of $\Delta V = 3$ MV, a wavelength $\lambda = 10$ cm, and a power $P = 5$ MW, corresponds to a magnetic field strength $H = 6$ kG, a quite practical value for the time. This scheme, which Schwinger and Alvarez dubbed the microtron,[1] was realized, but it was rather immediately superseded by the more practical synchrotron, with its fixed electron orbit, although there are still reportedly microtrons in operation in Germany and Russia. Nevertheless, we will here

---

[1] Apparently, it was independently invented by Veksler. For a bit more on the conventional history of the microtron, see [18].

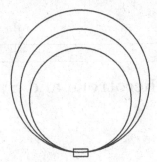

Fig. 14.1. Sketch of the microtron. The box represents the microwave cavity, and the successive circles, the orbits of the electron moving in a magnetic field perpendicular to the plane of the drawing

recount some of the theory behind the device, believing that the analysis still retains pedagogical value.

14.1.1 Cavity Resonators

Consider a circular cylindrical cavity of radius a, height b, and operating in the lowest E mode, which is the lowest mode of the cavity if $b/a < 2.03$ (see Problem 14.1). The fields are [cf. (8.45e), (8.45d), with $I \to V/i\zeta$]

$$E_z = J_0(\gamma r)\frac{V}{b}, \quad i\zeta H_\phi = J_1(\gamma r)\frac{V}{b}, \tag{14.2a}$$

$$k = \gamma = \frac{2.405}{a}, \quad \lambda = 2.613a. \tag{14.2b}$$

Here z-lies along the symmetry axis of the cylinder, and r is the radial coordinate. The average stored energy is,

$$E = \frac{1}{4}\int (\mathrm{dr})\left(\varepsilon|E_z|^2 + \mu|H_\phi|^2\right) = \frac{\varepsilon}{2}\int_0^a \mathrm{dr}\, r\, 2\pi b \frac{V^2}{b^2} J_0^2(\gamma r)$$

$$= \frac{\varepsilon}{2}\frac{V^2}{b}\pi a^2 J_1^2(\gamma a), \tag{14.3}$$

which uses (8.8a) and (8.41). The average power dissipated in skin effect losses is from (13.119)

$$P_{\mathrm{diss}} = \frac{k\delta}{4}\zeta\int \mathrm{dS}\,|H_\phi|^2 = \frac{k\delta}{4}\zeta\left(\frac{V}{\zeta b}\right)^2\left[J_1^2(\gamma a)2\pi ab + 2J_1^2(\gamma a)\pi a^2\right]$$

$$= \frac{k\delta}{4}\eta\frac{V^2}{b^2}2\pi a^2 J_1^2(\gamma a)\left(1 + \frac{b}{a}\right) \equiv \frac{\omega}{Q_0}E, \tag{14.4}$$

where the unloaded Q value is

$$Q_0 = \frac{b}{\delta}\frac{1}{1+\frac{b}{a}} . \tag{14.5}$$

The skin depth for a nonferromagnetic conductor is from (13.118)

$$\delta = \sqrt{\frac{2}{k\zeta\sigma}} , \quad \zeta = 120\pi\,\Omega , \tag{14.6}$$

where the last number is approximate – see (6.41). For Ag at room temperature, $\sigma = 6.14 \times 10^5\,\frac{\text{mho}}{\text{cm}}$, whence $\delta = 3.71\sqrt{\lambda} \times 10^{-5}$ cm, with the wavelength λ measured in cm. The loaded Q for a matched cavity is

$$Q = \frac{1}{2}Q_0 = \frac{b}{2\delta}\frac{1}{1+\frac{b}{a}} , \tag{14.7}$$

for the circular cylinder. The manner in which the electric field in the cavity builds up in time is indicated by

$$\mathbf{E}(t) = \mathbf{E}\left(1 - e^{-\omega t/2Q}\right) , \quad E = \frac{Q}{\omega}P . \tag{14.8}$$

Therefore, the build-up time, defined for our purposes as the time required for the field to reach 97% of the final value, is given by $\omega T_{\text{bu}}/2Q = 3.5$, or, for the cylindrical cavity, using the Ag values

$$T_{\text{bu}} = \frac{\lambda}{c}\frac{3.5}{2\pi}\frac{\lambda}{\delta}\frac{b}{\lambda}\frac{1}{1+\frac{b}{a}} = 0.250\lambda^{3/2}\frac{2b}{\lambda}\frac{1}{1+\frac{b}{a}}\,\mu\text{s} , \quad \lambda \text{ in cm} . \tag{14.9}$$

The voltage across the cavity for a given input power, under matched conditions, is determined by (14.3)

$$\frac{Q}{\omega}P = E = \frac{\varepsilon}{2}\frac{V^2}{b}\pi a^2 J_1^2(\gamma a) , \tag{14.10}$$

or

$$P = \frac{V^2}{2R} , \quad V = \sqrt{2RP} , \tag{14.11}$$

which defines the shunt resistance

$$R = \zeta\frac{Q}{\pi}\frac{b}{a}\frac{1}{\gamma a J_1^2(\gamma a)} = \frac{120}{2.405(0.5191)^2}\frac{b}{a}Q\,\Omega = 185\frac{b}{a}Q\,\Omega , \tag{14.12}$$

or recalling from (14.2b) that $\lambda = 2.613a$,

$$R = 1.63\sqrt{\lambda}\frac{\left(\frac{2b}{\lambda}\right)^2}{1+\frac{b}{a}}\,\text{M}\Omega , \quad \lambda \text{ in cm} . \tag{14.13}$$

Therefore, from (14.11)

$$V = 1.81 \frac{\frac{2b}{\lambda}}{\sqrt{1 + \frac{b}{a}}} \lambda^{1/4} P^{1/2} \, \text{MV} , \quad \lambda \text{ in cm and } P \text{ in MW} . \tag{14.14}$$

The maximum voltage transferred to an electron moving with the speed of light through the cavity is less than V, for with $t = z/c$,

$$\Delta V = \int_{-b/2}^{b/2} dz \, \frac{V}{b} \cos \omega t = \frac{V}{b} \int_{-b/2}^{b/2} dz \cos kz$$

$$= V \frac{\sin \frac{kb}{2}}{\frac{kb}{2}} = V \frac{\sin \frac{\pi b}{\lambda}}{\frac{\pi b}{\lambda}} . \tag{14.15}$$

Hence

$$\Delta V = 1.15 \frac{\sin \frac{\pi}{2} \frac{2b}{\lambda}}{\sqrt{1 + 1.306 \frac{2b}{\lambda}}} \lambda^{1/4} P^{1/2} \, \text{MV} , \tag{14.16}$$

and for a given λ and P, there is an optimum value of $2b/\lambda$, obtained by maximizing

$$\frac{\sin x}{\sqrt{1 + 0.8317x}} , \quad x = \frac{\pi}{2} \frac{2b}{\lambda} , \tag{14.17}$$

which is given by the solution of

$$\cot x = \frac{0.4158}{1 + 0.8317x} \Rightarrow x = 1.38 , \tag{14.18}$$

or $b/\lambda = 0.439$, which yields

$$\Delta V = 0.770 \lambda^{1/4} P^{1/2} \, \text{MV} . \tag{14.19}$$

The corresponding values of the quality factor, the build-up time, and the resistance are

$$Q = 2.76 \times 10^3 \sqrt{\lambda} , \quad Q_0 = 5.51 \times 10^3 \sqrt{\lambda} , \tag{14.20a}$$

$$T_{\text{bu}} = 0.102 \lambda^{3/2} \, \mu\text{s} , \tag{14.20b}$$

$$R = 0.586 \sqrt{\lambda} \, \text{M}\Omega , \tag{14.20c}$$

$$\mathcal{R} = 0.296 \sqrt{\lambda} \, \text{M}\Omega , \tag{14.20d}$$

where $\mathcal{R}$ is the effective shunt resistance,

$$\mathcal{R} = R \left(\frac{\sin \frac{\pi b}{\lambda}}{\frac{\pi b}{\lambda}} \right)^2 , \quad P = \frac{(\Delta V)^2}{2\mathcal{R}} . \tag{14.21}$$

The physically important quantities are T_{bu} and ΔV. A quantity of importance in connection with field emission is the maximum electric field strength in the cavity $\mathcal{E}$, a lower limit to the true maximum field:

$$\mathcal{E} = \frac{V}{b} = \frac{1}{0.711}\frac{\Delta V}{b} = 2.46\lambda^{-3/4}P^{1/2}\,\text{MV/cm}\,. \tag{14.22}$$

The magnetic field required to establish resonance conditions is determined by

$$\lambda = \frac{2\pi\Delta E}{eHc} = \frac{2\pi}{0.3}\frac{\Delta V}{H}\,, \tag{14.23}$$

for λ in cm, ΔV in MV, H in kG, or from (14.19)

$$H = \frac{2\pi}{0.3}\frac{\Delta V}{\lambda} = 16.1\lambda^{-3/4}P^{1/2}\,\text{kG}\,. \tag{14.24}$$

Since this can be written as

$$H = \frac{2\pi}{0.3}\frac{b}{\lambda}\frac{\Delta V}{b}\,, \tag{14.25}$$

H is related to the maximum field strength (14.22) by

$$H = 6.55\,\mathcal{E}\,\text{kG}\,, \quad \mathcal{E}\,\text{in MV/cm}\,. \tag{14.26}$$

14.1.2 Elementary Theory

Suppose the electron is injected at the proper phase with energy E_0; the energy after the first transit through the accelerating cavity is $E_0 + \Delta E$, and the resulting time for revolution is

$$T_1 = \frac{2\pi}{eHc^2}(E_0 + \Delta E) = n_1\tau\,, \tag{14.27}$$

if we require the revolution time to be a multiple, n_1, of the period of the cavity τ. After the second transit, the energy is $E_0 + 2\Delta E$, and the revolution time now is

$$T_2 = \frac{2\pi}{eHc^2}(E_0 + 2\Delta E) = n_2\tau\,. \tag{14.28}$$

In general, after n transits,

$$T_n = \frac{2\pi}{eHc^2}(E_0 + n\Delta E)\,. \tag{14.29}$$

Hence, the resonance conditions are

$$\frac{2\pi\Delta E}{eHc^2} = \tau, 2\tau \ldots\,, \quad \lambda = \frac{2\pi\Delta E}{eHc} = \frac{2\pi}{0.3}\frac{\Delta V}{H}\,, \tag{14.30a}$$

and assuming that the first option is satisfied,

$$\frac{2\pi E_0}{eHc^2} = n_0\tau\,, \quad \text{or} \quad E_0 = n_0\Delta E\,. \tag{14.30b}$$

On the elementary theory, $n_0 = 0$ is possible, that is, the electron starts from rest, and

$$T_n = n\tau , \quad E_n = n\Delta E . \tag{14.31}$$

The circumference of the nth orbit is $n\lambda$ and the diameter is $n\lambda/\pi$. Hence successive orbits are equally spaced on the symmetry axis. If the radius of the magnet is R, the maximum number of transits is N:

$$2R = N\frac{\lambda}{\pi} , \quad N = \frac{2\pi R}{\lambda} . \tag{14.32}$$

The final energy is

$$N\Delta E = \frac{2\pi R}{\lambda}\frac{eHc\lambda}{2\pi} = ecHR , \tag{14.33}$$

as it should be. The total time taken to reach this energy is

$$T = \sum_{n=1}^{\infty} T_n \approx \frac{N^2}{2}\tau , \tag{14.34}$$

and the total distance traveled by the electron is $N^2\lambda/2$.

If, for example, we consider $N = 50$ and $\lambda = 10$ cm, we find $R = 0.796$ m. If the acceleration is supplied by two resonators, each of power $P = 5$ MW, according to (14.19),

$$\Delta E = 0.77\lambda^{1/4}(2P)^{1/2} \,\text{MeV} = 4.33 \,\text{MeV} , \tag{14.35}$$

and the energy attained is 2.17×10^8 eV, at a magnetic field of 9.09 kG. The total distance traveled by the electron is 0.125 km, and the time consumed is $T = 0.417\,\mu$s.

The actual voltage available will be less than that considered, in consequence of the cavity loading by the electrons. If I is the output current during the acceleration period and V is the final voltage obtained, the actual available power is $P - IV$, or in terms of the efficiency of the device, $\varepsilon = IV/P$, the available power is $(1 - \epsilon)P$. If we are willing to tolerate a 10% reduction in voltage, ϵ can be 20% and $I = \frac{1}{5}\frac{P}{V}$. With $P = 5$ MW, and $V = 200$ MV, $I = 5$ mA. If pulsed at the ratio of 1000 to 1 so that the average power is 5 kW, the average output current is $5\,\mu$A. If say, 1% of a cycle is available for accelerating electrons from rest, the input current must be 100×5 mA $= 0.5$ A.

14.1.3 Vertical Defocusing

Vertical defocusing arises from the initial transverse velocities, and the electron will move with constant momentum perpendicular to the plane of the orbit, in terms of the relativistic mass:

$$mv_z = m_0v_{z0} . \tag{14.36}$$

Hence the vertical displacement acquired on the n orbit is

$$\Delta z = T_n v_n = \frac{2\pi E_n}{eHc^2} v_n = \frac{2\pi m_n}{eH} v_n = \frac{2\pi m_0 c^2}{eHc} \frac{v_{z0}}{c}, \tag{14.37}$$

from which follows the displacement after N orbits, according to (14.30a),

$$\Delta z = \frac{2\pi m_0 c^2}{eHc} \frac{v_{z0}}{c} N = \lambda \frac{m_0 c^2}{\Delta E} \frac{v_{z0}}{c} N. \tag{14.38}$$

For example, with $\Delta E = 8 m_0 c^2 \approx 4$ MeV, $N = 50$, and $v_{z0}/c = 10^{-3}$, we find $\Delta z = 0.6$ mm.

14.1.4 Radiation Losses

Radiation losses become important when the energy loss is sufficient to change the phase at which the electron returns to the cavity by say 5°, or when

$$\delta T = \frac{2\pi \delta E}{eHc^2} = \tau \frac{\delta E}{\Delta E} < \frac{5}{360} \tau. \tag{14.39}$$

Hence, we require the energy loss to satisfy

$$\delta E < \frac{\Delta E}{70}. \tag{14.40}$$

But we know the energy loss per revolution for an electron moving in a circle of radius R is given by (15.25),

$$\delta E = \frac{1}{3\varepsilon_0} \frac{e^2}{R} \left(\frac{E}{m_0 c^2} \right)^4 = \frac{1}{3\varepsilon_0} \frac{e^2}{E} \left(\frac{E}{m_0 c^2} \right)^4 eHc$$
$$= \frac{8\pi^2}{3} \frac{e^2/4\pi\varepsilon_0 m_0 c^2}{\lambda} \left(\frac{E}{m_0 c^2} \right)^3 \Delta E, \tag{14.41}$$

according to (14.33) and (14.30a), so by comparison with (14.40) we see that the energy must satisfy

$$\frac{8\pi^2}{3} \frac{r_0}{\lambda} \left(\frac{E}{m_0 c^2} \right)^3 < 10^{-2}, \tag{14.42}$$

or putting in the numbers ($r_0 = e^2/4\pi\varepsilon_0 m_0 c^2 = 2.818 \times 10^{-13}$ cm)

$$\frac{E}{m_0 c^2} < 1 \times 10^3 \lambda^{1/3}. \tag{14.43}$$

For example, for $\lambda = 10$ cm, the energy achieved cannot exceed about 1 GeV.

14.1.5 Phase Focusing

Here we consider a simplified model in which the acceleration region occupies a negligible portion of the orbit.

Let E_0 be the energy at injection, ϕ_n the phase at the nth transit, E_n the energy after the nth transit, and T_n the time for a circumnavigation after the nth transit:

$$T_n = \frac{2\pi E_n}{eHc^2} .$$ (14.44)

In terms of ΔV, the voltage gain per transit,

$$E_1 = E_0 + e\Delta V \cos\phi_1 ,\dots , E_n = E_{n-1} + e\Delta V \cos\phi_n ,$$ (14.45)

whence

$$T_n - T_{n-1} = \frac{2\pi\Delta V}{Hc^2} \cos\phi_n .$$ (14.46)

Now the phase advance in the nth cycle is

$$\phi_{n+1} = \phi_n + \omega T_n - 2\pi(n+1) ,$$ (14.47)

if we subtract a suitable multiple of 2π in order that $|\phi_n| \ll 1$, if resonance is maintained (it is assumed that the resonance increase in the period of revolution is one period of the accelerating field). Hence we have the difference equation

$$\phi_{n+1} - 2\phi_n + \phi_{n-1} = \omega(T_n - T_{n-1}) - 2\pi = \frac{2\pi\omega\Delta V}{Hc^2} \cos\phi_n - 2\pi .$$ (14.48)

A necessary condition for resonance at the phase $\phi_n = \phi$ is

$$\omega\Delta V \cos\phi = Hc^2 , \quad \text{or} \quad 2\pi\Delta V \cos\phi = H\lambda c .$$ (14.49)

Putting $\phi_n = \phi + \psi_n$, and assuming $|\psi_n| \ll 1$, we find

$$\psi_{n+1} - 2\psi_n + \psi_{n+1} = 2\pi\left(\frac{\cos\phi_n}{\cos\phi} - 1\right) = 2\pi\left(\cos\psi_n - 1 - \tan\phi\sin\psi_n\right)$$
$$\approx -2\pi\tan\phi\,\psi_n ,$$ (14.50)

which describes a stable oscillation if $\tan\phi > 0$, which implies that $0 < \phi < \pi/2$, if H and ΔV are positive. The solution is of the form

$$\psi_n = e^{i\gamma n} ,$$ (14.51)

with

$$e^{i\gamma} + e^{-i\gamma} - 2 = -2(1 - \cos\gamma) = -2\pi\tan\phi ,$$ (14.52)

or

$$2\sin\frac{\gamma}{2} = \sqrt{2\pi\tan\phi} .$$ (14.53)

However, in order that the oscillation be stable, not only must $\tan \phi$ be positive, but it must be sufficiently small that $\sin \frac{\gamma}{2} < 1$. Hence the stability requirement on ΔV, say, is from (14.49)

$$0.8436 < \frac{H\lambda c}{2\pi \Delta V} < 1 \,. \tag{14.54}$$

To find the complete solution for ψ_n, we add the initial conditions,

$$\psi_1 = \phi_1 - \phi \,, \tag{14.55a}$$

$$\psi_2 - \psi_1 = \omega T_1 - 4\pi = \frac{2\pi\omega}{eHc^2} \left(E_0 + e\Delta V \cos \phi_1 \right) - 4\pi$$

$$\approx 2\pi \left(\frac{\omega E_0}{eHc^2} - 1 \right) - 2\pi \tan \phi \, \psi_1 \,, \tag{14.55b}$$

using (14.49), to

$$\psi_n = A \sin \gamma(n - 1) + B \cos \gamma(n - 1) \,, \tag{14.56}$$

whence

$$B = \phi_1 - \phi = \psi_1 \tag{14.57}$$

and using (14.49) yet again,

$$A \sin \gamma + B(\cos \gamma - 1) = A \sin \gamma - 2\psi_1 \sin^2 \frac{\gamma}{2}$$

$$= 2\pi \left(\frac{E_0}{e\Delta V \cos \phi} - 1 \right) - 2\pi \tan \phi \, \psi_1 \,, \tag{14.58}$$

so from (14.53)

$$A \sin \gamma = 2\pi \left(\frac{E_0}{e\Delta V \cos \phi} - 1 \right) - \pi \tan \phi \, \psi_1 \,. \tag{14.59}$$

It is clear that to minimize the amplitude of the phase oscillation resulting from an incorrect value of the injection energy, $\sin \gamma$ should be unity, or $\gamma = \frac{\pi}{2}$; $\gamma = 0, \pi$ correspond to the stability limits; $\gamma = \pi/2$ corresponds to $\tan \phi = 1/\pi$ or $\phi = 0.308$.

14.2 Excitation of a Cavity by Electrons

Here we consider the excitation of a microwave cavity by electrons passing through it, assumed to travel at speed c. In the Lorenz gauge, the vector potential satisfies [see (1.102b), here in SI]

$$\left(\nabla^2 - \frac{1}{c^2} \frac{\partial^2}{\partial t^2} \right) \mathbf{A} = -\mathbf{J} \,. \tag{14.60}$$

Expand the potential in normal modes,

$$\mathbf{A}(\mathbf{r}, t) = \sum q(t) \mathbf{A}(\mathbf{r}) , \tag{14.61}$$

where the mode functions are eigenfunctions of the Laplace operator,

$$\nabla^2 \mathbf{A}(\mathbf{r}) = -k^2 \mathbf{A}(\mathbf{r}) , \tag{14.62}$$

and where they are normalized by

$$\int (\mathrm{d}\mathbf{r}) \mathbf{A}(\mathbf{r})^2 = 1 . \tag{14.63}$$

Thus the time dependence is governed by

$$\left(k^2 + \frac{1}{c^2} \frac{\mathrm{d}^2}{\mathrm{d}t^2} \right) q(t) = \int (\mathrm{d}\mathbf{r}) \, \mathbf{J}(\mathbf{r}, t) \cdot \mathbf{A}(\mathbf{r}) . \tag{14.64}$$

For a single electron entering at time t_0 and moving along the axis of a simple cavity of length b in the z-direction

$$\left(k^2 + \frac{1}{c^2} \frac{\mathrm{d}^2}{\mathrm{d}t^2} \right) q(t) = \begin{cases} 0 & t > t_0 + b/c , \\ ecA_z(0, 0, c(t - t_0)) , & t_0 < t < t_0 + b/c , \\ 0 , & t < t_0 . \end{cases} \tag{14.65}$$

In the following, we consider only excitation of the lowest E mode, where A_z does not depend on z. Consider the Fourier transform

$$\int_{-\infty}^{\infty} \mathrm{d}t \, e^{-i\zeta t} q(t) = q(\zeta) , \tag{14.66}$$

which is regular in the lower half plane. The transform satisfies

$$\left(k^2 - \frac{\zeta^2}{c^2} \right) q(\zeta) = ecA_z \int_{t_0}^{t_0 + b/c} \mathrm{d}t \, e^{-i\zeta t} = ecA_z e^{-i\zeta t_0} \frac{1 - e^{-i\zeta b/c}}{i\zeta} , \tag{14.67}$$

so

$$\begin{aligned} q(t) &= \frac{1}{2\pi} \int_{-\infty}^{\infty} \mathrm{d}\zeta \, e^{i\zeta t} \frac{ecA_z e^{-i\zeta t_0}}{i\zeta} \frac{1 - e^{-i\zeta b/c}}{k^2 - \zeta^2/c^2} \\ &= -\frac{ecA_z c^2}{2\pi i} \int_{-\infty}^{\infty} \mathrm{d}\zeta \, \frac{e^{i\zeta(t - t_0)}}{\zeta^2 - \omega^2} \frac{1 - e^{-i\zeta b/c}}{\zeta} . \end{aligned} \tag{14.68}$$

We resolve the singularity at $\zeta = \omega$ by requiring that $q(t) = 0$ for $t < t_0$, where the contour can be closed in the lower half plane, so both poles must be in the upper half plane. So after the electron has exited the cavity, $t > t_0 + b/c$, the contour can be closed in the upper half plane, and

$$q(t) = -ecA_z c^2 \text{Re}\, \frac{e^{i\omega(t-t_0)}}{\omega^2} \left(1 - e^{-i\omega b/c}\right)$$

$$= -\frac{ecA_z}{k^2} \left[\cos\omega(t - t_0) - \cos\omega(t - t_0 - b/c)\right] . \tag{14.69}$$

The optimal conditions occur when the two terms here reinforce each other, or

$$kb = \pi , \quad \text{or} \quad b = \frac{1}{2}\lambda , \tag{14.70}$$

in which case

$$q(t) = -\frac{2ecA_z}{k^2} \cos\omega(t - t_0) . \tag{14.71}$$

Hence the energy in the cavity after the passage of the electron is

$$E = \int (\mathrm{d}\mathbf{r})\, \mu_0 \overline{H^2} = \mu_0 k^2 \overline{q^2(t)} = \mu_0 \frac{2e^2 c^2 A_z^2}{k^2} . \tag{14.72}$$

For a rectangular cavity, with transverse sides of length a, the E_{110} mode is described by the vector potential (see Problem 14.2),

$$A_z(x, y, z) = \frac{\cos\pi x/a \, \cos\pi y/a}{\sqrt{b\frac{a}{2}\frac{a}{2}}} , \tag{14.73}$$

so

$$A_z^2(0, 0, z) = \frac{4}{a^2 b} = \frac{16}{\lambda^3} , \tag{14.74}$$

because from (14.70), $b = \lambda/2$, while the eigenvalue condition is $a = \lambda/\sqrt{2}$. Therefore

$$E = \frac{8}{\pi^2\varepsilon_0} \frac{e^2}{\lambda} \rightarrow \frac{32}{\pi} \frac{e^2}{\lambda} , \tag{14.75}$$

where the latter transformation marks the passage from SI units to Gaussian ones, so then e is in esu (see Appendix). If n electrons enter spaced over a time interval small compared to b/c, the energy of excitation is

$$E = \frac{32}{\pi} \frac{e^2}{\lambda} n^2 . \tag{14.76}$$

Thus, for $\lambda = 10$ cm, $n = 10^7$, we get $E = 2.3 \times 10^{-5}$ ergs or 1.5×10^7 eV, or roughly 1 eV per particle. Note that the minimum energy detectable is $kT = 0.025$ eV, so the smallest number of electrons detectable is about 400.

Suppose now that a succession of such pulses, spaced by the resonant period, enter. What is the energy stored in the cavity in the steady state, allowing for dissipation? We must solve

$$\left(k^2 + \frac{k}{Qc}\frac{\mathrm{d}}{\mathrm{d}t} + \frac{1}{c^2}\frac{\mathrm{d}^2}{\mathrm{d}t^2}\right) q(t) = ecn \sum_j A_z(0, 0, c(t - t_j)) , \tag{14.77}$$

which corresponds to the Fourier-transformed equation [see (14.67)]

$$\left(k^2 + \frac{ik}{Qc}\zeta - \frac{\zeta^2}{c^2}\right) q(\zeta) = ecA_z n \frac{1 - e^{-i\zeta b/c}}{i\zeta} \sum_j e^{-i\zeta t_j} . \tag{14.78}$$

If $t_j = j\tau$, $j = 0, 1, \ldots,$

$$\sum_j e^{-i\zeta t_j} = \frac{1}{1 - e^{-i\zeta\tau}} , \tag{14.79}$$

and so

$$q(\zeta) = -\frac{nec^3 A_z}{i\zeta} \frac{1 - e^{-i\zeta b/c}}{1 - e^{-i\zeta\tau}} \frac{1}{\zeta^2 - \frac{i\omega}{Q}\zeta - \omega^2} , \tag{14.80}$$

so the time dependence is

$$q(t) = -\frac{nec^3 A_z}{2\pi i} \int_{-\infty}^{\infty} d\zeta \frac{e^{i\zeta t}}{\zeta} \frac{1 - e^{-i\zeta b/c}}{1 - e^{-i\zeta\tau}} \frac{1}{\zeta^2 - \frac{i\omega}{Q}\zeta - \omega^2} . \tag{14.81}$$

The only important contribution to the steady state is the pole of $1 - e^{-i\zeta\tau}$, corresponding to $\zeta = \pm\omega$, that is, $\zeta = \pm 2\pi/\tau$ (the other pole leads to damping):

$$q(t) = -2nec^3 A_z \text{Re}\left[\frac{e^{i\omega t}}{\omega} \frac{1 - e^{-i\omega b/c}}{\tau} \frac{1}{\frac{\omega^2}{Q}}\right]$$

$$= -\frac{2Q}{\pi} \frac{nec}{k^2} A_z \cos\omega t , \tag{14.82}$$

where in the last line we used $\omega b/c = kb = \pi$. That is, comparing with (14.71), we see that Q/π pulses effectively contribute and the stored energy will be that produced by a single electron times $(nQ/\pi)^2$, that is,

$$E = \frac{n^2 Q^2}{\pi^2} \frac{32}{\pi} \frac{e^2}{\lambda} , \quad e \text{ in esu} . \tag{14.83}$$

This might be expressed in terms of the average input power P,

$$P = \frac{\omega}{Q} E = \frac{64}{\pi^2} c \frac{e^2}{\lambda^2} n^2 Q \text{ erg/s} . \tag{14.84}$$

If $\lambda = 10$ cm, $n = 10^7$, $Q = 5000$, we obtain a power of about 20 W. Even for one electron, $P \approx 2 \times 10^{-13}$ W, which is detectable.

Quantum effects on the excitation of a cavity by a transiting electron will be considered in Chap. 17.

14.3 Microwave Synchrotron

14.3.1 Accelerating Cavities

Here, for simplicity, we suppose there are two accelerating cavities, one at azimuthal angle $\varphi = 0$ and one at $\varphi = \pi$. Then the azimuthal electric field is

$$E_\phi = -\frac{V}{2R} [\delta(\varphi) - \delta(\varphi - \pi)] \sin \omega t . \tag{14.85}$$

If we resolve the delta functions in cosine series,

$$\delta(\varphi) = \frac{1}{2\pi} \sum_{m=-\infty}^{\infty} \cos m\varphi , \quad \delta(\varphi - \pi) = \frac{1}{2\pi} \sum_{m=-\infty}^{\infty} (-1)^m \cos m\varphi , \tag{14.86}$$

the quantity appearing here is

$$\delta(\varphi) - \delta(\varphi - \pi) = \frac{2}{\pi} \sum_{m=1,3,5,\ldots} \cos m\varphi . \tag{14.87}$$

Then the accelerating field is

$$
\begin{aligned}
E_\phi &= -\frac{V}{\pi R} \sum_{m=1,3,\ldots} \cos m\varphi \sin \omega t \\
&= \frac{V}{2\pi R} \sum_{m=\pm1,\pm3,\ldots} \sin(m\varphi - \omega t) \\
&= \frac{V}{2\pi R} \sum_{m=1,3,\ldots} \{\sin[(m-1)\omega t + m\phi] - \sin[(m+1)\omega t + m\phi]\} ,
\end{aligned}
\tag{14.88}
$$

where $\varphi - \omega t = \phi$. The term here explicitly independent of t is

$$E_\phi = \frac{V}{2\pi R} \sin \phi + \cdots = \mathcal{E} \sin \phi + \cdots . \tag{14.89}$$

14.3.2 Motion of Electron

We consider the motion of a relativistic electron ($v \approx c$) in a circular orbit of radius r in a uniform magnetic field directed perpendicularly to the plane of the orbit. The equations of motion are, in terms of the relativistic mass $m = m_0(1 - v^2/c^2)^{-1/2}$, in Heaviside–Lorentz or Gaussian units

$$\frac{d}{dt}(m\dot{r}) = \frac{mc^2}{r} - eH , \tag{14.90a}$$

$$\frac{d}{dt}(mc^2) = e\mathcal{E}c \sin(\mu\varphi - \omega t) + \frac{e}{2\pi r}\dot{\Phi} , \tag{14.90b}$$

$$r\frac{d}{dt}\varphi = c , \tag{14.90c}$$

where Φ is the magnetic flux contained within the orbit,

$$\Phi = 2\pi \int_0^r \mathrm{d}r' \, r' \, H(r') \,. \tag{14.91}$$

From (14.90a) we write

$$mc^2 = eHr + r\frac{\mathrm{d}}{\mathrm{d}t}(m\dot{r}) \,, \tag{14.92}$$

so if we look for small deviation from the equilibrium orbit, $r = R + x$, $|x|/R \ll 1$, and assume a power-law profile for the magnetic field,

$$H = H_0(t) \left(\frac{r}{R}\right)^{-n} \,, \tag{14.93}$$

we have (here and in the following we assume radial derivatives are evaluated at the equilibrium radius R)

$$Hr = H_0 R + x \left(H_0 + R\frac{\partial H}{\partial r}\right) = H_0 R + x(1 - n)H_0 \,. \tag{14.94}$$

Then (14.92) becomes

$$mc^2 = eH_0 R + x(1 - n)eH_0 + R\frac{\mathrm{d}}{\mathrm{d}t}\left(\frac{eH_0 R}{c^2}\frac{\mathrm{d}x}{\mathrm{d}t}\right) \,. \tag{14.95}$$

Now we can write, encompassing the situation discussed in the previous subsection,

$$\phi = \mu\varphi - \omega t \,, \quad \frac{\mathrm{d}\phi}{\mathrm{d}t} = \mu\frac{\mathrm{d}\varphi}{\mathrm{d}t} - \omega = \frac{\mu c}{r} - \omega \,, \tag{14.96}$$

or approximately,

$$\frac{\mathrm{d}\phi}{\mathrm{d}t} = \frac{\mu c}{R} - \omega - x\frac{\mu c}{R^2} = -x\frac{\omega}{R} \,, \tag{14.97}$$

where we have used

$$\mu = \frac{\omega R}{c} = \frac{2\pi R}{\lambda} \,. \tag{14.98}$$

Therefore, we can write (14.95) as

$$mc^2 = eH_0 R - (1 - n)eH_0 R\frac{1}{\omega}\frac{\mathrm{d}\phi}{\mathrm{d}t} - \frac{1}{\omega}\frac{\mathrm{d}}{\mathrm{d}t}\left(\frac{eH_0 R}{\omega_0^2}\frac{\mathrm{d}^2\phi}{\mathrm{d}t^2}\right) \,, \tag{14.99}$$

where we have written $\omega_0 = c/R$, $\omega = \mu\omega_0$.

We can do a similar expansion for the magnetic flux,

$$\frac{\Phi}{r} = \frac{\Phi_0}{R} + x\left(-\frac{\Phi_0}{R^2} + 2\pi H_0\right) \,, \tag{14.100}$$

and for the electric field

$$\mathcal{E} = \mathcal{E}_0 + x\frac{\partial \mathcal{E}}{\partial r} \ . \tag{14.101}$$

Then we deduce from (14.90b)

$$\frac{d}{dt}mc^2 = \frac{e}{2\pi R}\dot{\Phi}_0 + ex\left(\dot{H}_0 - \frac{\dot{\Phi}_0}{2\pi R^2}\right) + e\mathcal{E}_0 c\sin\phi + xec\frac{\partial \mathcal{E}}{\partial r}\sin\phi \ . \tag{14.102}$$

This can be rewritten as, from (14.99),

$$eR\frac{d}{dt}H_0 - (1-n)eR\frac{d}{dt}H_0\frac{1}{\omega}\frac{d\phi}{dt} - (1-n)eH_0R\frac{1}{\omega}\frac{d^2\phi}{dt^2}$$
$$- \frac{1}{\omega}\frac{d^2}{dt^2}\left(\frac{eH_0R}{\omega_0^2}\frac{d^2\phi}{dt^2}\right)$$
$$= \frac{e}{2\pi R}\frac{d}{dt}\Phi_0 - \frac{eR}{\omega}\frac{d\phi}{dt}\left(\frac{dH_0}{dt} - \frac{1}{2\pi R^2}\frac{d\Phi_0}{dt}\right) + e\mathcal{E}_0 c\sin\phi$$
$$- \frac{eRc}{\omega}\frac{\partial \mathcal{E}}{\partial r}\frac{d\phi}{dt}\sin\phi \ , \tag{14.103}$$

or

$$\frac{1}{\omega}\frac{d^2}{dt^2}\left(\frac{eH_0R}{\omega_0^2}\frac{d^2\phi}{dt^2}\right) + (1-n)\frac{eR}{\omega}\frac{d}{dt}\left(H_0\frac{d\phi}{dt}\right)$$
$$- \frac{eR}{\omega}\frac{d\phi}{dt}\left(\frac{dH_0}{dt} - \frac{1}{2\pi R^2}\frac{d\Phi_0}{dt}\right) - \frac{eRc}{\omega}\frac{\partial \mathcal{E}}{\partial r}\frac{d\phi}{dt}\sin\phi + e\mathcal{E}_0 c\sin\phi$$
$$= eR\left(\frac{dH_0}{dt} - \frac{1}{2\pi R^2}\frac{d\Phi_0}{dt}\right) \ . \tag{14.104}$$

Finally we have

$$\frac{1}{\omega_0^2}\frac{d^2}{dt^2}\left(H_0\frac{d^2\phi}{dt^2}\right) + (1-n)\frac{d}{dt}\left(H_0\frac{d\phi}{dt}\right) - \frac{d\phi}{dt}\left(\frac{dH_0}{dt} - \frac{1}{2\pi R^2}\frac{d\Phi_0}{dt}\right)$$
$$- c\frac{\partial \mathcal{E}}{\partial r}\frac{d\phi}{dt}\sin\phi + \omega\omega_0\mathcal{E}_0\sin\phi = \omega\left(\frac{dH_0}{dt} - \frac{1}{2\pi R^2}\frac{d\Phi_0}{dt}\right) \ . \tag{14.105}$$

We now briefly mention some special cases.

14.3.3 Betatron Regime: $\mathcal{E}_0 = 0, \Phi_0 = 2\pi R^2 H_0(t)$

The oscillation equation (14.105) reduces to

$$\frac{d}{dt}\left(H_0\frac{dx}{dt}\right) + (1-n)\omega_0^2 H_0 x = 0 \ , \tag{14.106}$$

the solution of which is

$$x \sim \frac{1}{\sqrt{H_0}} \sin(\omega_r t + \text{const.}) , \quad \omega_r = \omega_0 \sqrt{1-n} , \tag{14.107}$$

because if we treat H_0 as slowly varying, (14.106) is approximately

$$\frac{d^2}{dt^2} \sqrt{H_0}\, x + \omega_r^2 \sqrt{H_0}\, x = 0 . \tag{14.108}$$

14.3.4 Betatron Regime and Constant H_0

Here we neglect the variation of H_0, and we obtain

$$\frac{1}{\omega_0^2}\frac{d^4\phi}{dt^4} + (1-n)\frac{d^2\phi}{dt^2} - \frac{c}{H_0}\frac{\partial \mathcal{E}}{\partial r}\frac{d\phi}{dt}\sin\phi + \frac{\omega\omega_0\mathcal{E}_0}{H_0}\sin\phi = 0 . \tag{14.109}$$

Consider small oscillations about $\phi = 0$: $\sin\phi \approx \phi$, $\frac{d\phi}{dt}\phi \approx 0$. Then the above equation simplifies to

$$\frac{d^4\phi}{dt^4} + (1-n)\omega_0^2\frac{d^2\phi}{dt^2} + \omega\omega_0^3\frac{\mathcal{E}_0}{H_0}\phi = 0 , \tag{14.110}$$

so if we seek a solution of the form $\phi = e^{i\nu t}$, ν is given by

$$\nu^2 = \frac{1}{2}\left[(1-n)\omega_0^2 \pm \sqrt{(1-n)^2\omega_0^4 - 4\omega\omega_0^3\frac{\mathcal{E}_0}{H_0}}\,\right] . \tag{14.111}$$

Stable oscillations require

$$\omega\frac{\mathcal{E}_0}{H_0} < \omega_0\left(\frac{1-n}{2}\right)^2 , \tag{14.112}$$

or, with $V = 2\pi R\mathcal{E}_0$,

$$\frac{V}{\lambda H_0} < \left(\frac{1-n}{2}\right)^2 . \tag{14.113}$$

Thus with $\lambda = 10$ cm, $H_0 = 10^3$ G, $n = 1/2$, the limit is $V < 2 \times 10^5$ V (1 stat volt equals 300 V), while if $\lambda = 600$ cm, $H_0 = 10^2$ G, $n = 1/2$, the limit rises to a million volts. If the electric field is small, the two roots are:

$$\omega\frac{\mathcal{E}_0}{H_0} \ll \omega_0\left(\frac{1-n}{2}\right)^2 : \quad \nu^2 = (1-n)\omega_0^2 , \quad \nu^2 = \frac{\omega\omega_0}{1-n}\frac{\mathcal{E}_0}{H_0} . \tag{14.114}$$

There is a high-frequency oscillation determined by the magnetic field, and a low-frequency oscillation caused by the electric field. In treating the latter, the fourth derivative term in (14.110) is neglected.

For further details, particularly the inclusion of radiation damping, see the unpublished paper by Saxon and Schwinger, "Electron Orbits in the Synchrotron," printed in Part II.[2]

---

[2] Refers to the hardcover edition which includes in addition the reprints of seminal papers by J. Schwinger on these topics.

14.4 Modern Developments

The present discussion is, of course, not reflective of the current high development of the field of accelerator design. For a brief status report of current accelerators and their design parameters, the interested reader is referred to [19]. For recent treatises on the subject, we offer [20–22]. Our hope here in resurrecting these old notes on the subject by Schwinger is that techniques he invented in specific limited contexts may have applications elsewhere.

14.5 Problems for Chap. 14

1. Construct the E and H modes for a cylindrical cavity, and demonstrate that the lowest E mode has the lowest frequency if $b/a < 2.03$.
2. Calculate the E modes in a rectangular cavity of sides a, b, c. These are the modes that have no z-component of the magnetic field. Show that the eigenfrequencies have the expected form,

$$k^2 = \left(\frac{l\pi}{a}\right)^2 + \left(\frac{m\pi}{b}\right)^2 + \left(\frac{n\pi}{c}\right)^2 . \tag{14.115}$$

Show that the transverse components of **H** are derivable from a vector potential which has only a z-component, A_z. What is the corresponding scalar potential so that **E** is derivable from the two potentials? For the lowest mode when $c < a, b$, $l = m = 1$, $n = 0$, show that only a vector potential is present, which for $a = b$ here is exhibited in (14.73).
3. Investigate stability of electron orbits in the betatron regime with $\mathcal{E}_0$ varying. Here regard H_0 as constant, and consider low-frequency oscillations. Derive the governing equation from (14.105):

$$(1-n)\frac{\mathrm{d}^2\phi}{\mathrm{d}t^2} - \frac{c}{H_0}\frac{\partial \mathcal{E}}{\partial r}\sin\phi\frac{\mathrm{d}\phi}{\mathrm{d}t} + \frac{\omega\omega_0\mathcal{E}_0}{H_0}\sin\phi = 0 . \tag{14.116}$$

What constraints are placed on the spatial variation of $\mathcal{E}_0$ by stability requirements?

Synchrotron Radiation

An accelerated particle radiates. This familiar fact became a concern to physicists after the War as they contemplated building high-energy accelerators, where such radiation might provide a limitation to the operation of such devices. Nowadays, of course, dedicated synchrotron light sources have become valuable tools for research. Here we present the elementary theory of synchrotron radiation.

15.1 Relativistic Larmor Formula

In the first chapter, we showed that an accelerated charged particle radiates, and that therefore radiation necessarily exerts a reactive force on the charged particle. Accordingly, we should amend the Lorentz force law for a nonrelativistic particle by including this radiation reaction, inferred from the first term on the right side of (1.188), in Heaviside–Lorentz units,

$$m\frac{d\mathbf{v}}{dt} = e\left(\mathbf{E} + \frac{\mathbf{v}}{c} \times \mathbf{H}\right) + \frac{2}{3}\frac{1}{4\pi}\frac{e^2}{c^3}\frac{d^2\mathbf{v}}{dt^2} , \qquad (15.1)$$

or

$$\frac{d}{dt}\mathbf{p} = e\left(\mathbf{E} + \frac{\mathbf{v}}{c} \times \mathbf{H}\right) , \qquad (15.2)$$

where

$$\mathbf{p} = m\mathbf{v} - \frac{2}{3}\frac{1}{4\pi}\frac{e^2}{c^3}\frac{d\mathbf{v}}{dt} . \qquad (15.3)$$

The statement of energy conservation is obtained from (15.1) by multiplying by $\mathbf{v}$:

$$\frac{d}{dt}\mathcal{E} = e\mathbf{E} \cdot \mathbf{v} - \frac{2}{3}\frac{1}{4\pi}\frac{e^2}{c^3}\left(\frac{d\mathbf{v}}{dt}\right)^2 , \qquad (15.4)$$

where the energy is

$$\mathcal{E} = \frac{1}{2}mv^2 - \frac{2}{3}\frac{1}{4\pi}\frac{e^2}{c^3}\frac{d}{dt}\frac{v^2}{2} . \qquad (15.5)$$

Here the first term on the right side of (15.4) is the rate at which the applied electric field does work on the charged particle, and the second term on the right is the rate at which energy is radiated, corresponding to the Larmor (dipole radiation) formula (1.189).

Now we must recast this in covariant form so we can describe a relativistic particle. Recall the construction of the electromagnetic field strength tensor in Chap. 3,

$$F_{\mu\nu} = -F_{\nu\mu} , \quad F_{0i} = -E_i , \quad F_{ij} = \epsilon_{ijk}H_k , \tag{15.6}$$

and that position and momentum four-vectors are related to the corresponding three-dimensional quantities by

$$x^\mu = (ct, \mathbf{r}) , \qquad p^\mu = (\mathcal{E}/c, \mathbf{p}) . \tag{15.7}$$

So the energy equation (15.4) may be written as

$$\frac{d}{dt}\frac{\mathcal{E}}{c} = \frac{e}{c}F^{0i}\frac{dx_i}{dt} - \frac{2}{3}\frac{1}{4\pi}\frac{e^2}{c^4}\left(\frac{d^2x_i}{dt^2}\right)^2 . \tag{15.8}$$

Recalling the relation between coordinate time and proper time intervals (3.40), it is clear that (15.8) is the nonrelativistic version of the time component of the covariant equation of motion

$$\frac{d}{d\tau}p^\mu = \frac{e}{c}F^{\mu\nu}\frac{dx_\nu}{d\tau} - \frac{2}{3}\frac{1}{4\pi}\frac{e^2}{c^5}\frac{d^2x_\nu}{d\tau^2}\frac{d^2x^\nu}{d\tau^2}\frac{dx^\mu}{d\tau} . \tag{15.9}$$

The energy (15.5) divided by c generalizes to the time component of

$$p^\mu = m_0\frac{dx^\mu}{d\tau} - \frac{2}{3}\frac{1}{4\pi}\frac{e^2}{c^3}\frac{d^2x^\mu}{d\tau^2} , \tag{15.10}$$

because

$$\frac{d^2}{d\tau^2}ct = \frac{d}{d\tau}\frac{c}{\sqrt{1 - v^2/c^2}} \approx c\frac{d}{dt}\frac{1}{2}\frac{v^2}{c^2} . \tag{15.11}$$

Evidently, the space components of (15.10) generalize $\mathbf{p}$ in (15.3), and satisfy an equation of motion (15.9) which differs from the Lorentz force form (15.2) by v^2/c^2 corrections.

We now substitute the four-momentum (15.10) into (15.9) to obtain the third-order equation

$$m_0\frac{d^2x^\mu}{d\tau^2} = \frac{e}{c}F^{\mu\nu}\frac{dx_\nu}{d\tau} + \frac{2}{3}\frac{1}{4\pi}\frac{e^2}{c^3}\left[\frac{d^3x^\mu}{d\tau^3} - \frac{1}{c^2}\frac{dx^\mu}{d\tau}\left(\frac{d^2x_\nu}{d\tau^2}\frac{d^2x^\nu}{d\tau^2}\right)\right] , \tag{15.12}$$

or

$$m_0\frac{d^2x^\mu}{d\tau^2} = \frac{e}{c}(F^{\mu\nu} + f^{\mu\nu})\frac{dx_\nu}{d\tau} , \tag{15.13}$$

where $f^{\mu\nu}$ is the "dissipative part of the electron's self field," first discovered by Dirac,[1] which upon using

---

[1] Dirac [23] was referring to half the difference between the retarded and the advanced field of the point charge.

$$\frac{\mathrm{d}x^\nu}{\mathrm{d}\tau}\frac{\mathrm{d}x_\nu}{\mathrm{d}\tau} = -c^2 \,, \tag{15.14}$$

and its consequence

$$\frac{\mathrm{d}^2 x^\nu}{\mathrm{d}\tau^2}\frac{\mathrm{d}x_\nu}{\mathrm{d}\tau} = 0 \,, \tag{15.15}$$

has the manifestly antisymmetrical form

$$f^{\mu\nu} = -\frac{2}{3}\frac{1}{4\pi}\frac{e}{c^4}\left(\frac{\mathrm{d}^3 x^\mu}{\mathrm{d}\tau^3}\frac{\mathrm{d}x^\nu}{\mathrm{d}\tau} - \frac{\mathrm{d}^3 x^\nu}{\mathrm{d}\tau^3}\frac{\mathrm{d}x^\mu}{\mathrm{d}\tau}\right) \,. \tag{15.16}$$

Note that the antisymmetry of the $F^{\mu\nu} + f^{\mu\nu}$ structure in (15.13) leads to (15.15) upon multiplying by $\mathrm{d}x_\mu/\mathrm{d}\tau$.

Let us return to (15.9), take its time component, and drop the radiation reaction term in the energy and momentum:

$$\frac{\mathrm{d}\mathcal{E}}{\mathrm{d}t} = e\mathbf{E}\cdot\mathbf{v} - \frac{2}{3}\frac{1}{4\pi}\frac{e^2 c}{(m_0 c^2)^2}\left(\frac{\mathcal{E}}{m_0 c^2}\right)^2\left[\left(\frac{\mathrm{d}\mathbf{p}}{\mathrm{d}t}\right)^2 - \frac{1}{c^2}\left(\frac{\mathrm{d}\mathcal{E}}{\mathrm{d}t}\right)^2\right] \,. \tag{15.17}$$

We identify the second term on the right as the power radiated, so we infer the relativistic generalization of the Larmor formula (15.4),

$$P = \left(-\frac{\mathrm{d}\mathcal{E}}{\mathrm{d}t}\right)_{\mathrm{rad}} = \frac{2}{3}\frac{1}{4\pi}\frac{e^2}{m_0^2 c^3}\left(\frac{\mathcal{E}}{m_0 c^2}\right)^2\left[\left(\frac{\mathrm{d}\mathbf{p}}{\mathrm{d}t}\right)^2 - \frac{1}{c^2}\left(\frac{\mathrm{d}\mathcal{E}}{\mathrm{d}t}\right)^2\right] \,. \tag{15.18}$$

15.2 Energy Loss by a Synchrotron

Let us now apply this to a synchrotron, where a particle of charge e moves in a circular orbit of radius R with velocity $\mathbf{v}$; the angular frequency of revolution is $\omega_0 = v/R$. The circular motion is due to magnetic field $\mathbf{B}$ perpendicular to the plane of the orbit, and so, neglecting radiation reaction and the accelerating electric fields,

$$\frac{\mathrm{d}\mathbf{p}}{\mathrm{d}t} = \frac{e}{c}\mathbf{v}\times\mathbf{B} \,, \tag{15.19}$$

where the particle's momentum and energy are related by $\mathbf{p} = \mathcal{E}\mathbf{v}/c^2$. The magnetic field does no work, so the energy is conserved, and hence we conclude that

$$\frac{\mathrm{d}\mathbf{v}}{\mathrm{d}t} = -\frac{ec}{\mathcal{E}}\mathbf{B}\times\mathbf{v} \,, \tag{15.20}$$

that is, the angular velocity of precession of $\mathbf{v}$ is

$$\omega_0 = -\frac{ec}{\mathcal{E}}\mathbf{B} \,, \tag{15.21}$$

the sign of which depends on that of the charge. The momentum of the particle is given by the radius of the orbit,

$$p = \frac{|e|}{c}BR \ . \tag{15.22}$$

Now we may insert (15.19), or

$$\left(\frac{d\mathbf{p}}{dt}\right)^2 = \omega_0 \beta^3 \frac{\mathcal{E}^2}{Rc} \ , \tag{15.23}$$

into the relativistic Larmor formula (15.18), to obtain the formula for the total power radiated in synchrotron radiation,

$$P_{\text{rad}} = \frac{2}{3}\frac{1}{4\pi}\omega_0 \frac{e^2}{R}\left(\frac{\mathcal{E}}{m_0 c^2}\right)^4 \beta^3 \ . \tag{15.24}$$

Since the period of revolution is $\tau = 2\pi/\omega_0$, the energy lost per revolution is $(\beta \approx 1)$

$$\Delta E = \tau P = \frac{1}{3}\frac{e^2}{R}\left(\frac{\mathcal{E}}{m_0 c^2}\right)^4 \ . \tag{15.25}$$

For an electron synchrotron this reads

$$\Delta E_e(\text{keV}) = 88.4\frac{\mathcal{E}^4(\text{GeV})}{R(\text{m})} \ , \tag{15.26}$$

which gives a practical upper limit to the usefulness of an electron synchrotron. For protons, radiation losses are much smaller:

$$\Delta E_p(\text{eV}) = 7.78\frac{\mathcal{E}^4(\text{TeV})}{R(\text{km})} \ ; \tag{15.27}$$

so at the Large Hadron Collider (LHC), where the design beam energy is $\mathcal{E} = 7$ TeV, and the radius is 4.2 km, the loss per cycle is only 4.4 keV.

15.3 Spectrum of Radiation Emitted by Synchrotron

Here we will consider in more detail the rate at which the accelerated electron does work on the field. We assume the following charge distribution in cylindrical coordinates:

$$\rho(\mathbf{r}, t) = e\delta(\mathbf{r} - \mathbf{r}_e(t)) = \frac{e}{R}\delta(r - R)\delta(z)\delta(\varphi - \varphi_0 - \omega_0 t) \ , \tag{15.28}$$

where the delta function of the azimuthal coordinate is to be understood 'as periodic, with period 2π, so [see (7.44)]

$$\delta(\phi) = \frac{1}{2\pi} \sum_{m=-\infty}^{\infty} e^{im\varphi} . \tag{15.29}$$

Thus the time dependence of the charge density is seen to be given in the form of a Fourier series:

$$\rho(\mathbf{r}, t) = \sum_{m=-\infty}^{\infty} e^{-im\omega_0 t} \rho_m(\mathbf{r}) , \tag{15.30}$$

where

$$\rho_m(\mathbf{r}) = \frac{e}{R} \delta(r - R)\delta(z)\frac{1}{2\pi} e^{im(\varphi - \varphi_0)} . \tag{15.31}$$

The current density is given similarly:

$$\mathbf{j}(\mathbf{r}, t) = \rho(\mathbf{r}, t)\mathbf{v}(t) = \hat{\varphi}v\rho = \sum_{m=-\infty}^{\infty} e^{-im\omega_0 t} \mathbf{j}_m(\mathbf{r}) , \tag{15.32}$$

where

$$\mathbf{j}_m(\mathbf{r}) = \hat{\varphi}v\frac{e}{R} \delta(r - R)\delta(z)\frac{1}{2\pi} e^{im(\varphi - \varphi_0)} , \tag{15.33}$$

where $\hat{\varphi}$ is the tangent vector to the circular orbit.

Now in the Lorenz gauge, the vector and scalar potentials are given in terms of these currents by (1.117) and (1.116),

$$\begin{pmatrix} \mathbf{A} \\ \phi \end{pmatrix} (\mathbf{r}, t) = \int (d\mathbf{r}') \frac{1}{4\pi|\mathbf{r} - \mathbf{r}'|} \begin{pmatrix} \frac{\mathbf{j}}{c} \\ \rho \end{pmatrix} \left(\mathbf{r}', t - \frac{|\mathbf{r} - \mathbf{r}'|}{c} \right) , \tag{15.34}$$

so in terms of the above Fourier series coefficients

$$\begin{pmatrix} \mathbf{A}_m \\ \phi_m \end{pmatrix} (\mathbf{r}) = \int (d\mathbf{r}') \frac{1}{4\pi|\mathbf{r} - \mathbf{r}'|} e^{im\omega_0|\mathbf{r}-\mathbf{r}'|/c} \begin{pmatrix} \frac{\mathbf{j}_m}{c} \\ \rho_m \end{pmatrix} (\mathbf{r}') . \tag{15.35}$$

The electric field, given by (1.48), or

$$\mathbf{E}(\mathbf{r}, t) = -\frac{1}{c}\frac{\partial}{\partial t}\mathbf{A}(\mathbf{r}, t) - \nabla\phi(\mathbf{r}, t) , \tag{15.36}$$

then has as its Fourier component

$$\mathbf{E}_m(\mathbf{r}) = \frac{im\omega_0}{c}\mathbf{A}_m(\mathbf{r}) - \nabla\phi_m(\mathbf{r}) . \tag{15.37}$$

Inserting the Fourier series representation for the electric current and the electric field, we can write the power dissipated by the electron, averaged over one cycle, as

$$P = -\int (d\mathbf{r})\overline{\mathbf{j} \cdot \mathbf{E}} = -2\text{Re} \sum_{m=1}^{\infty} \int (d\mathbf{r})\mathbf{j}_m(\mathbf{r}) \cdot \mathbf{E}_m^*(\mathbf{r}) , \tag{15.38}$$

since no energy is carried by the $m = 0$ mode. Therefore, we can identify the power radiated in the mth harmonic of the fundamental frequency, $\omega = m\omega_0$, by

$$P = \sum_{m=1}^{\infty} P_m , \quad P_m = -2\mathrm{Re} \int (d\mathbf{r}) \mathbf{j}_m(\mathbf{r}) \cdot \mathbf{E}_m^*(\mathbf{r}) . \tag{15.39}$$

We simplify the latter successively:

$$P_m = -2\mathrm{Re} \int (d\mathbf{r}) \mathbf{j}_m \cdot \left(-\boldsymbol{\nabla}\phi_m^* - \frac{im\omega_0}{c} \mathbf{A}_m^* \right)$$

$$= 2\mathrm{Re}\, im\omega_0 \int (d\mathbf{r}) \left(\rho_m^* \phi_m - \frac{1}{c} \mathbf{j}_m^* \cdot \mathbf{A}_m \right) , \tag{15.40}$$

which uses the equation for current conservation (1.14). Now inserting the construction (15.35) and the explicit expressions for the charge and current densities (15.31) and (15.33), we write this as

$$P_m = 2\mathrm{Re}\, im\omega_0 \int (d\mathbf{r})(d\mathbf{r}') \frac{e^{im\omega_0|\mathbf{r}-\mathbf{r}'|/c}}{4\pi|\mathbf{r}-\mathbf{r}'|} \left[\rho_m^*(\mathbf{r})\rho_m(\mathbf{r}') - \frac{1}{c}\mathbf{j}_m^*(\mathbf{r}) \cdot \frac{1}{c}\mathbf{j}_m(\mathbf{r}') \right]$$

$$= \mathrm{Re}\, im\omega_0 \frac{e^2}{4\pi R} \int \frac{d\varphi\, d\varphi'}{2\pi\, 2\pi} \frac{e^{2im\omega_0(R/c)|\sin(\varphi-\varphi')/2|}}{|\sin(\varphi-\varphi')/2|}$$

$$\times \left[1 - \beta^2 \cos(\varphi - \varphi') \right] e^{-im(\varphi-\varphi')} . \tag{15.41}$$

Here we have used the following obvious geometric facts:

$$|\mathbf{r} - \mathbf{r}'| = 2R|\sin(\varphi - \varphi')/2| , \quad \hat{\boldsymbol{\varphi}} \cdot \hat{\boldsymbol{\varphi}}' = \cos(\varphi - \varphi') . \tag{15.42}$$

Using the symmetry of the φ integration, we can write the power radiated in the mth harmonic as

$$P_m = -m\omega_0 \frac{e^2}{4\pi R} \int_{-\pi}^{\pi} \frac{d\varphi}{2\pi} \frac{\sin(2m\beta \sin \varphi/2)}{\sin \varphi/2} (1 - \beta^2 \cos \varphi) \cos m\varphi$$

$$= -\frac{m\omega_0 e^2}{4\pi R} \left[(1 - \beta^2) \int_{-\pi}^{\pi} \frac{d\varphi}{2\pi} \frac{\sin(2m\beta \sin \varphi/2)}{\sin \varphi/2} \cos m\varphi \right.$$

$$\left. + 2\beta^2 \int_{-\pi}^{\pi} \frac{d\varphi}{2\pi} \sin(2m\beta \sin \varphi/2) \sin \varphi/2 \cos m\varphi \right] . \tag{15.43}$$

The integrals occurring here are immediately recognized as Bessel functions – see [9], p. 408. Thus the result for the power emitted at frequency $m\omega_0$, $m = 1, 2, 3, \ldots$, is

$$P_m = m\omega_0 \frac{e^2}{4\pi R} \left[2\beta^2 J_{2m}'(2m\beta) - (1 - \beta^2) \int_0^{2m\beta} dx\, J_{2m}(x) \right] . \tag{15.44}$$

A synchrotron works in the ultrarelativistic regime, where β is very close to one. Therefore, one might naively think that in that circumstance we could write

$$P_m \sim \frac{m\omega_0 e^2}{4\pi R} 2J_{2m}(2m) , \quad \beta \to 1 . \tag{15.45}$$

However, it is easy to prove ([9], p. 419) that for high harmonic numbers

$$J'_{2m}(2m) \sim \frac{3^{1/6}}{2\pi} \Gamma\left(\frac{2}{3}\right) m^{-2/3} , \quad m \gg 1 , \tag{15.46}$$

so then the total power radiated, obtained by summing P_m over all positive m, will diverge. If one artificially puts a cutoff in the summation at m_c, we would have

$$P = \sum_{m=1}^{\infty} P_m \sim \sum_{m}^{\sim m_c} m^{1/3} \sim m_c^{4/3} , \quad m_c \gg 1 , \tag{15.47}$$

which upon comparison with the total power (15.24) says that

$$m_c \sim \left(\frac{\mathcal{E}}{m_0 c^2}\right)^3 . \tag{15.48}$$

The corresponding wavelength is in the x-ray range (~ 1 nm) for a synchrotron of 1 m radius and 1 GeV energy.

The resolution of this conundrum is to recognize that although $1 - \beta$ may be small, it is never zero, and eventually the largeness of m will compensate for its smallness. Indeed, for $m \gg 1$,

$$J'_{2m}(2m\beta) \sim \begin{cases} (2\pi)^{-1} 3^{1/6} \Gamma(2/3) m^{-2/3} , & m(1-\beta^2)^{3/2} \ll 1 , \\ (4\pi m)^{-1/2} (1-\beta^2)^{1/4} e^{-\frac{2}{3} m(1-\beta^2)^{3/2}} , & m(1-\beta^2)^{3/2} \gg 1 , \end{cases} \tag{15.49}$$

and indeed the peak of the spectrum occurs for a large harmonic number characterized by

$$m_c = (1-\beta^2)^{-3/2} = \left(\frac{\mathcal{E}}{m_0 c^2}\right)^3 \tag{15.50}$$

and for $m \gg m_c$ the power decreases exponentially. For a 1 GeV electron synchrotron, $m_c = 7.5 \times 10^9$. For a detailed derivation of (15.49), see [9], Chap. 40.

A representation useful for verifying these assertions is the following:

$$J_{2m}(2m\beta) \sim \frac{1}{\sqrt{3}\pi} \left(\frac{3}{2m}\right)^{1/3} \xi^{1/3} K_{1/3}(\xi) , \quad m \gg 1 , \tag{15.51}$$

where $\xi = (2m/3)(m_0 c^2 / \mathcal{E})^3$, so

$$J'_{2m}(2m\beta) \sim \frac{1}{\sqrt{3}\pi} \left(\frac{3}{2m}\right)^{2/3} \xi^{2/3} K_{2/3}(\xi) , \quad m \gg 1 . \tag{15.52}$$

Hence, an asymptotic formula for the power radiated in the mth harmonic is

$$P_m \sim \frac{\omega_0}{4\pi} \frac{e^2}{R} \left(\frac{2m}{3}\right)^{1/3} \frac{\sqrt{3}}{2\pi} \xi^{2/3} \left[2K_{2/3}(\xi) - \int_\xi^\infty dx\, K_{1/3}(x)\right], \quad m \gg 1.$$

(15.53)

The modified Bessel function here are otherwise called Airy functions:

$$K_{1/3}(\xi) = \pi\sqrt{\frac{3}{z}}\mathrm{Ai}(z) = \sqrt{\frac{3}{z}} \int_0^\infty dt\, \cos\left(\frac{1}{3}t^3 + zt\right), \qquad (15.54a)$$

$$K_{2/3}(\xi) = -\pi\frac{\sqrt{3}}{z}\mathrm{Ai}'(z) = \frac{\sqrt{3}}{z} \int_0^\infty dt\, t \sin\left(\frac{1}{3}t^3 + zt\right), \quad (15.54b)$$

where $z = (3\xi/2)^{2/3}$.

The results given in (15.49) now follow immediately from the properties of the Bessel function of imaginary argument, the Macdonald function. Because of the dominance of very large harmonic numbers, we may use the representation (15.53) to reproduce the total power radiated (15.24) using the integral

$$\int_0^\infty dt\, t^{\mu-1}\, K_\nu(t) = 2^{\mu-2}\Gamma\left(\frac{\mu-\nu}{2}\right)\Gamma\left(\frac{\mu+\nu}{2}\right), \qquad (15.55)$$

that is,

$$
\begin{aligned}
P = \sum_{m=1}^\infty P_m &\approx \frac{3}{2}\left(\frac{\mathcal{E}}{m_0c^2}\right)^3 \int_0^\infty d\xi\, P_m \\
&= \omega_0 \frac{e^2}{4\pi R}\left(\frac{\mathcal{E}}{m_0c^2}\right)^4 \frac{3\sqrt{3}}{4\pi} \int_0^\infty \xi\, d\xi \left[2K_{2/3}(\xi) - \int_\xi^\infty dx\, K_{1/3}(x)\right] \\
&= \omega_0 \frac{e^2}{4\pi R}\left(\frac{\mathcal{E}}{m_0c^2}\right)^4 \frac{3\sqrt{3}}{4\pi}\left[2\int_0^\infty \xi\, d\xi\, K_{2/3}(\xi) - \frac{1}{2}\int_0^\infty \xi^2\, d\xi\, K_{1/3}(\xi)\right] \\
&= \frac{2}{3}\omega_0 \frac{e^2}{4\pi R}\left(\frac{\mathcal{E}}{m_0c^2}\right)^4.
\end{aligned}
$$

(15.56)

Numerically, it is interesting to compare the asymptotic formula (15.53) with the exact result for the spectral distribution (15.44). In fact for a very modest β of 0.999, the asymptotic formula is high by only 16% at $m = 1$, and the discrepancy drops rapidly to $1 - \beta^3$ for larger harmonic numbers. (The error at $m = 100$ is only 0.7%.) It is much faster to evaluate (15.53) than the exact (15.44). In Fig. 15.1, we show the power spectrum for $\beta = 0.999$, corresponding to an electron energy of 11 MeV, for which $m_c = 11,200$. Note that the error in using (15.53) is 3% at the lowest harmonic number.

15.4 Angular Distribution of Radiated Power

To study the angular distribution of the radiation emitted by an electron in a synchrotron, we may return to expression (15.35) and apply it to the region

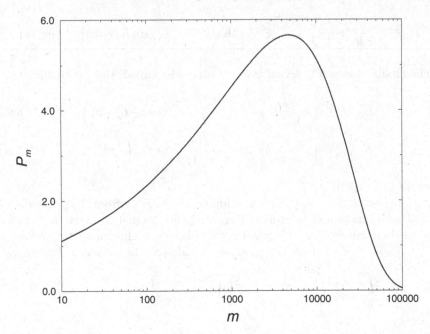

Fig. 15.1. Graph of the asymptotic power radiated, divided by $\omega_0 e^2/4\pi R$, given by (15.53), for $\beta = 0.999$

far from the synchrotron, where $r \gg R$. Thus we may approximate, in the exponent,

$$|\mathbf{r} - \mathbf{r}'| \approx r - \hat{\mathbf{n}} \cdot \mathbf{r}', \quad \hat{\mathbf{n}} = \frac{\mathbf{r}}{r}. \tag{15.57}$$

Then the vector potential in the radiation zone is, according to (15.33)

$$
\begin{aligned}
\mathbf{A}_m(\mathbf{r}) &\sim \frac{1}{4\pi rc} e^{i\omega_0 mr/c} \int (d\mathbf{r}') \, \mathbf{j}_m(\mathbf{r}') \, e^{-im\omega_0 \hat{\mathbf{n}} \cdot \mathbf{r}'/c} \\
&= \frac{1}{4\pi} \frac{\beta e}{2\pi r} e^{im\omega_0 r/c} \int_{-\pi}^{\pi} d\varphi' e^{im(\varphi'-\varphi_0)} \hat{\varphi}' e^{-im\beta \sin\theta \cos(\varphi-\varphi')}.
\end{aligned}
\tag{15.58}
$$

Now we write the azimuthal unit vector in terms of fixed basis vectors:

$$\hat{\varphi}' = \hat{\varphi}\cos(\varphi - \varphi') + \hat{\rho}\sin(\varphi - \varphi'). \tag{15.59}$$

The resulting angular integrals can be expressed in terms of the generating function for the Bessel functions (8.19), or

$$i^{-m} J_m(z) = \int_{-\pi}^{\pi} \frac{d\varphi'}{2\pi} e^{-iz\cos\varphi'} e^{im\varphi'}, \tag{15.60}$$

as follows:

$$\mathbf{A}_m \sim \frac{e\beta}{4\pi r} e^{im\omega_o r/c} e^{im(\varphi - \varphi_0 - \pi/2)} \left[i\hat{\varphi} J_m'(m\beta \sin\theta) + \hat{\rho} \frac{J_m(m\beta \sin\theta)}{\beta \sin\theta} \right] ,$$

(15.61)

which uses (8.7a). This implies

$$|\mathbf{A}_m|^2 = \frac{e^2\beta^2}{(4\pi)^2 r^2} \left[(J_m'(x))^2 + \frac{J_m^2(x)}{\beta^2 \sin^2\theta} \right] ,$$

(15.62a)

$$|\hat{\mathbf{n}} \cdot \mathbf{A}_m|^2 = \frac{e^2}{(4\pi)^2 r^2} J_m^2(x) ,$$

(15.62b)

with $x = m\beta \sin\theta$.

From the potential, we can calculate the Fourier components of the electric and magnetic fields, because in the radiation zone

$$\nabla \mathbf{A}_m \sim \frac{im\omega_0}{c} \hat{\mathbf{n}} \mathbf{A}_m :$$

(15.63)

$$\mathbf{E}_m = i\frac{m\omega_0}{c} \mathbf{A}_m - \frac{1}{im\omega_c/c} \nabla(\nabla \cdot \mathbf{A}_m)$$

$$\sim -i\frac{m\omega_0}{c} \hat{\mathbf{n}} \times (\hat{\mathbf{n}} \times \mathbf{A}_m) ,$$

(15.64a)

$$\mathbf{H}_m = \nabla \times \mathbf{A}_m \sim i\frac{m\omega_0}{c} \hat{\mathbf{n}} \times \mathbf{A}_m .$$

(15.64b)

Then the time-averaged Poynting vector (1.20b) is

$$\mathbf{S} = c\overline{\mathbf{E} \times \mathbf{H}} = 2c \operatorname{Re} \sum_{m=1}^{\infty} \mathbf{E}_m^* \times \mathbf{H}_m = \sum_{m=1}^{\infty} \mathbf{S}_m ,$$

(15.65)

so the power radiated per unit area into the mth harmonic is

$$\mathbf{S}_m = 2\hat{\mathbf{n}} \frac{m^2\omega_0^2}{c} \left(|\mathbf{A}_m|^2 - |\hat{\mathbf{n}} \cdot \mathbf{A}_m|^2 \right) ,$$

(15.66)

which involves just the combinations given in (15.62a) and (15.62b). Taking the radial component of this, and multiplying by r^2, we immediately obtain the power radiated into a given solid angle at frequency $m\omega_0$:

$$\frac{dP_m}{d\Omega} = \frac{\omega_0}{2\pi} \frac{e^2}{4\pi R} \beta^3 m^2 \left[J_m'^2(m\beta \sin\theta) + \frac{\cot^2\theta}{\beta^2} J_m^2(m\beta \sin\theta) \right].$$

(15.67)

It turns out that the two terms here have physical significance ([9], Chap. 39): They correspond to the power radiated in the polarization state corresponding to the electric field in the plane of the orbit, characterized by a unit vector $\mathbf{e}_\parallel$ perpendicular to the plane defined by $\hat{\mathbf{z}}$ and $\hat{\mathbf{n}}$, and by the electric field in the plane defined by the normal to the orbit and the direction of observation, characterized by a unit vector $\mathbf{e}_\perp$ in that plane, respectively. (Both unit vectors are necessarily perpendicular to $\hat{\mathbf{n}}$.) That is,

$$\left(\frac{\mathrm{d}P_m}{\mathrm{d}\Omega}\right)_\parallel = \frac{\omega_0}{2\pi}\frac{e^2}{4\pi R}\beta^3 m^2 \left[J_m'(m\beta\sin\theta)\right]^2 , \qquad (15.68a)$$

$$\left(\frac{\mathrm{d}P_m}{\mathrm{d}\Omega}\right)_\perp = \frac{\omega_0}{2\pi}\frac{e^2}{4\pi R}\beta^3 m^2 \left[\frac{J_m(m\beta\sin\theta)}{\beta\tan\theta}\right]^2 . \qquad (15.68b)$$

The total power radiated into these two polarizations are

$$P_\parallel = \frac{6+\beta^2}{8}P , \qquad P_\perp = \frac{2-\beta^2}{8}P , \qquad (15.69)$$

which means that the synchrotron radiation is predominantly polarized in the plane of the orbit, the ratio of the power in the two polarizations being 7 in the ultrarelativistic limit, and already 3 nonrelativistically.

The angular distribution of the high harmonics is of principal interest, for we have seen that little energy is radiated in the longer wavelengths. The general character of the radiation pattern is easily seen, for Bessel functions of high order are very small if the argument is appreciably less than the order, and therefore the radiation intensity is negligible unless $\sin\theta$ is close to unity. Hence the radiation is closely confined to the plane of the orbit. For a more precise analysis, we employ the approximate Bessel function representation already described in (15.51) and (15.52). Introducing the angle $\psi = \pi/2 - \theta$ between the point of observation and the plane of the orbit, which we suppose to be small, we write the power radiated into a unit angular range about the angle ψ, in the mth harmonic, as

$$\frac{\mathrm{d}P_m}{\mathrm{d}\Omega}(\psi) \sim \frac{\omega_0}{2\pi}\frac{e^2}{4\pi R}\frac{3}{\pi^2}\left(\frac{m}{3}\right)^{2/3}\left[\xi^{4/3}K_{2/3}^2(\xi) + \psi^2\left(\frac{m}{3}\right)^{2/3}\xi^{2/3}K_{1/3}^2(\xi)\right] , \qquad (15.70)$$

for $m \gg 1$ and $\psi \ll 1$, where

$$\xi = \frac{m}{3}(1 - \beta^2\sin^2\theta)^{3/2} \approx \frac{m}{3}\left[\psi^2 + \left(\frac{m_0 c^2}{\mathcal{E}}\right)^2\right]^{3/2} . \qquad (15.71)$$

In consequence of the properties of the cylinder functions, the radiation intensity decreases rapidly when ξ becomes appreciably greater than unity. Hence, for the modes of importance, $m \sim (\mathcal{E}/m_0 c^2)^3$, the angular range within which the energy is sensibly confined is of the order $m^{-1/3}$. Within this angular range, the intensity per unit angle is essentially independent of ψ and varies with m as $m^{2/3}$, which is consistent with the $m^{1/3}$ variation of the total power in a given harmonic. In virtue of the concentration of the radiation at the higher harmonics, $m \sim m_c$, it is clear that the mean angular range for the total radiation will be $\sim m_c^{-1/3} = m_0 c^2/\mathcal{E}$, which is approximately 0.03° for an electron at $\mathcal{E} = 1$ GeV. A plot of the angular distribution, for $\beta = 0.999$ and two different values of m, is shown in Fig. 15.2, illustrating the above points.

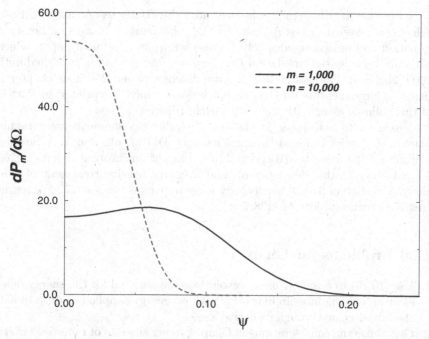

Fig. 15.2. The angular distribution of synchrotron radiation about the plane of the orbit calculated from (15.70) for $\beta = 0.999$, for $m = 1000$, and $m = 10,000$. Recall that $m_c = 11,200$ characterizes the peak of the spectrum in this case. Plotted is the power relative to $(\omega_0/2\pi)(e^2/4\pi R)$

15.5 Historical Note

As the Second World War wound down, many physicists at the Rad Lab and elsewhere started thinking again of fundamental physics, and of the necessity of overcoming the nonrelativistic limit of Lawrence's cyclotron. The betatron, which used a time-varying magnetic field to produce an accelerating voltage on electrons, was the first machine built on new principles, but devices using microwave cavities which had been developed so highly were on many people's minds. We described Schwinger's ideas in this regard in Chap. 14.

It appears to have been a conversation, in late 1944 or early 1945, with Marcel Schein of Chicago that sparked Schwinger's consideration of the radiation produced by the electrons in a betatron. The thinking at the time was that there would be little radiation because of destructive interference between the different electrons. Schwinger finally had the time to examine the problem in July 1945. He quickly developed the theory described here, and immediately communicated the salient features to Schein [24]. The conclusions were strikingly contrary to expectation: "(1) There is no interference, and the electrons radiate independently; (2) The energy is radiated, not in the micro-wave, but rather the infra-red region."

Later that month, actually the week after the Trinity test on July 16, 1945, Schwinger traveled to Los Alamos, and gave his famous lectures on betatron radiation and on waveguides [24]. He also wrote up a detailed paper, which had only very limited circulation as a preprint, and was never published until 2000 [2]. He did publish in 1949 a quite different paper [25]. This chapter is based in large measure on the original version, which is reprinted in Part II of this volume, along with the 1949 Physical Review article.

Unknown to Schwinger at the time, parallel developments were taking place in the Soviet Union at the same moment. D. Ivanenko and A. A. Sokolov worked out the angular and spectral properties of synchrotron radiation [26]; for a history of this development, and a comprehensive treatment of synchrotron radiation from a particularly Russian perspective see [27]. For recent treatises on the subject, see [28, 29].

15.6 Problems for Chap. 15

1. Use (15.25) to find how many revolutions are required for the energy of an electron moving in a circular orbit without energy supplied to drop to half the initial value through radiative loss.
2. Generalize the considerations in Chap. 1 to get the rate of radiative energy loss:

$$-\frac{dE}{dt}\bigg|_{\text{rad}} = \frac{2}{3}\frac{e^2}{4\pi c^3}\frac{1}{(1-\beta^2)^3}\left[\left(\frac{d\mathbf{v}}{dt}\right)^2 - \left(\frac{\mathbf{v}}{c}\times\frac{d\mathbf{v}}{dt}\right)^2\right]. \qquad (15.72)$$

Consider a charge accelerated from rest by a constant electric field $\mathbf{E}$, and show that the rate of radiation does *not* change with increasing energy. This is one reason that the successor to the LHC is projected to be a linear collider.

Problems 15.3 through 15.6 refer to what is called the *Wiggler*, which beside accelerator applications, has application to the free electron laser – see Problem 3.28.

3. Consider a helical magnetic field, for $0 < z < L = N\lambda_w$, where N is an integer, $(\lambda = \lambda/2\pi)$

$$B_x = B\cos\left(\frac{z}{\lambda_w}\right), \qquad B_y = B\sin\left(\frac{z}{\lambda_w}\right). \qquad (15.73)$$

At $t = 0$, an electron enters the helical magnetic field at $z = 0$ with $v_z = v$, $v_x = v_y = 0$. Find the leading approximations for small displacements to $v_x(t)$, $v_y(t)$, $v_z(t)$, and also $z(t)$, for $0 < t < L/v$.
4. Now apply the radiated energy formula for a point charge, $(L = N\lambda_\omega)$

$$\frac{dE}{d\Omega\,d\omega} = \frac{\omega^2}{16\pi^3 c^3}\left|\mathbf{n}\times\int_0^{L/v} dt\,e^{i\omega t}e^{-i\omega\mathbf{n}\cdot\mathbf{r}(t)/c}e\mathbf{v}(t)\right|^2, \qquad (15.74)$$

to radiation along the z-axis ($n_z = 1$, $n_x = n_y = 0$) under the assumption that $1 - \beta^2 \ll 1$. At what value of ω (ϖ) is the radiation a *maximum*? State that radiated energy, per unit solid angle and frequency, as a multiple of N^2.

5. What is the *smallest* fractional deviation of ω (positive or negative), $\delta\omega/\varpi$, that reduces the radiated energy to zero? This is a rough measure of the width of the spectral line centered around ϖ. Under the assumption that $\delta\omega/\varpi \ll 1$ (what does that say about N?) carry out the integration over ω to get $dE/d\Omega$. Check that your result has the dimensions of *energy*.

6. Now suppose that the radiation is emitted at a very small angle ($\theta \ll 1$) relative to the z-axis. What is the frequency? How must θ be restricted so that the frequency is changed by less than some given small fraction? After integrating $dE/d\Omega$ over that small angular range, how does the resulting radiated energy, E_{rad}, vary with the relativistic energy of the electron?

7. The asymptotic formula for the power radiated by a synchrotron into the mth harmonic is given in (15.53), where the modified Bessel function combination given there can be alternatively written as

$$2K_{2/3}(\xi) - \int_\xi^\infty d\eta \, K_{1/3}(\eta) = \int_\xi^\infty d\eta \, K_{5/3}(\eta) \, . \qquad (15.75)$$

Show that when we are considering parallel polarization, this combination is replaced by

$$\frac{1}{2}\left[3K_{2/3}(\xi) - \int_\xi^\infty d\eta \, K_{1/3}(\eta) \right] = \frac{1}{4} \int_\xi^\infty d\eta \, [3K_{5/3}(\eta) + K_{1/3}(\eta)] \, , \qquad (15.76)$$

while the remainder describes perpendicular polarization,

$$\frac{1}{2}\left[K_{2/3}(\xi) - \int_\xi^\infty d\eta \, K_{1/3}(\eta) \right] = \frac{1}{3} \int_\xi^\infty \frac{d\eta}{\eta} K_{2/3}(\eta)$$

$$= \frac{1}{4} \int_\xi^\infty d\eta \, [K_{5/3}(\eta) - K_{1/3}(\eta)] \, . \qquad (15.77)$$

Show that the 7:1 ratio seen in (15.69) emerges on performing the frequency integral.

16

Diffraction

The wave nature of light is manifested by diffraction. Such problems are notoriously difficult to solve. Here we indicate two approaches – the first, based on a variational principle, and the second, valid for short wavelengths, based on the exact solution found by Sommerfeld in 1896 for the diffraction by a straight edge.

16.1 Variational Principle for Scattering

In this chapter we will principally consider essentially two-dimensional situations, of scattering of waves by a thin conductor lying in the $x = 0$ plane, with translational symmetry along the y-axis. Consider a scattering by an H mode ($\perp$-mode), that is, the electric field of the incident electromagnetic wave points along the y-axis, and is therefore tangential to the conducting surface.[1] The geometry is sketched in Fig. 16.1. In terms of the current density in the conductor, which then necessarily also points in the y-direction, the scattered electric field intensity is, in Heaviside–Lorentz units with $k = \omega/c$,

$$E_{\text{scatt}} = \frac{ik}{4\pi c} \int (d\mathbf{r}') \frac{e^{ik|\mathbf{r}-\mathbf{r}'|}}{|\mathbf{r} - \mathbf{r}'|} J(\mathbf{r}') , \qquad (16.1)$$

which follows from the first line of (1.117). Since the conductor is infinitesimally thin, we can immediately carry out the integral over x',

$$\int dx' J(x', z') = K(z') , \qquad (16.2)$$

in terms of the surface current density. Because the latter does not depend on y', we can carry out the corresponding integral in terms of the definition of the Hankel function:

---

[1] This notation is that of the analogous situation discussed in Chap. 9. However, in view of the translational invariance along the y-direction, we could equally well refer to this as an E mode, and indeed Schwinger did so in his original notes.

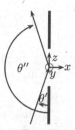

Fig. 16.1. Coordinate system adopted to describe scattering by a thin conducting screen containing an aperture. Grazing angles of the incident and scattered waves, θ' and θ'', respectively, are shown

$$\pi i H_0^{(1)}(k\rho) = \int_{-\infty}^{\infty} dy' \frac{e^{ik\sqrt{\rho^2+(y'-y)^2}}}{\sqrt{\rho^2+(y'-y)^2}} . \tag{16.3}$$

Therefore, in terms of the incident electric field E_{inc}, we have the following expression for the electric field amplitude,

$$E = E_{inc} + \frac{ik}{4\pi c} \int_{-\infty}^{\infty} dz' \, \pi i H_0^{(1)} \left(k\sqrt{x^2+(z-z')^2}\right) K(z') . \tag{16.4}$$

Of course, the z' integration is extended only over that portion of the $x = 0$ plane occupied by conductors. This expression is actually an integral equation, since the current in the conductor is determined by the electric field itself. On the conductor, $E = 0$, so we have

$$E_{inc}(0, z) = \frac{k}{4c} \int_{cond} dz' \, H_0^{(1)}(k|z - z'|)K(z') , \tag{16.5}$$

an integral equation for $K(z)$. There is only one situation in which this equation has been solved exactly, that for a straight edge, where the conductor extends along the negative z-axis, that is, $K(y) = 0$ for $z > 0$. This was first worked out by Sommerfeld [30]; Schwinger's solution of this problem is recounted in Chap. 48 of [9].

In general, we will have to treat this problem approximately, and make some intelligent guesses as to the form of the current distribution. For this purpose, it will be convenient to recast the problem in the form of a variational principle in the current density, so that a first-order error in the current translates into a second-order error in the scattering amplitude.

Let us consider a plane E-mode wave incident, with a propagation vector making an angle θ' with respect to the surface of the plane $x = 0$. The integral equation (16.5) becomes

$$e^{ikz\cos\theta'} = \frac{k}{4c} \int_{cond} dz' \, H_0^{(1)}(k|z - z'|)K(z') , \tag{16.6}$$

for z on the conductor. From this follows, for arbitrary K',

$$\int_{\text{cond}} \mathrm{d}z\, \mathrm{e}^{\mathrm{i}kz\cos\theta'} K'(z) = \frac{k}{4c} \int_{\text{cond}} \mathrm{d}z\, \mathrm{d}z'\, K'(z) H_0^{(1)}(k|z-z'|) K(z') \ . \quad (16.7)$$

We are interested in the scattering amplitude describing fields far away from the conductor, so we may use in (16.4) the asymptotic form of the Hankel function:

$$H_0^{(1)}\left(k\sqrt{x^2+(z-z')^2}\right) \sim \sqrt{\frac{2}{\pi \mathrm{i}k}} \frac{\mathrm{e}^{\mathrm{i}k[x^2+(y-y')^2]^{1/2}}}{[x^2+(y-y')^2]^{1/4}}, \quad kx \gg 1 \ . \quad (16.8)$$

Furthermore, since z' is a coordinate in the conductor, asymptotically far away,

$$\sqrt{x^2+(z-z')^2} \sim \rho - \frac{z}{\rho}z' = \rho + z'\cos\theta'' \ , \quad (16.9)$$

where $\rho = \sqrt{x^2+z^2} \gg |z'|$, and θ'' is the angle of the scattered wave again relative to the plane of the conductor. Thus, the Hankel function, in the asymptotic region, may be replaced by

$$H_0^{(1)}\left(k\sqrt{x^2+(z-z')^2}\right) \sim \sqrt{\frac{2}{\pi \mathrm{i}k\rho}} \mathrm{e}^{\mathrm{i}k\rho} \mathrm{e}^{\mathrm{i}kz'\cos\theta''} \ . \quad (16.10)$$

We now define the scattering amplitude by comparison with (16.4),

$$A(\theta',\theta'') = \frac{\mathrm{i}k}{c} \int_{\text{cond}} \mathrm{d}z'\mathrm{e}^{\mathrm{i}kz'\cos\theta''} K(z') \ , \quad (16.11)$$

which has been normalized so that the differential cross section is obtained from it by

$$\sigma(\theta',\theta'') = \frac{1}{8\pi k}|A(\theta',\theta'')|^2 \ , \quad (16.12)$$

and the total cross section is related to the imaginary part of the forward scattering amplitude by the optical theorem,

$$\sigma(\theta') = \frac{1}{k}\mathrm{Im}\, A(\theta',\theta'+\pi) \ . \quad (16.13)$$

In view of the integral equation (16.6), which determines the surface current for the wave incident on the conductor with grazing angle θ' (call that current $K_{\theta'}$), and the analogously defined

$$\mathrm{e}^{\mathrm{i}kz\cos\theta''} = \frac{k}{4c} \int_{\text{cond}} \mathrm{d}z'\, H_0^{(1)}(k|z-z'|) K_{\theta''}(z') \ , \quad (16.14)$$

we have from (16.7) the following equation, homogeneous in the two current densities, for the scattering amplitude (16.11):

$$\frac{1}{A(\theta',\theta'')} = -\frac{\mathrm{i}}{4} \frac{\int \mathrm{d}z\, \mathrm{d}z'\, K_{\theta'}(z) H_0^{(1)}(k|z-z'|) K_{\theta''}(z')}{\int \mathrm{d}z\, \mathrm{e}^{\mathrm{i}kz\cos\theta''} K_{\theta'}(z) \int \mathrm{d}z'\, \mathrm{e}^{\mathrm{i}kz'\cos\theta'} K_{\theta''}(z')} \ . \quad (16.15)$$

This form now is stationary with respect to both $K_{\theta'}$ and $K_{\theta''}$. That is, if we make independent small variations in both of these currents, and use the integral equation (16.14), as well as the analogous equation obtained by interchanging θ' and θ'', we get a vanishing result in first order. This means that a relatively crude estimate for the surface current should lead to a considerably more accurate estimate for the scattering amplitude, and for the cross section.

16.2 Scattering by a Strip

As an example of the use of this method, we consider a metallic strip of width δ of negligible thickness and infinite length. The scattering amplitude for an E-mode wave incident at an angle θ' relative to the plane of the strip and scattering at angle θ'' is given by (16.15), with coordinates illustrated in Fig. 16.1. If $\theta' = \theta''$ we are describing backscattering, which leads to the "radar" cross section (so called because only backscattering is detected by a radar receiver), while if $\pi + \theta' = \theta''$ we describe forward scattering, which by the optical theorem [see (16.13)] represents the total cross section.

16.2.1 Normal Incidence

As a first example consider normal incidence, backscattering, $\theta' = \theta'' = \frac{\pi}{2}$. Assume $K_{\pi/2}(z) = $ constant. Then we have from (16.15) for the radar amplitude

$$\frac{1}{R} = -\frac{i}{4\delta^2} \int dz\, dz'\, H_0^{(1)}(k|z - z'|) \,. \tag{16.16}$$

To evaluate this integral, introduce a rotation of $\pi/4$ by defining new variables

$$\xi = \frac{z - z'}{\sqrt{2}}, \qquad \eta = \frac{z + z'}{\sqrt{2}}, \tag{16.17}$$

so (16.16) becomes

$$\begin{aligned}
\frac{1}{R} &= -\frac{i}{4\delta^2} \int_{-\delta/\sqrt{2}}^{\delta/\sqrt{2}} d\xi \int_{-(\delta/\sqrt{2}-|\xi|)}^{\delta/\sqrt{2}-|\xi|} d\eta\, H_0^{(1)}(k\sqrt{2}|\xi|) \\
&= -\frac{i}{\delta^2} \int_0^{\delta/\sqrt{2}} d\xi \left(\frac{\delta}{\sqrt{2}} - \xi\right) H_0^{(1)}(k\sqrt{2}\xi) \\
&= -\frac{i}{2} \int_0^1 dx\, (1 - x) H_0^{(1)}(kx\delta) \,,
\end{aligned} \tag{16.18}$$

where we have introduced $\xi = x\delta/\sqrt{2}$. Now remember from (8.8a) that

$$\frac{d}{dx}\left[x H_1^{(1)}(\lambda x)\right] = \lambda x H_0^{(1)}(\lambda x) \,, \tag{16.19}$$

so

$$\int_0^1 \mathrm{d}x \, x \, H_0^{(1)}(\lambda x) = \frac{1}{\lambda}\left[H_1^{(1)}(\lambda) + \frac{2\mathrm{i}}{\pi \lambda}\right] , \tag{16.20}$$

which uses (8.30a). The remaining integral can be evaluated in terms of Struve functions. The fundamental such function is defined by

$$S_0(x) = \frac{2}{\pi}\int_0^{\pi/2} \mathrm{d}\vartheta \sin(x\cos\vartheta) , \tag{16.21}$$

from which we deduce

$$\frac{\mathrm{d}S_0(x)}{\mathrm{d}x} = \frac{2}{\pi}\int_0^{\pi/2} \mathrm{d}\vartheta \cos\vartheta \cos(x\cos\vartheta)$$

$$= \frac{2}{\pi} - \frac{2}{\pi}x\int_0^{\pi/2} \mathrm{d}\vartheta \sin^2\vartheta \sin(x\cos\vartheta) , \tag{16.22}$$

after integration by parts. $S_1(x)$ is defined by

$$S_1(x) = \frac{2x}{\pi}\int_0^{\pi/2} \mathrm{d}\vartheta \sin^2\vartheta \sin(x\cos\vartheta) , \tag{16.23}$$

so the differential equation for $S_1(x)$ is

$$\frac{\mathrm{d}}{\mathrm{d}x}(xS_1(x)) = \frac{4x}{\pi}\int_0^{\pi/2} \mathrm{d}\vartheta \sin^2\vartheta \sin(x\cos\vartheta)$$

$$+ \frac{2x^2}{\pi}\int_0^{\pi/2} \mathrm{d}\vartheta \cos\vartheta \sin^2\vartheta \cos(x\cos\vartheta)$$

$$= -\frac{2x}{\pi}\int_0^{\pi/2} \mathrm{d}[\sin(x\cos\vartheta)]\sin\vartheta\cos\vartheta$$

$$+ \frac{4x}{\pi}\int_0^{\pi/2} \mathrm{d}\vartheta \sin^2\vartheta \sin(x\cos\vartheta)$$

$$= \frac{2x}{\pi}\int_0^{\pi/2} \mathrm{d}\vartheta \sin(x\cos\vartheta) = xS_0(x) , \tag{16.24}$$

after integrating by parts. Thus we have from (16.22) and (16.24)

$$\frac{\mathrm{d}S_0}{\mathrm{d}x} = \frac{2}{\pi} - S_1 , \qquad \frac{\mathrm{d}}{\mathrm{d}x}(xS_1) = xS_0 . \tag{16.25}$$

If the constant were not present in the first of these relations, these equations would be the same as the Bessel function recurrence relations (8.8a) and (8.8b) with $\mu = 1$ and 0, respectively. Therefore, the 0th Struve function satisfies the differential equation

$$\left(\frac{\mathrm{d}^2}{\mathrm{d}x^2} + \frac{1}{x}\frac{\mathrm{d}}{\mathrm{d}x} + 1\right)S_0(x) = \frac{2}{\pi x} , \tag{16.26}$$

which is to be compared with the equation satisfied by the 0th Hankel function:

$$\left(\frac{\mathrm{d}^2}{\mathrm{d}x^2} + \frac{1}{x}\frac{\mathrm{d}}{\mathrm{d}x} + 1\right) H_0^{(1)}(x) = 0 .\tag{16.27}$$

Multiply the first of these equations by $xH_0^{(1)}$ and the second by $x\mathcal{S}_0$, and subtract. Then, the left-hand side is a total differential,

$$\frac{\mathrm{d}}{\mathrm{d}x}\left[x\left(H_0^{(1)}(x)\frac{\mathrm{d}}{\mathrm{d}x}\mathcal{S}_0(x) - \mathcal{S}_0(x)\frac{\mathrm{d}}{\mathrm{d}x}H_0^{(1)}(x)\right)\right] = \frac{2}{\pi}H_0^{(1)}(x) ,\tag{16.28}$$

and hence if we integrate this equation, we obtain

$$\frac{2}{\pi}\int_0^x \mathrm{d}t\, H_0^{(1)}(t) = x\left\{H_0^{(1)}(x)\left[\frac{2}{\pi} - \mathcal{S}_1(x)\right] + H_1^{(1)}(x)\mathcal{S}_0(x)\right\} .\tag{16.29}$$

Now returning to (16.18), we find the backscattering amplitude to be, from (16.20) and (16.29),

$$\frac{1}{R} = -\frac{i}{2k\delta}\left[\int_0^{k\delta} \mathrm{d}x\, H_0^{(1)}(x) - H_1^{(1)}(k\delta) - \frac{2i}{\pi k\delta}\right]$$

$$= -\frac{i}{2k\delta}\left[k\delta H_0^{(1)}(k\delta) - H_1^{(1)}(k\delta)\right.$$

$$\left. + \frac{\pi}{2}k\delta\left(H_1^{(1)}(k\delta)\mathcal{S}_0(k\delta) - H_0^{(1)}(k\delta)\mathcal{S}_1(k\delta)\right) - \frac{2i}{\pi k\delta}\right] .\tag{16.30}$$

Consider first the short-wavelength limit, $k\delta \gg 1$. Then, from the representation[2]

$$H_0^{(1)}(|x|) = \frac{1}{\pi}\int_{-\infty}^{\infty} \mathrm{d}t\frac{e^{ixt}}{\sqrt{1-t^2}} ,\tag{16.31}$$

we find

$$\int_0^{\infty} \mathrm{d}x\, H_0^{(1)}(x) = \int_{-\infty}^{\infty} \mathrm{d}t\frac{\delta(t)}{\sqrt{1-t^2}} = 1 ,\tag{16.32}$$

which is equivalent to

$$\int_0^{\infty} \mathrm{d}x\, J_0(x) = 1 , \qquad \int_0^{\infty} \mathrm{d}x\, N_0(x) = 0 .\tag{16.33}$$

From (16.32) we get the geometrical limit entirely from the first term in the first line of (16.30),

$$R = 2ik\delta ,\tag{16.34}$$

and the backscattering cross section per unit angle is

---

[2] The branch point is interpreted by setting in the square root $1 \to 1 + i\epsilon$. See [9], p. 513, for example, for a derivation.

$$\sigma\left(\frac{\pi}{2},\frac{\pi}{2}\right) = \frac{1}{8\pi k}|R|^2 = \frac{k\delta^2}{2\pi} = \frac{\delta^2}{\lambda} . \tag{16.35}$$

In the same limit, the total cross section is given by the optical theorem (16.13),

$$\sigma\left(\frac{\pi}{2}\right) = \frac{1}{k}\text{Im}\,(2ik\delta) = 2\delta . \tag{16.36}$$

We use here the fact that as much is scattered forward as backward. The reason for the factor of δ/λ in the backscattering cross section is that radiation is scattering in a narrow width (specular reflection). The reason for the factor of 2 in the total cross section is that there is a cross section δ for forward scattering to make the shadow, and the backward scattering gives another δ. We can easily get corrections to these results by using asymptotic forms of the Hankel functions.

In the opposite limit, $k\delta \ll 1$, it is easiest to go back to (16.18), where from (8.30b):

$$\frac{1}{R} \approx -\frac{i}{2}\int_0^1 dx\,(1-x)\left[1 + \frac{2i}{\pi}\log\left(\frac{\gamma}{2}xk\delta\right)\right]$$
$$= -\frac{i}{4}\left[1 + \frac{2i}{\pi}\left(\log\frac{\gamma k\delta}{2} - \frac{3}{2}\right)\right] . \tag{16.37}$$

Here we see the appearance of the Euler constant,

$$C = \ln\gamma = 0.577216\ldots . \tag{16.38}$$

The correct answer differs from this by the replacement of the factor multiplying $\frac{2i}{\pi}$ by $\log\gamma k\delta/4$, which is to say that in place of 4 in the logarithm we have here $2e^{3/2}$, an error in the ratio of 0.45, or an additive error in that coefficient of 0.81. The backscattering cross section per unit angle is

$$\sigma\left(\frac{\pi}{2},\frac{\pi}{2}\right) = \frac{1}{8\pi k}\frac{16}{1 + \frac{4}{\pi^2}\left(\log\frac{\gamma k\delta}{2} - \frac{3}{2}\right)^2} , \tag{16.39}$$

and from the optical theorem, the total cross section is

$$\sigma\left(\frac{\pi}{2}\right) = \frac{1}{k}\frac{4}{1 + \frac{4}{\pi^2}\left(\log\frac{\gamma k\delta}{2} - \frac{3}{2}\right)^2} , \tag{16.40}$$

which is $2\pi\sigma\left(\frac{\pi}{2},\frac{\pi}{2}\right)$, which indicates that the scattering is uniform as is to be expected in long-wavelength limit.

16.2.2 Grazing Incidence

Next, consider the problem of the wave incident at angle $\theta' = 0$. Let us assume now that the surface currents are

$$K_0(z) = e^{ikz} , \qquad K_\pi(z) = e^{-ikz} , \tag{16.41}$$

since K must be constant in the long-wavelength limit, and must be a plane wave for short wavelength. The backscattering amplitude is given by (16.15), or

$$\frac{1}{A(0,0)} = -\frac{i}{4} \frac{\int_{-\delta/2}^{\delta/2} dz\, dz'\, H_0^{(1)}(k|z - z'|)e^{ik(z+z')}}{4\left(\frac{\sin k\delta}{2k}\right)^2} . \tag{16.42}$$

Now we do the integral appearing in the numerator using the variables (16.17)

$$\int_{-\delta/2}^{\delta/2} dz\, dz' e^{ik(z+z')} H_0^{(1)}(k|z - z'|) = \int_{-\delta/\sqrt{2}}^{\delta/\sqrt{2}} d\xi \int_{-(\delta/\sqrt{2}-|\xi|)}^{\delta/\sqrt{2}-|\xi|} d\eta\, e^{ik\sqrt{2}\eta}$$
$$\times H_0^{(1)}(k\sqrt{2}|\xi|)$$
$$= \frac{2}{k^2} \int_0^{k\delta} dt\, H_0^{(1)}(t) \sin(k\delta - t) . \tag{16.43}$$

Now define a function F by

$$F(x) = \int_0^x dt\, H_0^{(1)}(t) \sin(x - t) . \tag{16.44}$$

Its derivative is

$$F'(x) = \int_0^x dt\, H_0^{(1)}(t) \cos(x - t) , \tag{16.45}$$

and its second derivative is

$$F''(x) = -\int_0^x dt\, H_0^{(1)}(t) \sin(x - t) + H_0^{(1)}(x) , \tag{16.46}$$

leading to the differential equation

$$F(x) + F''(x) = H_0^{(1)}(x) , \tag{16.47}$$

so a particular solution is, according to (8.9) and (8.8a),

$$F(x) = xH_1^{(1)}(x) . \tag{16.48}$$

To get the general solution, add $A\cos x + B\sin x$ as a solution to the homogeneous equation. Imposing the boundary condition $F(0) = 0$, we deduce from (8.30a) that $A = 2i/\pi$, and $F'(0) = 0$ implies $B = 0$. Thus,

$$F(x) = xH_1^{(1)}(x) + \frac{2i}{\pi} \cos x , \tag{16.49}$$

The backscattering amplitude (16.42) therefore has the form

$$\frac{1}{A(0,0)} = -\frac{i}{2}\frac{k\delta\, J_1(k\delta) + i\left(k\delta\, N_1(k\delta) + \frac{2}{\pi}\cos k\delta\right)}{\sin^2 k\delta}, \tag{16.50}$$

so the cross section per unit angle for backscattering is

$$\sigma_{\text{radar}} = \frac{1}{8\pi k}\frac{4\sin^4 k\delta}{[k\delta\, J_1(k\delta)]^2 + \left[k\delta\, N_1(k\delta) + \frac{2}{\pi}\cos k\delta\right]^2}. \tag{16.51}$$

For short wavelengths, we recall from (8.33a) and (8.33b) the asymptotic behavior of the Bessel functions, $k\delta \gg 1$:

$$J_1(k\delta) \sim \sqrt{\frac{2}{\pi k\delta}}\cos\left(k\delta - \frac{3\pi}{4}\right), \tag{16.52a}$$

$$N_1(k\delta) \sim \sqrt{\frac{2}{\pi k\delta}}\sin\left(k\delta - \frac{3\pi}{4}\right). \tag{16.52b}$$

Thus the backscattering cross section becomes in this limit

$$k\delta \gg 1: \quad \sigma_{\text{radar}} = \frac{1}{2\pi k}\frac{\sin^4 k\delta}{2k\delta/\pi} = \frac{\sin^2 k\delta}{4k^2\delta}, \tag{16.53}$$

so it is very small for short wavelength, as is to be expected.

For long wavelengths $k\delta \ll 1$, we get the same answer (16.39) as for normal incidence, so that as already anticipated the scattering is independent of the angle in the static limit.

To obtain the total cross section, we need the forward scattering amplitude,

$$\frac{1}{A(0,\pi)} = -\frac{i}{4\delta^2}\int_{-\delta/2}^{\delta/2} dz\, dz\, e^{ik(z-z')} H_0^{(1)}(k|z-z'|)$$

$$= -\frac{i}{4\delta^2} 2\delta^2 \int_0^1 dx\,(1-x)\cos kx\delta\, H_0^{(1)}(kx\delta). \tag{16.54}$$

First, we observe that

$$\int_0^x dt \cos t\, H_0^{(1)}(t) = F(x)\sin x + F'(x)\cos x$$

$$= x[H_0^{(1)}(x)\cos x + H_1^{(1)}(x)\sin x], \tag{16.55}$$

which follows from (16.49) and its derivative. To obtain the second integral in (16.54) consider

$$G(x) = \int_0^x dt\, t\, H_0^{(1)}(t)\sin(x-t). \tag{16.56}$$

As before, G satisfies

$$G''(x) + G(x) = x H_0^{(1)}(x) = [x H_1^{(1)}(x)]', \tag{16.57}$$

the general solution of which has the form

$$G(x) = \frac{x^2}{3}H_1^{(1)}(x) + A\cos x + B\sin x . \tag{16.58}$$

The boundary condition $G(0) = 0$ implies $A = 0$, while $G'(0) = 0$ implies $B = 2\mathrm{i}/3\pi$, that is,

$$G(x) = \frac{x^2}{3}H_1^{(1)}(x) + \frac{2\mathrm{i}}{3\pi}\sin x . \tag{16.59}$$

As a result we can conclude

$$\int_0^x \mathrm{d}t\, t\cos t\, H_0^{(1)}(t) = G(x)\sin x + G'(x)\cos x$$

$$= \frac{x^2}{3}\left(H_0^{(1)}(x)\cos x + H_1^{(1)}(x)\sin x\right)$$

$$+ \frac{x}{3}H_1^{(1)}(x)\cos x + \frac{2\mathrm{i}}{3\pi} . \tag{16.60}$$

From this we read off the forward scattering amplitude, with $x = k\delta$,

$$A(0,\pi) = \frac{3\mathrm{i}}{H_0^{(1)}(x)\cos x + H_1^{(1)}(x)\sin x - \frac{1}{2}\frac{\cos x}{x}H_1^{(1)}(x) - \frac{\mathrm{i}}{\pi x^2}} . \tag{16.61}$$

By expanding this for small x, we find the static limit is the same as before, (16.37). For $x \gg 1$, from (8.31a)

$$A(0,\pi) \approx \frac{3\mathrm{i}}{\sqrt{\frac{2}{\pi x}}\mathrm{e}^{\mathrm{i}(x-\pi/4)}(\cos x - \mathrm{i}\sin x)}$$

$$\approx \sqrt{\frac{\pi x}{2}}3\mathrm{i}\,\mathrm{e}^{\pi \mathrm{i}/4} \approx \frac{3}{2}\sqrt{\pi k\delta}(\mathrm{i} - 1) . \tag{16.62}$$

The total cross section is obtained from this by taking the imaginary part and dividing by k:

$$\sigma(0) = \frac{3}{2}\sqrt{\frac{\pi\delta}{k}} = \frac{3}{2}\sqrt{\frac{\lambda\delta}{2}} . \tag{16.63}$$

This is the first diffraction result. The geometrical limit is zero, of course,

$$\lim_{\delta\to\infty}\frac{\sigma}{\delta} = 0 . \tag{16.64}$$

16.2.3 General Incident Angle

Now consider the general situation where the wave is incident at angle θ', and correspondingly assume $K_{\theta'}(z) = \mathrm{e}^{\mathrm{i}k\cos\theta' z}$. Now we have for forward scattering

$$\frac{1}{A(\theta', \theta' + \pi)} = -\frac{i}{4}\frac{1}{\delta^2} \int_{-\delta/2}^{\delta/2} dz\, dz'\, e^{ik\cos\theta'|z-z'|} H_0^{(1)}(k|z - z'|)$$

$$= -\frac{i}{2} \int_0^1 dx\, (1 - x) \cos(kx\delta \cos\theta') H_0^{(1)}(kx\delta)\,. \quad (16.65)$$

This cannot be evaluated in general. We can extract the high and low frequency limits, however. Let us change variables to $t = kx\delta$. Then

$$A(\theta', \theta' + \pi) = \frac{2ik\delta}{\int_0^{k\delta} dt\,(1 - \frac{t}{k\delta}) \cos(t \cos\theta') H_0^{(1)}(t)}\,. \quad (16.66)$$

For low frequency, we can get the result (16.37) by expanding in powers of $k\delta$. To get the high frequency limit, we want to discuss

$$f(x) = \int_0^x dt\,(x - t) \cos\lambda t\, H_0^{(1)}(t)\,, \quad (16.67)$$

which satisfies

$$f''(x) = \cos\lambda x H_0^{(1)}(x)\,. \quad (16.68)$$

Since we want the answer for large x, we need the asymptotic forms and integral representations. We start from (16.31) or

$$H_0^{(1)}(|x|) = \frac{1}{\pi} \int_{-\infty}^{\infty} dt\,\frac{e^{ixt}}{\sqrt{1 - t^2}}\,. \quad (16.69)$$

We choose the branch lines to run along the real axis, from ± 1 to $\pm\infty$ and the path of integration to lie below the branch line on the positive real axis, and above that on the negative real axis. Then, by Jordan's lemma, we can rotate the left-hand contour through the upper half plane, and obtain the contour shown in Fig. 16.2. Hence we have the representation

$$\sqrt{1 - t^2} = -i\sqrt{t^2 - 1}$$
$$\sqrt{1 - t^2} = i\sqrt{t^2 - 1}$$

Fig. 16.2. Contour for defining the 0th Hankel function

$$H_0^{(1)}(x) = \frac{2}{\pi i} \int_1^{\infty} dt\,\frac{e^{ixt}}{\sqrt{t^2 - 1}}\,, \quad (16.70)$$

from which follows

$$J_0(x) = \frac{2}{\pi} \int_1^{\infty} dt\,\frac{\sin xt}{\sqrt{t^2 - 1}} = \frac{2}{\pi} \int_0^{\infty} d\vartheta \sin(x \cosh\vartheta)\,, \quad (16.71a)$$

$$N_0(x) = -\frac{2}{\pi} \int_1^{\infty} dt\,\frac{\cos xt}{\sqrt{t^2 - 1}} = -\frac{2}{\pi} \int_0^{\infty} d\vartheta \cos(x \cosh\vartheta)\,. \quad (16.71b)$$

Now let $t = 1 + \xi$:

$$H_0^{(1)}(x) = \frac{2}{\pi i} e^{ix} \int_0^\infty d\xi \frac{e^{ix\xi}}{\sqrt{\xi(2+\xi)}} . \tag{16.72}$$

For large x, the main contribution for ξ comes from the region of ξ near zero. Hence we get the asymptotic series by expanding the denominator: The first term in the series is, in terms of $x\xi = y$,

$$H_0^{(1)}(x) \sim \frac{2}{\pi i} \frac{e^{ix}}{\sqrt{2x}} \int_0^\infty dy \frac{e^{iy}}{\sqrt{y}} = \sqrt{\frac{2}{\pi x}} e^{ix - \pi i/4} , \quad x \gg 1 , \tag{16.73}$$

which is the known answer (8.31a). We can write the complete expression by noting

$$\xi e^{ix\xi} = \frac{d}{d(ix)} e^{ix\xi} , \tag{16.74}$$

and then, symbolically, from (16.72)

$$
\begin{aligned}
H_0^{(1)}(x) &= \sqrt{\frac{2}{\pi}} e^{ix - \pi i/4} \frac{1}{\sqrt{1 + \frac{1}{2i}\frac{d}{dx}}} \frac{1}{\sqrt{x}} \\
&= \sqrt{\frac{2}{\pi}} e^{ix - \pi i/4} \left(1 + \frac{i}{4}\frac{d}{dx} + \cdots \right) \frac{1}{\sqrt{x}} \\
&= \sqrt{\frac{2}{\pi}} e^{ix - \pi i/4} \left(1 - \frac{i}{8x} + \cdots \right) x^{-1/2} ,
\end{aligned}
\tag{16.75}
$$

etc., which reproduces the first correction embodied in (8.32b).

We will now use the same ideas for f in (16.67). From (16.68) and (16.70) we have

$$f''(x) = \frac{1}{\pi i} \int_1^\infty \frac{dt}{\sqrt{t^2 - 1}} \left[e^{ix(t+\lambda)} + e^{ix(t-\lambda)} \right] \tag{16.76}$$

and then

$$f(x) = \frac{i}{\pi} \int_1^\infty \frac{dt}{\sqrt{t^2 - 1}} \left[\frac{e^{ix(t+\lambda)}}{(t+\lambda)^2} + \frac{e^{ix(t-\lambda)}}{(t-\lambda)^2} \right] + A + Bx . \tag{16.77}$$

The boundary conditions are $f(0) = f'(0) = 0$. Therefore

$$
\begin{aligned}
B &= \frac{2}{\pi} \int_1^\infty \frac{dt}{\sqrt{t^2 - 1}} \frac{t}{t^2 - \lambda^2} \\
&= \frac{2}{\pi} \int_0^\infty \frac{du}{u^2 + 1 - \lambda^2} = \frac{1}{\sqrt{1 - \lambda^2}} .
\end{aligned}
\tag{16.78}
$$

The other constant is

$$A = -\frac{2i}{\pi} \int_1^\infty \frac{dt}{\sqrt{t^2-1}} \frac{t^2+\lambda^2}{(t^2-\lambda^2)^2}$$

$$= -\frac{2i}{\pi} \frac{1}{1-\lambda^2} \left(1 + \frac{\lambda}{\sqrt{1-\lambda^2}} \tan^{-1} \frac{\lambda}{\sqrt{1-\lambda^2}}\right). \qquad (16.79)$$

To get the rest of the answer, we expand t in the integrand of (16.77) about 1 for large x. The approximation is valid provided $x(1-\lambda) \gg 1$. So for the first approximation,

$$\int_0^x dt \left(1 - \frac{t}{x}\right) \cos \lambda t \, H_0^{(1)}(t) = \frac{1}{\sqrt{1-\lambda^2}} \left[1 - \frac{2i}{\pi x} \frac{1}{\sqrt{1-\lambda^2}} \right.$$

$$\times \left(1 + \frac{\lambda}{\sqrt{1-\lambda^2}} \tan^{-1} \frac{\lambda}{\sqrt{1-\lambda^2}}\right)$$

$$\left. - \frac{\sqrt{1-\lambda^2} \, e^{-i\pi/4}}{x\sqrt{2\pi x}} \left(\frac{e^{i(1+\lambda)x}}{(1+\lambda)^2} + \frac{e^{i(1-\lambda)x}}{(1-\lambda)^2}\right)\right]. \qquad (16.80)$$

If we insert this result in (16.66) and consider the $k\delta \gg 1$ limit with $\lambda = \cos \theta'$, we obtain from the optical theorem (16.13)

$$\sigma = 2\delta \sin \theta', \qquad (16.81)$$

the obvious geometrical generalization of (16.36). To a better approximations, since the correction to the real part of (16.80) comes only from the last term there,

$$\frac{\frac{1}{2}\sigma}{\delta \sin \theta'} = 1 + \frac{\sin \theta'}{4\sqrt{2\pi}(k\delta)^{3/2}} \left[\frac{\cos(2k\delta \cos^2 \theta'/2 - \pi/4)}{\cos^4 \theta'/2}\right.$$

$$\left. + \frac{\cos(2k\delta \sin^2 \theta'/2 - \pi/4)}{\sin^4 \theta'/2}\right]$$

$$\sim 1 + \frac{\cos(k\delta - \pi/4)}{2\pi^2(\delta/\lambda)^{3/2}}, \qquad \theta' = \frac{\pi}{2}. \qquad (16.82)$$

Stopping at this level of approximation is good for $\delta/\lambda = 1$ or even $1/2$.

This calculation can be extended to other scattering angles to compute $A(\theta', \theta'')$, but this is a tedious calculation, so we will move on to a complementary problem.

16.3 Diffraction by a Slit

We have thus far used a rigorous variational method, useful for short wavelengths, $\lambda \ll \delta$. We can also consider the opposite limit, $\lambda \gg \delta$. We will now develop a Fourier transform procedure, for $\delta \gg \lambda$. We will treat the exact half-plane solution [9, 30] as a basis – with a wide slit this gives a good answer.

Consider now a slit of width $2l = \delta$ in a conducting screen lying in the $x = 0$ plane. The geometry is sketched in Fig. 16.1. The condition for the convergence of the method to be described is that the wavelength be short, $\lambda < 4l$, and that we not have grazing incidence.

As usual, the scattering for H modes is described by (16.4), or

$$E = E_{\text{inc}} - \frac{k}{4c} \int dz' H_0^{(1)}(k\sqrt{(z - z')^2 + x^2})K(z') \,, \tag{16.83}$$

where now the integration is over the entire range of z' except for the slit, $-l < z' < l$. On the plane, $E(0, z) = E(z)$, and we have the integral equation (16.5), or

$$E(z) = E_{\text{inc}}(0, z) - \frac{k}{4c} \int dz' H_0^{(1)}(k|z - z'|)K(z') = 0, \quad |z| > l \,. \tag{16.84}$$

Take for the incident field that of an incoming plane wave with grazing angle θ', as sketched in Fig. 16.1:

$$E_{\text{inc}} = e^{ik(\cos \theta' z + \sin \theta' x)}, \qquad E_{\text{inc}}(0, z) = e^{ik \cos \theta' z} \,. \tag{16.85}$$

Introduce a convergence factor $e^{-\epsilon|z|}$ into E_{inc} in order to simplify and regulate the calculation. We will drop its explicit appearance later:

$$E_{\text{inc}}(0, z) = e^{ik \cos \theta' z} e^{-\epsilon|z|} \,. \tag{16.86}$$

Break up the surface current into two parts, one corresponding to each half plane:

$$K(z) = K_1(z) + K_2(z) \,, \quad K_1 = 0 \,, z > -l \,, \quad K_2 = 0 \,, z < l \,. \tag{16.87}$$

Introduce the Fourier transform of the field in the slit,

$$E(\zeta) = \int_{-l}^{l} dz \, e^{-i\zeta z} E(z) \,, \tag{16.88}$$

which is regular over all space. Examine the asymptotic forms, with $\zeta = \xi + i\eta$:

$$\eta > 0 \,, \quad |\zeta| \to \infty : \quad E(\zeta) \sim e^{-i\zeta l} \frac{1}{(-i\zeta)^{3/2}} \,, \tag{16.89a}$$

$$\eta < 0 \,, \quad |\zeta| \to \infty : \quad E(\zeta) \sim e^{i\zeta l} \frac{1}{(i\zeta)^{3/2}} \,, \tag{16.89b}$$

which follow from the behavior of the electric field near the edge ([9], Chap. 48),

$$E(z) \sim \sqrt{l - |z|} \,, \quad \text{as} \quad |z| \to l \,. \tag{16.90}$$

The factors of i in (16.89a) and (16.89b) are inserted for convenience. Conversely, we can see that $E(z) = 0$ for $|z| > l$ just from the behavior at infinity of $E(\zeta)$, from the inverse transform,

$$E(z) = \frac{1}{2\pi} \int_{-\infty}^{\infty} d\zeta\, e^{i\zeta z} E(\zeta)\,, \tag{16.91}$$

by closing the contour in the upper (or lower) half plane.

Now consider the transform of the current

$$K_2(\zeta) = \int_{l}^{\infty} dz\, e^{-i\zeta z} K_2(z)\,, \tag{16.92}$$

which is regular in the lower half plane, $\eta = \operatorname{Im}\zeta < 0$. Then using the same argument, because $K_2(z) \sim (z-l)^{-1/2}$ near the edge, and similarly for K_1, which is regular in the upper half plane, we see that

$$\eta < 0\,, \quad |\zeta| \to \infty: \quad K_2(\zeta) \sim e^{-i\zeta l}\frac{1}{\sqrt{i\zeta}}\,, \tag{16.93a}$$

$$\eta > 0\,, \quad |\zeta| \to \infty: \quad K_1(\zeta) \sim e^{i\zeta l}\frac{1}{\sqrt{-i\zeta}}\,. \tag{16.93b}$$

By the same argument as before we can show that K_1 and K_2 vanish properly in the slit just from the exponential factor.

Now from E_{inc} we see the transforms

$$\int_{z>0} dz\, e^{-i\zeta z} e^{-\epsilon z} \quad \text{exists for } \eta < \epsilon\,, \tag{16.94a}$$

$$\int_{z<0} dz\, e^{-i\zeta z} e^{\epsilon z} \quad \text{exists for } \eta > -\epsilon\,, \tag{16.94b}$$

so the transform of E_{inc} exists is the narrow band $-\epsilon < \eta < \epsilon$ about the real ζ axis. In the kernel of (16.83), we take k to be complex (lossy for the same reason as E_{inc}). Again the transform will exist in a narrow region,

$$|\eta| < \operatorname{Im} k\,. \tag{16.95}$$

Further $K_1(\zeta)$ is regular for $\eta > -\epsilon$, while $K_2(\zeta)$ is regular for $\eta < \epsilon$, since the ϵ factor in the field will also influence the currents.

Now we are ready to calculate all the transforms. First, the incident field:

$$E_{\text{inc}}(\zeta) = \int_{-\infty}^{0} dz\, e^{-i\zeta z} e^{ik\cos\theta' z + \epsilon z} + \int_{0}^{\infty} dz\, e^{-i\zeta z} e^{ik\cos\theta' z - \epsilon z}$$

$$= -\frac{1}{i(\zeta - k\cos\theta' + i\epsilon)} + \frac{1}{i(\zeta - k\cos\theta' - i\epsilon)}\,, \tag{16.96}$$

where the second term has a pole in the upper half plane, and the first in the lower half plane. Call

$$\zeta_- = k\cos\theta' - i\epsilon\,, \qquad \zeta_+ = k\cos\theta' + i\epsilon\,, \tag{16.97}$$

so

$$E_{\text{inc}}(\zeta) = i\left(\frac{1}{\zeta - \zeta_-} - \frac{1}{\zeta - \zeta_+}\right). \tag{16.98}$$

Note for ζ real, $\epsilon \to 0$,

$$E_{\text{inc}}(\zeta) = 2\pi\delta(\zeta - k\cos\theta'). \tag{16.99}$$

Next, we recognize that the transform of $H_0^{(1)}(kz)$ occurs in the integral representation, equivalent to (16.69),

$$H_0^{(1)}(k|z|) = \frac{1}{\pi}\int_{-\infty}^{\infty} d\zeta \frac{e^{i\zeta z}}{\sqrt{k^2 - \zeta^2}},, \tag{16.100}$$

and so therefore the transform of the integral equation (16.84) is

$$E(\zeta) = i\left(\frac{1}{\zeta - \zeta_-} - \frac{1}{\zeta - \zeta_+}\right) - \frac{k}{2c}\frac{K_1(\zeta) + K_2(\zeta)}{\sqrt{k^2 - \zeta^2}}. \tag{16.101}$$

Now we seek to separate this equation into parts regular in the upper and lower half planes, as in the case of the straight edge. Define

$$I_1(\zeta) = i\frac{k}{c}e^{-i\zeta l}K_1(\zeta), \tag{16.102a}$$

$$I_2(\zeta) = i\frac{k}{c}e^{i\zeta l}K_2(\zeta). \tag{16.102b}$$

In order to focus attention on say $K_2(\zeta)$ and consider $K_1(\zeta)$ as part of the integral equation (16.101), suppose we multiply the integral equation by $e^{i\zeta l}\sqrt{k + \zeta}$:

$$e^{i\zeta l}\sqrt{k + \zeta}E(\zeta) = ie^{i\zeta l}\frac{\sqrt{k + \zeta}}{\zeta - \zeta_-} - ie^{i\zeta l}\frac{\sqrt{k + \zeta}}{\zeta - \zeta_+}$$
$$+ \frac{i}{2}\frac{I_2(\zeta)}{\sqrt{k - \zeta}} + \frac{i}{2}\frac{e^{2i\zeta l}I_1(\zeta)}{\sqrt{k - \zeta}}. \tag{16.103}$$

The term on the left is regular for $\eta > -\text{Im}\,k$ or $\eta > -\epsilon$. (Henceforth we will choose $\text{Im}\,k > \epsilon$.) The first term on the right is regular for $\eta > -\epsilon$, while the third term on the right is regular for $\eta < \epsilon$. The remaining two terms have no particular character. We split the second term (times i)

$$\frac{e^{i\zeta l}\sqrt{k + \zeta}}{\zeta - \zeta_+} = \frac{e^{i\zeta l}\sqrt{k + \zeta} - e^{i\zeta_+ l}\sqrt{k + \zeta_+}}{\zeta - \zeta_+} + \frac{e^{i\zeta_+ l}\sqrt{k + \zeta_+}}{\zeta - \zeta_+}, \tag{16.104}$$

where the first term is regular for $\eta > -\epsilon$ while the second is for $\eta < \epsilon$. Next, we work on the unknown function appearing as the last term in (16.103). Suppose we have a function $F(\zeta)$ defined in a strip, $-\epsilon < \eta < \epsilon$, and suppose as $\xi \to \pm\infty$, $|\eta| < \epsilon$, $F(\zeta) \to 0$. Let the lines $\eta = \pm\epsilon$ be denoted $\sigma_\pm$ (actually choose them infinitesimally inside this strip). Then, by Cauchy's theorem, for ζ in the strip $|\eta| < \epsilon$,

$$F(\zeta) = \frac{1}{2\pi i} \int_{\sigma_+} dt \frac{F(t)}{t - \zeta} + \frac{1}{2\pi i} \int_{\sigma_-} dt \frac{F(t)}{t - \zeta} , \qquad (16.105)$$

where σ_- is traversed in a positive sense, while σ_+ is traversed in the negative sense. The first term is an analytic function of ζ in the lower half plane, $\eta < \epsilon$, while the second is analytic in the upper half plane, $\eta > -\epsilon$. Illustrate this for the function (16.104): the second term in (16.105) is then

$$\frac{1}{2\pi i} \int_{\sigma_-} \frac{dt}{t - \zeta} e^{itl} \frac{\sqrt{k + t}}{t - \zeta_+} ; \qquad (16.106)$$

for $\eta > -\epsilon$ we can close the contour in the upper half plane, and thereby obtain precisely the first term in (16.104). Similarly, for $\eta < \epsilon$ the first term in (16.105) gives the second term in (16.104).

In this way we can pick out the part of the last term in (16.103) which is regular in the upper half plane,

$$\frac{1}{2\pi i} \int_{\sigma_-} \frac{dt}{t - \zeta} \frac{i}{2} \frac{e^{2itl} I_1(t)}{\sqrt{k - t}} . \qquad (16.107)$$

We close the contour in the upper half plane; the branch point appears in the lower half plane. Similarly the σ_+ integral gives a regular function for the lower half plane. So the integral equation can be divided into a part that is regular above and a part regular below; everything being regular in a common strip. Thus writing this in the form

$$F_+ = F_- , \qquad (16.108)$$

we recognize that this defines an integral or entire function, regular everywhere. That is

$$\frac{i}{2} \frac{I_2(\zeta)}{\sqrt{k - \zeta}} + \frac{1}{4\pi} \int_{\sigma_+} \frac{dt}{t - \zeta} \frac{e^{2itl}}{\sqrt{k - t}} I_1(t) - i \frac{e^{i\zeta + l} \sqrt{k + \zeta_+}}{\zeta - \zeta_+} \qquad (16.109)$$

is an integral function, and similarly for the $-$ terms. We will show this is zero, by examining the asymptotic form. In fact, $I_2(\zeta) \sim \zeta^{-1/2}$, and then every term here behaves as $1/\zeta$, so the whole function vanishes at infinity, and therefore vanishes everywhere.

Now consider the σ_+ integral in (16.109). The integrand has a branch point at k, and vanishes at infinity in the upper half plane exponentially fast, and hence we can close the contour above, and then bend it back as we did for the Hankel function in Fig. 16.2. In that way that term becomes

$$\frac{i}{2\pi} \int_k^\infty \frac{dt}{t - \zeta} \frac{e^{2itl}}{\sqrt{t - k}} I_1(t) . \qquad (16.110)$$

Then we recast (16.109) into

$$\frac{I_2(\zeta)}{\sqrt{k-\zeta}} + \frac{1}{\pi}\int_k^\infty \frac{dt}{t-\zeta}\frac{e^{2itl}}{\sqrt{t-k}}I_1(t) = \frac{2\sqrt{k+\zeta_+}\,e^{i\zeta_+ l}}{\zeta-\zeta_+}. \tag{16.111}$$

Similarly for the other case, that is, where attention is focussed on K_1 and K_2 is considered unexciting, we get (see Problem 16.2)

$$\frac{I_1(\zeta)}{\sqrt{k+\zeta}} + \frac{1}{\pi}\int_k^\infty \frac{dt}{t+\zeta}\frac{e^{2itl}}{\sqrt{t-k}}I_2(-t) = -\frac{2\sqrt{k-\zeta_-}\,e^{-i\zeta_- l}}{\zeta-\zeta_-}, \tag{16.112}$$

and these lead to two simultaneous integral equations for I_1 and I_2.

16.3.1 Approximate Field

Before proceeding with them, let us look at the expression for $E(\zeta)$, which we obtain from the original transform equation (16.101) by substituting (16.111) and (16.112):

$$E(\zeta) = \frac{i}{\zeta-\zeta_-}\left(1 - e^{i(\zeta-\zeta_-)l}\sqrt{\frac{k-\zeta_-}{k-\zeta}}\right) - \frac{i}{\zeta-\zeta_+}\left(1 - e^{-i(\zeta-\zeta_+)l}\sqrt{\frac{k+\zeta_+}{k+\zeta}}\right)$$
$$-\frac{i}{2\pi}\left[\frac{e^{i\zeta l}}{\sqrt{k-\zeta}}\int_k^\infty \frac{dt}{t+\zeta}e^{2itl}\frac{I_2(-t)}{\sqrt{t-k}} + \frac{e^{-i\zeta l}}{\sqrt{k+\zeta}}\int_k^\infty \frac{dt}{t-\zeta}e^{2itl}\frac{I_1(t)}{\sqrt{t-k}}\right]. \tag{16.113}$$

The first term is regular, and hence there we can let $\epsilon \to 0$. Therefore, we can set $\zeta_- = \zeta_+ = \zeta_0 = k\cos\theta'$, so that the first line of (16.113) can be written as

$$E^{(0)}(\zeta) = \frac{i}{\zeta-\zeta_0}\left[e^{-i(\zeta-\zeta_0)l}\sqrt{\frac{k+\zeta_0}{k+\zeta}} - e^{i(\zeta-\zeta_0)l}\sqrt{\frac{k-\zeta_0}{k-\zeta}}\right]. \tag{16.114}$$

We will shortly show that this term is the first approximation to the answer.

16.3.2 Transform of Scattered Field

Now return to (16.83), and write it in terms of Fourier transforms. Using the integral representation derived in Problem 16.3 for $H_0^{(1)}$, the term involving that function is

$$-\frac{k}{4\pi c}\int dz'\,d\zeta\,\frac{e^{i\zeta(z-z')+i\sqrt{k^2-\zeta^2}|x|}}{\sqrt{k^2-\zeta^2}}K(z'). \tag{16.115}$$

Do the z' integral first, so (16.83) becomes

$$E = E_{\text{inc}} - \frac{k}{4\pi c} \int_{-\infty}^{\infty} d\zeta \frac{e^{i\zeta z + i\sqrt{k^2 - \zeta^2}|x|}}{\sqrt{k^2 - \zeta^2}} K(\zeta) .$$

(16.116)

Now recall the transformation equation (16.101) for ζ on the real axis:

$$-\frac{k}{2c} \frac{K(\zeta)}{\sqrt{k^2 - \zeta^2}} = E(\zeta) - 2\pi\delta(\zeta - k\cos\theta') .$$

(16.117)

Then we get

$$E(x,z) = e^{ik(\cos\theta' z + \sin\theta' x)}$$
$$+ \frac{1}{2\pi} \int d\zeta\, e^{i\zeta z + i\sqrt{k^2 - \zeta^2}|x|} \left[-2\pi\delta(\zeta - k\cos\theta') + E(\zeta) \right] ,$$

(16.118)

and so

$$E(x,z) = e^{ik(\cos\theta' z + \sin\theta' x)} - e^{ik(\cos\theta' z + \sin\theta'|x|)}$$
$$+ \frac{1}{2\pi} \int d\zeta\, e^{i\zeta z + i\sqrt{k^2 - \zeta^2}|x|} E(\zeta) .$$

(16.119)

Again using the representation (16.192), we can write the last term here as

$$-\frac{i}{2\pi} \frac{\partial}{\partial x} \int d\zeta \frac{e^{i\zeta z + i\sqrt{k^2 - \zeta^2}|x|}}{\sqrt{k^2 - \zeta^2}} E(\zeta) = -\frac{i}{2} \frac{\partial}{\partial x} \int dz'\, H_0^{(1)} \left(k\sqrt{(z-z')^2 + x^2} \right)$$
$$\times E(z') ,$$

(16.120)

so the above comes directly from the Green's function.

For $x > 0$ the first two terms in (16.119) cancel. For $\rho \gg \lambda$, using the asymptotic form (16.8) of the Hankel function, we get ($\theta = \theta'' - \pi$)

$$E(x,z) = -\frac{i}{2} \frac{\partial}{\partial x} \int dz' \sqrt{\frac{2}{\pi k\rho}} e^{ik\rho - ikz'\cos\theta - i\pi/4} E(z')$$
$$= \frac{k}{2} \sqrt{\frac{2}{\pi k\rho}} \sin\theta\, e^{ik\rho - i\pi/4} E(\zeta = k\cos\theta) .$$

(16.121)

16.3.3 Differential Cross Section

From this we can find the differential cross section per unit grazing angle θ for an incident wave at grazing angle θ' as

$$\sigma(\theta, \theta') = |E|^2 \rho = \frac{k}{2\pi} \sin^2\theta |E_{\theta'}(k\cos\theta)|^2 .$$

(16.122)

Instead of integrating over all angles to get the total "absorption" cross section, corresponding to the energy that passes through the slit rather than being scattered through a grazing angle less than π, we can obtain the same object by examining the scattered wave in the direction of incidence. We prove this by looking at the Poynting vector,

$$\mathbf{S} = cE^* \frac{\nabla}{ik} E \,, \qquad \nabla \cdot \operatorname{Re} \mathbf{S} = 0 \,. \tag{16.123}$$

At large distances [cf. (1.167)] (in the radiation zone, $\nabla \to ik\mathbf{n}$),

$$\mathbf{S} = c\mathbf{n}|E|^2 \,, \tag{16.124}$$

where $\mathbf{n}$ is a unit vector in the direction of propagation. Then, since with a unit plane wave incident the incident energy flux is c, the total cross section per unit length is

$$\sigma(\theta') = \int_{C'} ds \operatorname{Re} \mathbf{n} \cdot E^* \frac{\nabla E}{ik} \,, \tag{16.125}$$

where C' is a semicircle at infinity. But this is the same as the energy passing through the aperture since energy is conserved:

$$\sigma(\theta') = \int_{\text{aperture}} dz \operatorname{Re} \frac{1}{ik} E^* \frac{\partial}{\partial x} E \,. \tag{16.126}$$

But in the aperture, from (16.85)

$$\frac{\partial}{\partial x} E = \frac{\partial}{\partial x} E_{\text{inc}} = ik \sin \theta' e^{ikz \cos \theta'} \,, \tag{16.127}$$

because the magnetic field in the aperture is equal to the incident field.[3] Then

$$\sigma(\theta') = \operatorname{Re} \int dz \, E^* \sin \theta' e^{ikz \cos \theta'}$$

$$= \operatorname{Re} \sin \theta' \int dz \, e^{-ikz \cos \theta'} E(z)$$

$$= \sin \theta' \operatorname{Re} E(\zeta_0 = k \cos \theta') \,. \tag{16.128}$$

16.3.4 First Approximation

Now return to (16.114), which was the first term of our expression for the Fourier transform of the electric field. Let us use this as a first approximation. It is easy to show that

$$\operatorname{Re} E^{(0)}(\zeta_0) = 2l = \delta \,, \tag{16.129}$$

and hence from (16.128) the total cross section is

$$\sigma(\theta') = \delta \sin \theta' \,, \tag{16.130}$$

---

[3] B_z is proportional to $\partial E/\partial x$, and the x-derivative of $H_0^{(1)}(k\sqrt{x^2 + (z - z')^2})$ vanishes at $x = 0$.

as expected in the limit of geometrical optics, $k\delta \gg 1$. Next, consider the differential cross section (16.122):

$$\sigma(\theta,\theta') = \frac{1}{\pi k}\frac{1-\cos(\theta-\theta')+2\sin\theta\sin\theta'\sin^2[kl(\cos\theta-\cos\theta')]}{(\cos\theta-\cos\theta')^2} . \quad (16.131)$$

As $\theta \to \theta'$,

$$\sigma(\theta',\theta') \to \frac{2}{\pi k}(kl)^2 \sin^2\theta' . \quad (16.132)$$

Thus we see in comparison with the total cross section that the forward scattering is large, because kl is large. In the first lobe, $\theta - \theta'$ is small, but kl is large so the $\sin^2$ term in the differential cross section is not expandable. If we write

$$\cos\theta - \cos\theta' \approx -\sin\theta'(\theta-\theta') , \quad (16.133)$$

and call $\psi = \theta - \theta'$, the differential cross section is

$$\sigma(\theta,\theta') \approx \frac{2}{\pi k}\frac{\sin^2(kl\sin\theta'\psi)}{\psi^2} , \quad (16.134)$$

which is strongly peaked about $\psi = 0$. We integrate this over ψ as a check to get the total cross section (16.130):

$$\sigma \approx \frac{2}{\pi k}\int_{-\infty}^{\infty} d\psi\,\frac{\sin^2(kl\sin\theta'\psi)}{\psi^2} = \frac{2}{\pi}l\sin\theta'\int_{-\infty}^{\infty} dx\frac{\sin^2 x}{x^2} = 2l\sin\theta'. \quad (16.135)$$

Now let us explore the connection between this and the Kirchhoff theory, which assumes that the electric field in the aperture is the incident field,

$$E(z) = \begin{cases} e^{ikz\cos\theta'} = e^{i\zeta_0 z} , & |z| < l , \\ 0 , & |z| > l . \end{cases} \quad (16.136)$$

Hence the transformed field is

$$\begin{aligned} E(\zeta) &= \int_{-l}^{l} dz\,e^{-i\zeta z}e^{i\zeta_0 z} \\ &= 2\int_0^l dz\cos(\zeta-\zeta_0)z = 2\frac{\sin(\zeta-\zeta_0)l}{\zeta-\zeta_0} \\ &= \frac{i}{\zeta-\zeta_0}\left[e^{-i(\zeta-\zeta_0)l} - e^{i(\zeta-\zeta_0)l}\right] , \end{aligned} \quad (16.137)$$

so in comparison with our result (16.114), we see that the Kirchhoff theory is satisfactory when $\zeta \approx \zeta_0$, but not otherwise.

Note that in the long-wavelength limit, the Kirchhoff field (16.136) is constant in the aperture, which is a bad feature, but it has the good feature of vanishing on the metal. In contrast, our result has good behavior in the slit, behaving properly at the edges, but does not vanish on the metal. So our approximation is good only when the incident field does *not* come in at grazing incidence. (But see below.) If it does not, the field in the metal is very small and does not matter.

Let us calculate the electric field corresponding to our approximation. Inverting the transform,

$$E(z) = \frac{1}{2\pi} \int_{-\infty}^{\infty} d\zeta \, e^{i\zeta z} E(\zeta) \,. \tag{16.138}$$

Consider the first term in (16.114),

$$\frac{i}{2\pi} \int_{-\infty}^{\infty} \frac{d\zeta}{\zeta - \zeta_0} e^{i\zeta z} e^{-i(\zeta - \zeta_0)l} \sqrt{\frac{k + \zeta_0}{k + \zeta}} \,. \tag{16.139}$$

Each term in (16.114) has a pole at $\zeta = \zeta_0$, but together they do not. To handle each term separately, let us take ζ_0 to be below the real axis so then this term is

$$\frac{i}{2\pi} \sqrt{k + \zeta_0} \, e^{i\zeta_0 l} \int_{-\infty}^{\infty} \frac{d\zeta}{\sqrt{k + \zeta}} \frac{e^{i\zeta(z-l)}}{\zeta - \zeta_0} \,. \tag{16.140}$$

This vanishes for $z > l$ because we can close the contour in the upper half plane, and no singularities are contained inside. (Remember, $\operatorname{Im} k > 0$.) For $z < l$ we close the contour in the lower half plane, and encircle the branch line beginning at $-k$ and extending along the negative ζ axis. From the pole at ζ_0 we get $e^{i\zeta_0 z}$, while from the branch line, we get

$$2\sqrt{k + \zeta_0} \, e^{i\zeta_0 l} \frac{1}{2\pi} \int_{-\infty}^{-k} d\zeta \frac{e^{i\zeta(z-l)}}{\sqrt{-k - \zeta}\,(\zeta - \zeta_0)}$$
$$= -e^{i(k+\zeta_0)l} \frac{e^{-ikz}}{\pi} \int_{0}^{\infty} \frac{dx}{\sqrt{x}} \frac{e^{i(k+\zeta_0)(l-z)x}}{x + 1} \,, \tag{16.141}$$

where we have substituted

$$\zeta = -k - (k + \zeta_0)x \,. \tag{16.142}$$

To evaluate this, consider

$$F(\xi) = \int_{0}^{\infty} \frac{dx}{\sqrt{x}} \frac{e^{i\xi x}}{x + 1} \,, \tag{16.143}$$

which satisfies the differential equation

$$\left(\frac{d}{d\xi} + i\right) F(\xi) = i \int_{0}^{\infty} \frac{dx}{\sqrt{x}} e^{i\xi x} = i \frac{\sqrt{\pi} e^{\pi i/4}}{\sqrt{\xi}} \,. \tag{16.144}$$

To solve this equation, multiply by $e^{i\xi}$ and so

$$\frac{d}{d\xi}\left(e^{i\xi}F(\xi)\right) = \frac{i\sqrt{\pi}}{\sqrt{\xi}}e^{i\pi/4}e^{i\xi} \, , \tag{16.145}$$

and then, with a constant of integration C,

$$F(\xi)e^{i\xi} = -\sqrt{\pi}e^{-\pi i/4}\int_0^\xi dx\frac{e^{ix}}{\sqrt{x}} + C$$

$$= C - 2\sqrt{\pi}e^{-i\pi/4}\int_0^{\sqrt{\xi}} dx\,e^{ix^2} \, . \tag{16.146}$$

Then, putting all this together, we find for the first term in (16.140), for $z < l$,

$$e^{i\zeta_0 z} - \frac{e^{i\zeta_0 z}}{\pi}\left[C - 2\sqrt{\pi}e^{-\pi i/4}\int_0^{\sqrt{(k+\zeta_0)(l-z)}} dx\,e^{ix^2}\right] \, . \tag{16.147}$$

Since this term is zero for $z > l$, and it must be continuous at $z = l$, we determine the constant $C = \pi$, and we have, for $z < l$,

$$\frac{2}{\sqrt{\pi}}e^{i\zeta_0 z - \pi i/4}\int_0^{\sqrt{(k+\zeta_0)(l-z)}} dx\,e^{ix^2} \, . \tag{16.148}$$

Now we consider the transform (16.138) of the second term in (16.114),

$$-\frac{i}{2\pi}\sqrt{k-\zeta_0}\,e^{-i\zeta_0 l}\int_{-\infty}^\infty d\zeta\frac{e^{i\zeta(z+l)}}{(\zeta-\zeta_0)\sqrt{k-\zeta}} \, . \tag{16.149}$$

Now for $z + l < 0$, we close the contour below, and get $-e^{i\zeta_0 z}$ from the pole at ζ_0. For $z + l > 0$ we close above and get from the branch line extending along the positive real axis,

$$-\frac{i}{\pi}\sqrt{k-\zeta_0}\,e^{-i\zeta_0 l}\int_k^\infty d\zeta\frac{e^{i\zeta(z+l)}}{(\zeta-\zeta_0)i\sqrt{\zeta-k}}$$

$$= -\frac{1}{\pi}e^{i\zeta_0 z}e^{i(k-\zeta_0)(z+l)}\int_0^\infty \frac{dx}{\sqrt{x}}\frac{e^{i(k-\zeta_0)(z+l)x}}{1+x}$$

$$= -\frac{2}{\sqrt{\pi}}e^{i\zeta_0 z - \pi i/4}\int_{\sqrt{(k-\zeta_0)(z+l)}}^\infty dx\,e^{ix^2} \, , \tag{16.150}$$

where in the second line we changed variables, $\zeta = k + (k-\zeta_0)x$, and in the third line used (16.146) together with continuity at $z + l = 0$.

Now we combine the first and second terms, from (16.147) and (16.150), and obtain the following form for the electric field on the $x = 0$ plane:

$$E^{(0)}(z) = -\frac{2}{\sqrt{\pi}}e^{i\zeta_0 z - \pi i/4}\int_{\sqrt{(k-\zeta_0)(z+l)}}^\infty dx\,e^{ix^2} \, , \quad z > l, \tag{16.151a}$$

$$= -\frac{2}{\sqrt{\pi}} e^{i\zeta_0 z - \pi i/4} \int_{\sqrt{(k+\zeta_0)(l-z)}}^{\infty} dx\, e^{ix^2} , \quad z < -l \,, (16.151b)$$

$$= e^{i\zeta_0 z} - \frac{2}{\sqrt{\pi}} e^{i\zeta_0 z - \pi i/4} \left[\int_{\sqrt{(k-\zeta_0)(z+l)}}^{\infty} dx\, e^{ix^2} \right.$$

$$\left. + \int_{\sqrt{(k+\zeta_0)(l-z)}}^{\infty} dx\, e^{ix^2} \right], \quad |z| < l \,. \qquad (16.151c)$$

The first two terms give the (nonzero) value of the electric field on the metal, while the last gives the value in the aperture.

We now must investigate the values of these Fresnel integrals. We can do this by repeatedly integrating by parts:

$$\int_{\xi}^{\infty} dx\, e^{ix^2} = \int_{\xi}^{\infty} \frac{d(e^{ix^2})}{2ix} = \frac{i}{2\xi} e^{i\xi^2} + \int_{\xi}^{\infty} dx\, \frac{e^{ix^2}}{2ix^2}$$

$$= \frac{i}{2\xi} e^{i\xi^2} - \int_{\xi}^{\infty} \frac{d(e^{ix^2})}{4x^3} , \qquad (16.152)$$

and so on. Thus for large ξ,

$$\int_{\xi}^{\infty} dx\, e^{ix^2} = \frac{i}{2\xi} e^{i\xi^2} + \frac{1}{4\xi^3} e^{i\xi^2} + \cdots . \qquad (16.153)$$

To apply this for $z > l$ we must require $\sqrt{(k - \zeta_0)2l}$ must be large, so $k = \zeta_0 = k\cos\theta'$ is forbidden, meaning no grazing incidence. For normal incidence, $\zeta_0 = 0$, and then $2kl \gg 1$, or $2\pi\delta/\lambda \gg 1$, which is reasonably well satisfied even for $\delta = \lambda/2$. The electric field on the metal, $z > l$, in this approximation is given by

$$E^{(0)}(z) = -\frac{e^{\pi i/4}}{\sqrt{\pi}} \frac{e^{i(k-\zeta_0)l} e^{ikz}}{\sqrt{k - \zeta_0}\sqrt{z + l}} . \qquad (16.154)$$

At the edge,

$$|E^{(0)}|_{z=l} = \frac{1}{\sqrt{\pi(k - \zeta_0)2l}} \approx \frac{1}{\sqrt{2\pi^2\delta/\lambda}} . \qquad (16.155)$$

This should be small compared to unity. In fact, even for $\delta = \lambda$ this is only about 20% of the incident field.

16.3.5 Exact Electric Field

Recall that the transform of the electric field was given exactly by (16.113), or

$$E(\zeta) = E^{(0)}(\zeta) - \frac{i}{2\pi} \left[\frac{e^{i\zeta l}}{\sqrt{k - \zeta}} \int_{k}^{\infty} \frac{dt}{t + \zeta} e^{2itl} \frac{I_2(-t)}{\sqrt{t - k}} \right.$$

$$\left. + \frac{e^{-i\zeta l}}{\sqrt{k + \zeta}} \int_{k}^{\infty} \frac{dt}{t - \zeta} e^{2itl} \frac{I_1(t)}{\sqrt{t - k}} \right]. \qquad (16.156)$$

The electric field corresponding to this transform must vanish on the walls, and this is the integral equation [another sort from the type considered earlier in (16.111) and (16.112)]. Thus inverting this transform

$$E(z) = E^{(0)}(z) - \frac{i}{4\pi^2} \int_k^\infty \frac{dt}{\sqrt{t-k}} e^{2itl} I_2(-t) \int_{-\infty}^\infty \frac{d\zeta}{t+\zeta} \frac{e^{i\zeta(z+l)}}{\sqrt{k-\zeta}}$$
$$+ \frac{i}{4\pi^2} \int_k^\infty \frac{dt}{\sqrt{t-k}} e^{2itl} I_1(t) \int_{-\infty}^\infty \frac{d\zeta}{\zeta-t} \frac{e^{i\zeta(z-l)}}{\sqrt{k+\zeta}} . \qquad (16.157)$$

We want to evaluate the integrals over ζ. To this end, consider functions $F(z)$, $G(z)$, with transforms $F(\zeta)$, $G(\zeta)$, from which follows

$$\frac{1}{2\pi} \int_{-\infty}^\infty d\zeta\, e^{i\zeta z} F(\zeta) G(\zeta) = \int_{-\infty}^\infty dz'\, F(z-z') G(z') , \qquad (16.158)$$

the convolution theorem. Use this property to evaluate the above integrals, by taking $F(\zeta) = 1/\sqrt{k-\zeta}$, $G(\zeta) = 1/(\zeta+t)$, so

$$G(z) = \frac{1}{2\pi} \int_{-\infty}^\infty d\zeta \frac{e^{iz\zeta}}{\zeta+t} . \qquad (16.159)$$

Suppose t has a positive imaginary part. Then

$$G(z) = \begin{cases} 0 , & z > 0 , \\ -ie^{-itz} , & z < 0 . \end{cases} \qquad (16.160)$$

Similarly

$$F(z) = \frac{1}{2\pi} \int_{-\infty}^\infty d\zeta \frac{e^{i\zeta z}}{\sqrt{k-\zeta}} , \qquad (16.161)$$

where the integrand possesses a branch point at $\zeta = k$. Putting this, and the associated branch line above the positive axis, we obtain

$$F(z) = 0 , \quad z < 0 , \qquad (16.162a)$$
$$= \frac{1}{\pi i} \int_k^\infty d\zeta \frac{e^{i\zeta z}}{\sqrt{\zeta-k}} = \frac{e^{-\pi i/4}}{\sqrt{\pi}} \frac{e^{ikz}}{\sqrt{z}} , \quad z > 0. \qquad (16.162b)$$

Thus, with $z_<$ being the lesser of z and 0,

$$\frac{1}{2\pi} \int_{-\infty}^\infty d\zeta\, e^{i\zeta z} F(\zeta) G(\zeta) = - \int_{-\infty}^{z_<} dz'\, i \frac{e^{-\pi i/4}}{\sqrt{\pi}} \frac{e^{ik(z-z')}}{\sqrt{z-z'}} e^{-itz'}$$
$$= -i \frac{e^{-itz}}{\sqrt{k+t}} , \quad z < 0 , \qquad (16.163a)$$
$$= -2 \frac{e^{\pi i/4}}{\sqrt{\pi}} \frac{e^{-itz}}{\sqrt{k+t}} \int_{\sqrt{(k+t)z}}^\infty dx\, e^{ix^2} , \quad z > 0 . \qquad (16.163b)$$

We can similarly proceed with the second ζ integral in (16.157), with the result

$$\frac{1}{2\pi} \int_{-\infty}^{\infty} d\zeta \frac{e^{i\zeta z}}{\sqrt{k \pm \zeta}(\zeta \mp t)} = \pm i \frac{e^{\pm itz}}{\sqrt{k+t}} , \qquad \pm z > 0 , \qquad (16.164a)$$

$$= \pm 2 \frac{e^{\pi i/4}}{\sqrt{\pi}} \frac{e^{\pm itz}}{\sqrt{k+t}} \int_{\sqrt{\mp(k+t)z}}^{\infty} dx \, e^{ix^2} , \qquad \pm z < 0 . \qquad (16.164b)$$

Hence on the upper sheet, $z > l$,

$$E(z) = E^{(0)}(z) - \frac{1}{2\pi} \int_k^{\infty} \frac{dt}{\sqrt{t^2 - k^2}} e^{it(z+l)} I_1(t)$$

$$- \frac{1}{2\pi} \int_k^{\infty} \frac{dt}{\sqrt{t^2 - k^2}} e^{2itl} I_2(-t) \frac{e^{-\pi i/4}}{\sqrt{\pi}} e^{-it(z+l)} 2 \int_{\sqrt{(k+t)(z+l)}}^{\infty} dx \, e^{ix^2} ; \qquad (16.165)$$

this should be zero and hence is an integral equation. On the lower sheet, $z < -l$, by symmetry,

$$E(z) = E^{(0)}(z) - \frac{1}{2\pi} \int_k^{\infty} \frac{dt}{\sqrt{t^2 - k^2}} e^{it(l-z)} I_2(-t)$$

$$- \frac{1}{2\pi} \int_k^{\infty} \frac{dt}{\sqrt{t^2 - k^2}} e^{2itl} I_1(t) \frac{e^{-\pi i/4}}{\sqrt{\pi}} e^{-it(l-z)} 2 \int_{\sqrt{(k+t)(l-z)}}^{\infty} dx \, e^{ix^2} . \qquad (16.166)$$

This should also be zero and hence we have a pair of simultaneous integral equations. These are equivalent to the original pair of integral equations (16.111) and (16.112). We will use this equivalence later.

16.3.6 Approximate Surface Current

We will now proceed with the approximation procedure. For large l, the only rapidly varying functions under the integral signs for the original set of integral equations are e^{2itl} and $\sqrt{t-k}$, and the dominant value comes from t near k. Thus, at $\zeta = \pm k$, we have from (16.112) and (16.111), respectively,

$$\frac{I_1(k)}{\sqrt{2k}} + \frac{1}{\pi} \frac{I_2(-k)}{2k} \int_k^{\infty} dt \frac{e^{2itl}}{\sqrt{t-k}} = -2 \frac{e^{-i\zeta_- l} \sqrt{k - \zeta_-}}{k - \zeta_-} , \qquad (16.167a)$$

$$\frac{I_2(-k)}{\sqrt{2k}} + \frac{1}{\pi} \frac{I_1(k)}{2k} \int_k^{\infty} dt \frac{e^{2itl}}{\sqrt{t-k}} = -2 \frac{e^{i\zeta_+ l} \sqrt{k + \zeta_+}}{k + \zeta_+} . \qquad (16.167b)$$

The integrals are easily evaluated by introducing $u = t - k$ as a variable

$$e^{2ikl} \int_0^\infty du \frac{e^{2iul}}{\sqrt{u}} = \frac{e^{2ikl}}{\sqrt{2l}} \sqrt{\pi} e^{\pi i/4} . \tag{16.168}$$

. Now check to see how the electric field reacts to the same approximation and how nearly it comes to vanishing. In the Fresnel integral in (16.165) the lower limit is

$$\sqrt{(k+t)(z+l)} > \sqrt{4kl} , \tag{16.169}$$

and we treat this as large. Thus, remembering from (16.153) that

$$\int_\xi^\infty dx\, e^{ix^2} \sim \frac{i}{2\xi} e^{i\xi^2} , \tag{16.170}$$

for large ξ, we see that the z-dependence separates out in the $I_2(-t)$ term and the z-dependence is $\frac{e^{ikz}}{\sqrt{z+l}}$. We also showed in (16.154) that

$$E^{(0)} \sim -\frac{e^{\pi i/4}}{\sqrt{\pi}} \frac{e^{i(k-\zeta_0)l}}{\sqrt{k-\zeta_0}} \frac{e^{ikz}}{\sqrt{z+l}} , \tag{16.171}$$

valid for $|\zeta_0| < k$ and $l/\lambda \gg 1$. The remaining term in (16.165) is also approximately

$$-\frac{1}{2\pi} \int_k^\infty dt \frac{e^{it(z+l)}}{\sqrt{t-k}} \frac{I_1(k)}{\sqrt{2k}} \sim \frac{e^{ik(z+l)}}{\sqrt{z+l}} . \tag{16.172}$$

The requirement that $E(z)$ vanishes in this limit then leads back to the original (simultaneous) equations for the transform, (16.167a) and (16.167b). (See Problem 16.4.)

We will now show that the only essential assumption is in the use of the asymptotic form for the Fresnel integrals. If the Fresnel integral is replaced by its asymptotic form, then at any stage of approximation, the problem can be rigorously solved. The requirement is that $\sqrt{kl} \gg 1$ (but not too much), and this is the *only* approximation (no requirement is placed on the angle of incidence).

Recall (16.112) and (16.111), which may be written as

$$\frac{I_1(\zeta)}{\sqrt{k+\zeta}} + \frac{1}{\pi} \int_k^\infty \frac{dt}{t+\zeta} \frac{e^{2itl} I_2(-t)}{\sqrt{t-k}} = -\frac{2e^{-i\zeta_0 l}\sqrt{k-\zeta_0}}{\zeta - \zeta_0} , \tag{16.173a}$$

$$\frac{I_2(-\zeta)}{\sqrt{k+\zeta}} + \frac{1}{\pi} \int_k^\infty \frac{dt}{t+\zeta} \frac{e^{2itl} I_1(t)}{\sqrt{t-k}} = -\frac{2e^{i\zeta_0 l}\sqrt{k+\zeta_0}}{\zeta + \zeta_0} , \tag{16.173b}$$

where on the left-hand side we have, as was discussed in the sentence before (16.114), replaced $\zeta_\pm$ by ζ_0. The approximation in the Fresnel integral is equivalent to replacing ζ by k in the above integral. That is, it is equivalent to writing

$$\frac{1}{\zeta+t} = \frac{1}{k+t+(\zeta-k)} = \frac{1}{k+t}\left(1 - \frac{\zeta-k}{k+t} + \cdots\right) . \tag{16.174}$$

The first approximation to (16.173a) and (16.173b) is then obtained by setting $\zeta = k$, which is a correction to that exhibited in (16.167a) and (16.167b), or

$$\frac{I_1(\zeta)}{\sqrt{k+\zeta}} = -\frac{2e^{-i\zeta_0 l}\sqrt{k-\zeta_0}}{\zeta - \zeta_0} - \frac{C_2}{\pi}, \tag{16.175a}$$

$$\frac{I_2(-\zeta)}{\sqrt{k+\zeta}} = -\frac{2e^{i\zeta_0 l}\sqrt{k+\zeta_0}}{\zeta + \zeta_0} - \frac{C_1}{\pi}, \tag{16.175b}$$

where

$$C_2 = \int_k^\infty \frac{dt}{k+t} e^{2itl} \frac{I_2(-t)}{\sqrt{t-k}}, \tag{16.176a}$$

$$C_1 = \int_k^\infty \frac{dt}{k+t} e^{2itl} \frac{I_1(t)}{\sqrt{t-k}}. \tag{16.176b}$$

Eliminating $I_1(t)$ in the latter by using (16.175a) we have

$$C_1 = \int_k^\infty \frac{d\zeta}{\sqrt{\zeta-k}\sqrt{\zeta+k}} e^{2i\zeta l} \left(-\frac{2e^{-i\zeta_0 l}\sqrt{k-\zeta_0}}{\zeta - \zeta_0} - \frac{C_2}{\pi} \right), \tag{16.177}$$

or, using (16.70), ($\delta = 2l$)

$$C_1 + \frac{i}{2}H_0^{(1)}(k\delta)C_2 = -2\sqrt{k-\zeta_0} e^{-i\zeta_0 l} \int_k^\infty \frac{d\zeta}{\sqrt{\zeta^2-k^2}} \frac{e^{2i\zeta l}}{\zeta - \zeta_0}. \tag{16.178a}$$

Similarly,

$$C_2 + \frac{i}{2}H_0^{(1)}(k\delta)C_1 = -2\sqrt{k+\zeta_0} e^{i\zeta_0 l} \int_k^\infty \frac{d\zeta}{\sqrt{\zeta^2-k^2}} \frac{e^{2i\zeta l}}{\zeta + \zeta_0}. \tag{16.178b}$$

If we solve these two equations, we get C_2 and C_1. Next, calculate the transform of the electric field. We need to evaluate the integrals in (16.156). We need the values only for $\zeta = k\cos\theta$, where θ is the angle of scattering, that is, for $|\zeta| < k$. Using (16.175a), we find the second integral to be

$$-\frac{i}{2\pi}\frac{e^{-i\zeta l}}{\sqrt{k+\zeta}} \int_k^\infty \frac{dt}{t-\zeta} e^{2itl} \sqrt{\frac{t+k}{t-k}} \left[-2\frac{e^{-i\zeta_0 l}\sqrt{k-\zeta_0}}{t-\zeta_0} - \frac{C_2}{\pi} \right]. \tag{16.179}$$

So we are led to consider integrals of the type

$$\int_k^\infty \frac{dt}{t-\zeta} e^{it\delta} \sqrt{\frac{t+k}{t-k}} = \int_k^\infty \frac{dt}{t-\zeta} \frac{t-\zeta+k+\zeta}{\sqrt{t^2-k^2}} e^{it\delta}$$

$$= \frac{\pi i}{2}H_0^{(1)}(k\delta) + (k+\zeta) \int_k^\infty \frac{dt}{t-\zeta} \frac{e^{it\delta}}{\sqrt{t^2-k^2}}. \tag{16.180}$$

The latter integral is the same as that appearing in the equations (16.178a) and (16.178b) for C_1 and C_2. Let us denote this integral by

$$F(z) = \int_k^\infty \frac{dt}{\sqrt{t^2 - k^2}} \frac{e^{itz}}{t - \zeta} , \tag{16.181}$$

which satisfies the differential equation

$$\left(\frac{d}{dz} - i\zeta\right) F(z) = i \int_k^\infty dt \frac{e^{itz}}{\sqrt{t^2 - k^2}} = -\frac{\pi}{2} H_0^{(1)}(kz) . \tag{16.182}$$

Multiply both sides of this equation by the integrating factor $e^{-i\zeta z}$. Then

$$\frac{d}{dz} \left(e^{-i\zeta z} F(z)\right) = -\frac{\pi}{2} H_0^{(1)}(kz) e^{-i\zeta z} , \tag{16.183}$$

so

$$F(z) = \frac{\pi}{2} e^{i\zeta z} \int_z^\infty dz' e^{-i\zeta z'} H_0^{(1)}(kz') , \tag{16.184}$$

since $F(z) \to 0$ as $z \to \infty$. This integral cannot be done exactly, but may be evaluated asymptotically. Next, consider the remaining integral which appears in $E(\zeta)$, (16.179). For simplicity, let us consider $\zeta = \zeta_0$, so, by (16.128), we are describing total scattering only. Then the integral we need is

$$\int_k^\infty \frac{dt}{(t - \zeta)^2} e^{it\delta} \sqrt{\frac{t + k}{t - k}} = -\frac{d}{d\zeta} \int_k^\infty \frac{dt}{t - \zeta} e^{it\delta} \sqrt{\frac{t + k}{t - k}} . \tag{16.185}$$

Then, we obtain for this integral from (16.180) and (16.184)

$$\frac{\pi}{2} e^{i\zeta\delta} \int_\delta^\infty dz' e^{-i\zeta z'} H_0^{(1)}(kz') + \frac{\pi i}{2}(k + \zeta)\delta \, e^{i\zeta\delta} \int_\delta^\infty dz' e^{-i\zeta z'} H_0^{(1)}(kz')$$
$$- \frac{\pi i}{2}(k + \zeta) e^{i\zeta\delta} \int_\delta^\infty dz' \, z' e^{-i\zeta z'} H_0^{(1)}(kz') . \tag{16.186}$$

Again for simplicity take normal incidence, so $\zeta = \zeta_0 = 0$. For normal incidence we obtain for the electric field from (16.156), (16.186), and (16.19) [the two terms in square brackets in (16.156) give equal contributions in this case]

$$E(0) = E^{(0)}(0) + \frac{1}{k}\left\{i \int_{k\delta}^\infty dx \, H_0^{(1)}(x) - k\delta \int_{k\delta}^\infty dx \, H_0^{(1)}(x)\right.$$
$$\left. - k\delta \, H_1^{(1)}(k\delta) - \frac{i}{2} \frac{\int_{k\delta}^\infty dx \, H_0^{(1)}(x)}{1 + \frac{i}{2} H_0^{(1)}(k\delta)} \left[\int_{k\delta}^\infty dx H_0^{(1)}(x) + i H_0^{(1)}(k\delta)\right]\right\} . \tag{16.187}$$

Here we have used, for normal incidence, from (16.178b) and (16.184),

$$C_1 \left[1 + \frac{i}{2} H_0^{(1)}(k\delta) \right] = C_2 \left[1 + \frac{i}{2} H_0^{(1)}(k\delta) \right] = -\frac{\pi}{\sqrt{k}} \int_{k\delta}^{\infty} dx \, H_0^{(1)}(x) \,.$$

$$(16.188)$$

The first three correction terms in (16.187) are important here, so looking at those, and taking the real part, we get the total cross section according to (16.128) and (16.129),

$$\sigma = \delta \left[1 - \int_{k\delta}^{\infty} dx \, J_0(x) - J_1(k\delta) \right] - \frac{1}{k} \int_{k\delta}^{\infty} dx \, N_0(x) + \cdots$$

$$= \delta \left[1 - \int_{k\delta}^{\infty} \frac{dx}{x} J_1(x) \right] - \frac{1}{k} \int_{k\delta}^{\infty} dx \, N_0(x) + \cdots \,, \qquad (16.189)$$

where the remaining terms are small. For $k\delta$ large, we can use the asymptotic form, from (8.33a) and (8.33b),

$$\int_{k\delta}^{\infty} dx \, N_0(x) \sim \sqrt{\frac{2}{\pi k\delta}} \cos\left(k\delta - \frac{\pi}{4} \right) \,, \qquad (16.190a)$$

$$\int_{k\delta}^{\infty} \frac{dx}{x} J_1(x) \sim \frac{1}{k\delta} \sqrt{\frac{2}{\pi k\delta}} \cos\left(k\delta - \frac{\pi}{4} \right) \,. \qquad (16.190b)$$

Adding these terms, we find

$$\sigma = \delta - \frac{4}{k\sqrt{2\pi k\delta}} \cos\left(k\delta - \frac{\pi}{4} \right) \,. \qquad (16.191)$$

This is sketched in Fig. 16.3, and indicates the principal correction to the geometrical limit (16.130). Note that this result is very similar to the cross section for a strip, given in (16.82), at normal incidence, $\theta' = \frac{\pi}{2}$. For opposite polarizations, the two cross sections must agree (apart from a factor of two due to shadowing or forward scattering), according to Babinet's principle.

We close this chapter by noting that in comparison with rigorous results found by Morse and Rubenstein [31] (who incidentally used elliptic cylinder coordinates and Mathieu functions), the Fourier transform method is satisfactory down to $\delta < \lambda/4$, where static methods apply, so the approximation method works well over the whole range. We could now go on to discuss E (or H, depending on one's point of view) polarization, that is, where **H** is parallel to the slit, but we will leave such considerations for Harold.[4]

16.4 Problems for Chap. 16

1. Use the formalism developed in this chapter to prove the optical theorem relating the total cross section per unit length to the forward scattering amplitude (16.13).

---

[4] Harold, the "Hypothetical alert reader of limitless dedication," was the name of Schwinger's older brother. Harold in this guise made his first appearance in [12].

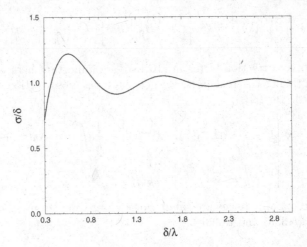

Fig. 16.3. Plot of the first correction to the total cross section per unit length for H-mode scattering by a slit of width δ in a perfectly conducting screen. What is plotted is the ratio of the cross section to the geometric cross section δ as a function of δ/λ, where λ is the wavelength of the radiation

2. Derive the relation (16.112).
3. Derive the Fourier representation for the Hankel function,

$$H_0(k\sqrt{z^2 + x^2}) = \frac{1}{\pi} \int_{-\infty}^{\infty} d\zeta \, e^{iz\zeta} \, \frac{e^{i\sqrt{k^2-\zeta^2}|x|}}{\sqrt{k^2-\zeta^2}} \, , \qquad (16.192)$$

by showing successively that

$$\int_{-\infty}^{\infty} dz \, e^{-i\zeta z} H_0(k\sqrt{z^2+x^2}) = \frac{1}{\pi i} \int_{-\infty}^{\infty} dz \int_{-\infty}^{\infty} dy \, e^{-i\zeta z} \, \frac{e^{ik\sqrt{x^2+y^2+z^2}}}{\sqrt{x^2+y^2+z^2}}$$

$$= \frac{4}{i} \int_{-\infty}^{\infty} dz \, e^{-i\zeta z} \int \frac{dk_x \, dk_z}{(2\pi)^2}$$

$$\times \frac{e^{ik_z z} e^{ik_x x}}{k_x^2 + k_z^2 - (k+i\epsilon)^2}$$

$$= 2\frac{e^{i\sqrt{k^2-\zeta^2}|x|}}{\sqrt{k^2-\zeta^2}} \, . \qquad (16.193)$$

4. Show that the condition that the electric field vanishes on the conducting surfaces is equivalent to the integral equations (16.167a) and (16.167b), in the approximation treated there.

5. The differential cross section for transmission through an aperture of size a in the circumstance $\lambda \ll a$ is

$$\frac{d\sigma}{d\Omega} = \left(\frac{k}{2\pi}\right)^2 \left| \int_{app} dS\, e^{-i\mathbf{k}_\perp \cdot \mathbf{r}_\perp} \right|^2 . \qquad (16.194)$$

Show that ($\cos\theta \approx 1$)

$$d\Omega \approx \frac{(d\mathbf{k}_\perp)}{k^2} . \qquad (16.195)$$

Using this, prove that the total cross section equals the area of the aperture, without regard to its shape.

6. An electromagnetic wave of reduced wavelength λ falls normally on a perfectly conducting flat disk of radius a and negligible thickness. In the center of the shadow behind the disk, at a distance x such that $a \ll x \ll a^2/\lambda$, there is a bright spot of intensity equal to that of the incident beam. Demonstrate this, by considering the appropriate approximation to the surface current on the disk. Then find the smallest displacement from the center that reduces the intensity of the bright spot to zero. (It involves a zero of a Bessel function, of course.) Choose numbers to illustrate the size of the central bright spot.

7. Two infinitely long parallel slits of width a and separation $b \gg a$ are cut in a perfectly conducting sheet of zero thickness. An electromagnetic wave of frequency ω with its electric vector parallel to the slit edges is normally incident on one side. Use physical reasoning to discuss the diffraction pattern of the transmitted wave, as observed at a distance x from the sheet, such that $\lambda \ll x \ll ab/\lambda$. What is the value of the total cross section? Then work out the differential cross section for $x \gg b^2/\lambda \gg ab/\lambda$. Locate the zeroes and maxima of the interference pattern that are attributable to the presence of both slits. Now what value do you find for the total cross section?

8. A plane wave is normally incident on a plane screen containing a large number of similar apertures arranged in a line, with constant spacing d. Show that the diffracted wave observed at a large distance from the screen is destroyed by interference, save in those directions specified by

$$d\cos\theta = n\lambda , \quad n = 0, \pm1, \pm2, \dots , \qquad (16.196)$$

where θ is the angle between the line of apertures and the direction of observation. (This is Bragg scattering.)

9. Babinet's principle states that the diffraction pattern of an aperture in a screen is identical, except in the direction of the incident wave, with that of the complementary situation obtained by replacing the aperture with an obstacle and removing the screen. Prove this by showing that the sum of the field quantities describing the diffracted fields in the two situations equals the incident field.

10. Determine the diffraction pattern for light normally incident on a half plane from the Fresnel–Kirchhoff theory, and compare the result with the rigorous Sommerfeld solution.

Quantum Limitations on Microwave Oscillators

This book has been entirely devoted to *classical* electrodynamics. However, we live in a quantum world, so it is fitting that we end this volume with a discussion of the quantum nature of the interaction of charged particles with the fields in a microwave cavity, represented here as a harmonic oscillator.

17.1 Introduction

The question of quantum limitations in microwave oscillators refers to an essentially classical domain where the conventional types of quantum mechanical description are inappropriate. For a dynamical system characterized by complementary canonical variables q and p, the usual description employs states for which the variables q, say, have definite values and, correspondingly, all values of p are equally probable. This is at the opposite pole from the classical picture in which the q and p variables are simultaneously known. The nearest equivalent to the latter is a quantum mechanical description in which neither the q nor the p variables are precisely specified, but rather both are determined to the optimum precision allowed by their incompatibility.

17.2 Coherent States

To characterize this new description, let A and B be operators symbolizing any two incompatible quantities. For the state symbolized by the unit vector Ψ, the expectation value and dispersion of property A is defined by

$$\langle A \rangle = \Psi^\dagger A \Psi \,, \tag{17.1a}$$

$$\Delta A = \langle (A - \langle A \rangle)^2 \rangle^{1/2} \,, \tag{17.1b}$$

and it is known that

$$\Delta A \, \Delta B \geq \frac{1}{2} |\langle C \rangle| \,, \tag{17.2}$$

where

$$C = \frac{1}{i}[A, B] \, . \tag{17.3}$$

The equality sign in (17.2) holds only when

$$(A - \langle A \rangle)\Psi = \lambda(B - \langle B \rangle)\Psi \tag{17.4}$$

and

$$\langle \{A - \langle A \rangle, B - \langle B \rangle\} \rangle = 0 \, . \tag{17.5}$$

These conditions imply that

$$(\lambda + \lambda^*)(\Delta B)^2 = 0 \, , \tag{17.6}$$

which requires that λ be pure imaginary,

$$\lambda = -i\gamma \, , \qquad \gamma = \gamma^* \, . \tag{17.7}$$

Thus, with $\Delta A \, \Delta B$ as a measure of the simultaneous specifiability of the two physical quantities, the optimum state with minimum $\Delta A \, \Delta B$ is such that

$$(A + i\gamma B)\Psi = (\langle A \rangle + i\gamma\langle B \rangle)\Psi \, , \tag{17.8}$$

which characterizes it as an eigenvector of the non-Hermitian operator $A + i\gamma B$, with the complex eigenvalue $\langle A \rangle + i\gamma\langle B \rangle$.

The magnitude of γ is determined by

$$(\Delta A)^2 = \gamma^2(\Delta B)^2 \, , \tag{17.9}$$

while the sign of γ is that of C. This follows from the equations

$$0 = \langle [A - \langle A \rangle + i\gamma(B - \langle B \rangle)]^2 \rangle = (\Delta A)^2 - \gamma^2(\Delta B)^2 \, , \tag{17.10a}$$

and

$$0 = \langle [A - \langle A \rangle - i\gamma(B - \langle B \rangle)][A - \langle A \rangle + i\gamma(B - \langle B \rangle)] \rangle$$
$$= (\Delta A)^2 + \gamma^2(\Delta B)^2 - \gamma\langle C \rangle \, . \tag{17.10b}$$

Thus

$$\gamma = \frac{(\Delta A)^2}{\frac{1}{2}\langle C \rangle} = \frac{\frac{1}{2}\langle C \rangle}{(\Delta B)^2} \, , \tag{17.11}$$

which, of course, is consistent with the equality in (17.2).

On placing $A = q$, $B = p$, and $C = -i[q, p] = \hbar$, we learn that the state with $\Delta q \, \Delta p = \frac{1}{2}\hbar$ is characterized by

$$(q_d + ip_d)\Psi = (\langle q_d \rangle + i\langle p_d \rangle)\Psi \, , \tag{17.12}$$

where

$$q_d = \frac{q}{\sqrt{2}\Delta q} , \qquad p_d = \frac{p}{\sqrt{2}\Delta p} \qquad (17.13)$$

are dimensionless operators obeying the commutation relations

$$[q_d, p_d] = \mathrm{i} . \qquad (17.14)$$

We are thus led to investigate the eigenvectors of the non-Hermitian operators (working now with the dimensionless operators)

$$y = 2^{-1/2}(q + \mathrm{i}p) , \qquad \mathrm{i}y^\dagger = 2^{-1/2}(p + \mathrm{i}q) , \qquad (17.15)$$

which obey the commutation property

$$[y, \mathrm{i}y^\dagger] = \mathrm{i} , \qquad (17.16)$$

characteristic of canonical variables. The possibility of introducing non-Hermitian canonical variables appears more generally in terms of the Lagrangian operator (written for an arbitrary number of degrees of freedom)

$$L = \sum_k \frac{1}{2} \left(p_k \cdot \frac{\mathrm{d}q_k}{\mathrm{d}t} - \frac{\mathrm{d}p_k}{\mathrm{d}t} \cdot q_k \right) - H , \qquad (17.17)$$

where the dot indicates symmetrized multiplication. On writing

$$q_k = 2^{-1/2}(y_k + y_k^\dagger) , \qquad p_k = -\mathrm{i}2^{-1/2}(y_k - y_k^\dagger) , \qquad (17.18)$$

we obtain

$$L = \sum_k \frac{1}{2} \left(\mathrm{i}y_k^\dagger \cdot \frac{\mathrm{d}y_k}{\mathrm{d}t} - \mathrm{i}\frac{\mathrm{d}y_k^\dagger}{\mathrm{d}t} \cdot y_k \right) - H , \qquad (17.19)$$

which exhibits the same form as (17.17) with the substitution $q \to y$, $p \to \mathrm{i}y^\dagger$. Hence, every property derived from the Lagrangian operator, the equations of motion, generators of infinitesimal transformations, and the commutation relations, maintain their form on introducing the non-Hermitian variables. What is not maintained, of course, are the Hermitian properties. In particular, the infinitesimal generators,

$$G_q = \sum_k p_k \, \delta q_k , \qquad G_p = -\sum_k \delta p_k \, q_k , \qquad (17.20)$$

with meanings indicated by

$$\frac{1}{\mathrm{i}}[F(q,p), G_q] = \sum_k \frac{\partial F}{\partial q_k} \delta q_k = \delta_q F \qquad (17.21)$$

and

$$\langle p'|iG_p = \sum_k \delta p'_k \frac{\partial}{\partial p_k}\langle p'| = \delta_p\langle p'| \,, \tag{17.22a}$$

$$\delta_q|q'\rangle = -iG_q|q'\rangle \tag{17.22b}$$

imply the infinitesimal generators

$$G_y = \sum_k iy_k^\dagger \delta y_k \,, \qquad G_{y^\dagger} = -\sum_k i\delta y_k^\dagger y_k \,. \tag{17.23}$$

The construction of the right eigenvectors of the operator y (returning to one degree of freedom for simplicity),

$$y|y'\rangle = y'|y'\rangle \,, \tag{17.24}$$

is equivalent to the construction of the transformation function $\langle q'|y'\rangle$. Now, the dependence of the latter on the eigenvalues is given by

$$\delta\langle q'|y'\rangle = i\langle q'|(G_q - G_y)|y'\rangle \,, \tag{17.25}$$

where

$$iG_q = ip\delta q' = (2^{1/2}y - q)\delta q' \,, \tag{17.26a}$$

$$-iG_y = y^\dagger \delta y' = (2^{1/2}q - y)\delta y' \,. \tag{17.26b}$$

Hence, since the operators act directly on the eigenvectors, we get

$$\delta\langle q'|y'\rangle = \delta\left[-\frac{1}{2}q'^2 + 2^{1/2}q'y' - \frac{1}{2}y'^2\right]\langle q'|y'\rangle \,, \tag{17.27}$$

and

$$\langle q'|y'\rangle = C\exp\left[-\frac{1}{2}q'^2 + 2^{1/2}q'y' - \frac{1}{2}y'^2\right] \,. \tag{17.28}$$

Now the adjoint of the right eigenvector equation for y (17.24) is the left eigenvector equation $y^\dagger$:

$$\langle y^{\dagger\prime}|y^\dagger = \langle y^{\dagger\prime}|y^{\dagger\prime} \,, \qquad y^{\dagger\prime} = y'^* \,. \tag{17.29}$$

Hence the complex conjugate of the transformation function $\langle q'|y'\rangle$ is

$$\langle y^{\dagger\prime}|q'\rangle = C^*\exp\left[-\frac{1}{2}q'^2 + 2^{1/2}y^{\dagger\prime}q' - \frac{1}{2}y^{\dagger\prime 2}\right] \,, \tag{17.30}$$

and we conclude that

$$\langle y^{\dagger\prime}|y''\rangle = \int_{-\infty}^{\infty}\langle y^{\dagger\prime}|q'\rangle dq'\langle q'|y''\rangle$$

$$= |C|^2\int_{-\infty}^{\infty}dq'\exp\left[-\left(q' - \frac{y^{\dagger\prime} + y''}{\sqrt{2}}\right)^2\right]e^{y^{\dagger\prime}y''}$$

$$= \sqrt{\pi}|C|^2 e^{y^{\dagger\prime}y''} \,. \tag{17.31}$$

In particular,
$$\langle y^{\dagger'}|y'\rangle = ||y'||^2 = \sqrt{\pi}|C|^2 e^{|y'|^2} , \qquad (17.32)$$
which is finite for arbitrary complex eigenvalues y'. The vector $|y'\rangle$ of minimum norm is the one with zero eigenvalue, and we adopt a normalization to unit length for this vector. Hence, with a conventional choice of phase,
$$C = \pi^{-1/4} , \qquad (17.33)$$
and
$$\langle y^{\dagger'}|y''\rangle = e^{y^{\dagger'}y''} , \qquad (17.34a)$$
$$\langle q'|y'\rangle = \pi^{-1/4} \exp\left[-\frac{1}{2}q'^2 + 2^{1/2}q'y' - \frac{1}{2}y'^2\right] . \qquad (17.34b)$$

We shall adopt a special notation for the vectors $|y'\rangle e^{-|y'|^2/2}$, which are of unit length. On writing
$$y' = \frac{1}{2^{1/2}}(q' + ip') , \qquad (17.35)$$
we define
$$|q'p'\rangle = e^{-\frac{1}{2}|y'|^2}|y'\rangle = e^{-\frac{1}{4}(p'^2+q'^2)}|y'\rangle . \qquad (17.36)$$
In this notation, the transformation functions we have evaluated read
$$\langle q'p'|q''p''\rangle = e^{-\frac{1}{2}|y'|^2}\langle y^{\dagger'}|y''\rangle e^{-\frac{1}{2}|y''|^2}$$
$$= e^{\frac{1}{2}(q'p''-p'q'')}e^{-\frac{1}{4}(q'-q'')^2-\frac{1}{4}(p'-p'')^2} \qquad (17.37a)$$
and
$$\langle q'|q''p''\rangle = \langle q'|y''\rangle e^{-\frac{1}{2}|y''|^2}$$
$$= \pi^{-1/4}e^{i(q'p''-\frac{1}{2}q''p'')}e^{-\frac{1}{2}(q'-q'')^2} . \qquad (17.37b)$$

The completeness property of the $|qp\rangle$ states can be inferred from the last result, namely
$$1 = \int |qp\rangle \frac{dq\,dp}{2\pi}\langle qp| , \qquad (17.38)$$
since we find by direct calculation that
$$\int \langle q'|qp\rangle \frac{dq\,dp}{2\pi}\langle qp|q''\rangle = \pi^{-1/2}\int_{-\infty}^{\infty} dq\, e^{-\frac{1}{2}(q'-q)^2-\frac{1}{2}(q''-q)^2} \int_{-\infty}^{\infty} \frac{dp}{2\pi}e^{i(q'-q'')p}$$
$$= \delta(q'-q'') . \qquad (17.39)$$

The usual probability interpretation would follow from this completeness property of the vectors $|qp\rangle$ if they were also linearly independent. However, the transformation function $\langle q'p'|q''p''\rangle$, which equals unity for $q' = q''$,

$p' = p''$, remains of this order for $|q' - q''| \leq 1$, $|p' - p''| \leq 1$, so that a displacement of q' or p' of the order of unity or less does not create a new state. On the other hand, the Gaussian factor in (17.37a) ensures that a change in q' or p' which is large compared with unity certainly does produce a new state. Thus in a sense there is one state associated with each eigenvalue interval $\Delta q' \Delta p' / 2\pi = 1$. The precise construction of the individual states is not necessary, however, if we are concerned only with a large number of states, or eigenvalue intervals $\Delta q' \Delta p' / 2\pi > 1$, which is appropriate to essentially classical measurements. Then we can assert that the wave function

$$\Psi(q'p') = \langle q'p'|\Psi \tag{17.40}$$

is a probability amplitude, giving, according to

$$\int_\Omega \frac{dq\,dp}{2\pi} |\Psi(qp)|^2 \,, \tag{17.41}$$

the total probability that q and p measurements, performed on the state Ψ with optimum compatibility, will give results lying within the region Ω of the qp phase space.

17.3 Harmonic Oscillator

The simplest dynamical application of these methods appears for the harmonic oscillator system described by the Hamiltonian

$$H = \omega \frac{1}{2}(p^2 + q^2) = \omega \left(y^\dagger y + \frac{1}{2} \right) \,, \tag{17.42}$$

which obeys the equation of motion

$$i\frac{dy}{dt} = \frac{\partial H}{\partial y^\dagger} = \omega y \,. \tag{17.43}$$

The dynamical situation can be described by the transformation function $\langle y^{\dagger\prime} t_1 | y'' t_2 \rangle$, which varies with t_1 in accordance with

$$i\frac{\partial}{\partial t_1} \langle y^{\dagger\prime} t_1 | y'' t_2 \rangle = \langle y^{\dagger\prime} t_1 | H | y'' t_2 \rangle \,. \tag{17.44}$$

But, on applying the explicit solution of the equations of motion (17.43)

$$y(t_1) = e^{-i\omega\tau} y(t_2) \,, \quad \tau = t_1 - t_2 \,, \tag{17.45}$$

we can write

$$H = \omega \left(y(t_1)^\dagger y(t_1) + \frac{1}{2} \right) = \omega \left(y(t_1)^\dagger y(t_2) e^{-i\omega\tau} + \frac{1}{2} \right) \,, \tag{17.46}$$

which gives

$$i\frac{\partial}{\partial t_1} \langle y^{\dagger\prime} t_1 | y^{\prime\prime} t_2 \rangle = \omega \left(y^{\dagger\prime} y^{\prime\prime} e^{-i\omega\tau} + \frac{1}{2} \right) \langle y^{\dagger\prime} t_2 | y^{\prime\prime} t_2 \rangle . \tag{17.47}$$

This has the solution

$$\langle y^{\dagger\prime} t_1 | y^{\prime\prime} t_2 \rangle = e^{-i\omega\tau/2} \exp\left[y^{\dagger\prime} y^{\prime\prime} e^{-i\omega\tau} \right] , \tag{17.48}$$

which correctly reduces to $\langle y^{\dagger\prime} | y^{\prime\prime} \rangle$, (17.34a), for equal times. From this result we derive

$$\langle q^\prime p^\prime t_1 | q^{\prime\prime} p^{\prime\prime} t_2 \rangle = e^{-i\omega\tau/2} e^{-\frac{1}{2}|y^\prime|^2} e^{-\frac{1}{2}|y^{\prime\prime}|^2} \exp\left[y^{\dagger\prime} y^{\prime\prime} e^{-i\omega\tau} \right] , \tag{17.49}$$

and

$$
\begin{aligned}
|\langle q^\prime p^\prime t_1 | q^{\prime\prime} p^{\prime\prime} t_2 \rangle|^2 &= \exp\left(-|y^\prime - y^{\prime\prime} e^{-i\omega\tau}|^2 \right) \\
&= \exp\left[-\frac{1}{2}(q^\prime - q^{\prime\prime}\cos\omega\tau - p^{\prime\prime}\sin\omega\tau)^2 \right. \\
&\qquad \left. -\frac{1}{2}(p^\prime - p^{\prime\prime}\cos\omega\tau + q^{\prime\prime}\sin\omega\tau)^2 \right] , \tag{17.50}
\end{aligned}
$$

which clearly show the limiting classical results and the differences.

The transformation function $\langle y^{\dagger\prime} t_1 | y^{\prime\prime} t_2 \rangle$ also supplies complete information about the state of definite energy, since

$$\langle y^{\dagger\prime} t_1 | = \langle y^{\dagger\prime} t_2 | e^{-iH\tau} \tag{17.51}$$

shows that

$$\langle y^{\dagger\prime} t_1 | y^{\prime\prime} t_2 \rangle = \langle y^{\dagger\prime} | e^{-iH\tau} | y^{\prime\prime} \rangle = \sum_E \langle y^{\dagger\prime} | E \rangle e^{-iE\tau} \langle E | y^{\prime\prime} \rangle . \tag{17.52}$$

Indeed, since (17.48) can be expanded as

$$\langle y^{\dagger\prime} t_1 | y^{\prime\prime} t_2 \rangle = \sum_{n=1}^{\infty} \frac{(y^{\dagger\prime})^n}{\sqrt{n!}} e^{-i(n+1/2)\omega\tau} \frac{(y^{\prime\prime})^n}{\sqrt{n!}} , \tag{17.53}$$

we infer the energy spectrum

$$E_n = \left(n + \frac{1}{2} \right) \omega , \quad n = 0, 1, \dots , \tag{17.54}$$

and the wavefunctions,

$$\langle y^{\dagger\prime} | n \rangle = \frac{(y^{\dagger\prime})^n}{\sqrt{n!}} , \qquad \langle n | y^{\prime\prime} \rangle = \frac{(y^{\prime\prime})^n}{\sqrt{n!}} . \tag{17.55}$$

Hence,

$$\langle q'p'|n\rangle = e^{-\frac{1}{2}|y^{\dagger\prime}|^2}\langle y^{\dagger\prime}|n\rangle = e^{-\frac{1}{4}(q'^2+p'^2)}\frac{(q'-ip')^n}{\sqrt{2^n n!}} \,, \tag{17.56}$$

which gives the probability density

$$\frac{dq'\,dp'}{2\pi}|\langle q'p'|n\rangle|^2 = \frac{dq'\,dp'}{2\pi}e^{-\frac{1}{2}(q'^2+p'^2)}\frac{\left(\frac{1}{2}(q'^2+p'^2)\right)^n}{n!}$$

$$= dE'\frac{(E')^n}{n!}e^{-E'} \,, \tag{17.57}$$

where the last form refers to the distribution of the "energy,"

$$E' = \frac{1}{2}(p'^2+q'^2) \,. \tag{17.58}$$

With increasing n, this Poisson distribution rapidly approaches the Gaussian distribution

$$\frac{dE'}{\sqrt{2\pi n}}e^{-(E'-n)^2/(2n)} \,, \tag{17.59}$$

characteristic of quantum phenomena.

17.4 Free Particle

For the dynamical system of a free particle, described by the Hamiltonian

$$H = \frac{p^2}{2m} = -\frac{(y-y^{\dagger})^2}{4m} \,, \tag{17.60}$$

the equations of motion are

$$i\frac{dy}{dt} = \frac{y-y^{\dagger}}{2m} \,, \qquad i\frac{dy^{\dagger}}{dt} = \frac{y-y^{\dagger}}{2m} \,, \tag{17.61}$$

which imply (writing $y_1 = y(t_1)$, etc.)

$$y_1 - y_1^{\dagger} = y_2 - y_2^{\dagger} \,, \tag{17.62a}$$

$$y_1 - y_2 = -i\frac{\tau}{2m}(y_1 - y_1^{\dagger}) \,. \tag{17.62b}$$

Hence

$$y_1 = \frac{\frac{i\tau}{2m}y_1^{\dagger} + y_2}{1 + \frac{i\tau}{2m}} \,, \qquad y_2^{\dagger} = \frac{y_1^{\dagger} + \frac{i\tau}{2m}y_2}{1 + \frac{i\tau}{2m}} \,, \tag{17.63}$$

so by commuting the first of these with $y_1^{\dagger}$, we infer that

$$[y_2, y_1^{\dagger}] = 1 + \frac{i\tau}{2m} \,. \tag{17.64}$$

Now we can write the Hamiltonian as

$$H = -\frac{(y_1 - y_1^\dagger)^2}{4m} = -\frac{1}{4m}\left(\frac{y_2 - y_1^\dagger}{1 + \frac{i\tau}{2m}}\right)^2$$

$$= -\frac{1}{4m}\frac{y_1^{\dagger 2} - 2y_1^\dagger y_2 + y_2^2}{\left(1 + \frac{i\tau}{2m}\right)^2} + \frac{1}{4m}\frac{1}{1 + \frac{i\tau}{2m}}, \qquad (17.65)$$

and therefore the Schrödinger equation is

$$i\frac{\partial}{\partial t}\langle y^{\dagger\prime}t_1 | y''t_2\rangle = \langle y^{\dagger\prime}t_1 | H | y''t_2\rangle$$

$$= \left[-\frac{1}{4m}\frac{(y^{\dagger\prime} - y'')^2}{\left(1 + \frac{i\tau}{2m}\right)^2} + \frac{1}{4m}\frac{1}{1 + \frac{i\tau}{2m}}\right]\langle y^{\dagger\prime}t_1 | y''t_2\rangle .$$

$$(17.66)$$

The solution is

$$\langle y^{\dagger\prime}t_1 | y''t_2\rangle = \frac{1}{\left(1 + \frac{i\tau}{2m}\right)^{1/2}} \exp\left[-\frac{1}{2}\frac{(y^{\dagger\prime} - y'')^2}{1 + \frac{i\tau}{2m}} + \frac{1}{2}(y^{\dagger\prime 2} + y''^2)\right], \quad (17.67)$$

where the time independent factor is determined by the initial condition. Alternatively, from (17.23) and (17.63), we observe that

$$\frac{\partial}{\partial y^{\dagger\prime}}\langle y^{\dagger\prime}t_1 | y''t_2\rangle = \langle y^{\dagger\prime}t_1 | y_1 | y''t_2\rangle$$

$$= \frac{\frac{i\tau}{2m}y^{\dagger\prime} + y''}{1 + \frac{i\tau}{2m}}\langle y^{\dagger\prime}t_1 | y''t_2\rangle = \left(y^{\dagger\prime} - \frac{y^{\dagger\prime} - y''}{1 + \frac{i\tau}{2m}}\right)\langle y^{\dagger\prime}t_1 | y''t_2\rangle ,$$

$$(17.68)$$

which supplies the complete dependence upon the eigenvalue $y^{\dagger\prime}$, and similarly for y''. From this result we infer the transformation function

$$\langle q'p't_1 | q''p''t_2\rangle = \left(1 + \frac{i\tau}{2m}\right)^{-1/2} \exp\left[\frac{i}{2}(q'p'' - p'q'')\right]$$

$$\times \exp\left(-\frac{1}{4}\frac{1}{1 + i\tau/2m}\left[q' - q'' - \frac{\tau}{2m}(p' + p'')\right]^2\right)$$

$$\times \exp\left[-\frac{1}{4}(p' - p'')^2\right]\exp\left[-\frac{i\tau}{2m}\left(\frac{p' + p''}{2}\right)^2\right] ,$$

$$(17.69)$$

and the probability distribution

$$|\langle q'p't_1 | q''p''t_2\rangle|^2 = \frac{1}{\left[1 + \left(\frac{\tau}{2m}\right)^2\right]^{1/2}} \exp\left[-\frac{1}{2}(p' - p'')^2\right]$$

$$\times \exp\left\{-\frac{1}{2}\frac{1}{1+\left(\frac{\tau}{2m}\right)^2}\left[q'-q''-\frac{\tau}{2m}(p'+p'')\right]^2\right\} .$$

$$(17.70)$$

Here again we recognize the limiting classical results, and a quantum difference that grows in time, for the coordinate distribution.

On using the identity

$$\int_{-\infty}^{\infty}dx\,e^{-ax^2+i\sqrt{2}xy} = \sqrt{\frac{\pi}{a}}e^{-y^2/2a} ,$$

$$(17.71)$$

the above transformation function (17.67) becomes

$$\langle y^{\dagger\prime}t_1|y''t_2\rangle = \frac{1}{\sqrt{\pi}}e^{\frac{1}{2}y^{\dagger\prime2}}\int_{-\infty}^{\infty}dp\exp\left[-\left(1+\frac{i\tau}{2m}\right)p^2+i\sqrt{2}p(y^{\dagger\prime}-y'')\right]$$
$$\times e^{\frac{1}{2}y''^2} ,$$

$$(17.72)$$

from which we identify from (17.52) the energy spectrum for a free particle,

$$E = \frac{p^2}{2m} , \quad -\infty < p < \infty ,$$

$$(17.73)$$

and then the wavefunction,

$$\langle y^{\dagger\prime}|p\rangle = \pi^{-1/4}\exp\left(\frac{1}{2}y^{\dagger\prime2}+i\sqrt{2}py^{\dagger\prime}-\frac{1}{2}p^2\right) ,$$

$$(17.74a)$$

$$\langle p|y''\rangle = \pi^{-1/4}\exp\left(\frac{1}{2}y''^2-i\sqrt{2}py''-\frac{1}{2}p^2\right) .$$

$$(17.74b)$$

The implied probability distribution for an energy state occupying the momentum range dp is, essentially,

$$\frac{dq'\,dp'}{2\pi}\left|\langle q'p'|p\rangle[dp]^{1/2}\right|^2 = \frac{dq'}{2\pi/dp}\frac{dp'}{\sqrt{\pi}}e^{-(p'-p)^2} ,$$

$$(17.75)$$

where the normalization factor for the q' distribution indicates that all values of q' are equally probable with a range given by $2\pi/dp$. The various formulae for the free particle are converted to conventional units by the substitution of

$$q \to \frac{q}{\sqrt{2}\Delta q} , \quad p \to \frac{p}{\sqrt{2}\Delta p} , \quad m \to m\frac{\Delta q}{\Delta p} .$$

$$(17.76)$$

17.5 Electron Interacting with an Oscillator

Now let us consider the dynamical problem described by

$$H = \omega y^\dagger y + \frac{p^2}{2m} + \lambda(y + y^\dagger)\mu(q) , \tag{17.77}$$

which is an idealization of an electron interacting with an electromagnetic oscillator. On calling the non-Hermitian operators for the particle, defined by q and p, by z and $z^\dagger$, this Hamiltonian becomes

$$H = \omega y^\dagger y - \frac{(z - z^\dagger)^2}{4m} + \lambda(y + y^\dagger)\mu\left(\frac{z + z^\dagger}{\sqrt{2}}\right) . \tag{17.78}$$

We shall make use of an approximate method, based upon the general differential property of transformation functions, the quantum action principle [8],

$$\delta_H \langle t_1 | t_2 \rangle = -i \langle t_1 | \int_{t_2}^{t_1} dt\, \delta H | t_2 \rangle . \tag{17.79}$$

Thus with $H = H_0 + \lambda H_1$, we have

$$\frac{\partial}{\partial \lambda} \langle t_1 | t_2 \rangle = -i \langle t_1 | \int_{t_2}^{t_1} dt\, H_1 | t_2 \rangle , \tag{17.80}$$

and a first approximation is obtained by using the equation of motion for $\lambda = 0$ to replace the operator $H_1(t)$ by numerical eigenvalues, say $\langle H_1(t) \rangle$. Then, the differential equation becomes

$$\frac{\partial}{\partial \lambda} \langle t_1 | t_2 \rangle = -i \int_{t_2}^{t_1} dt\, \langle H_1(t) \rangle \langle t_1 | t_2 \rangle , \tag{17.81}$$

and the solution is then

$$\langle t_1 | t_2 \rangle = \langle t_1 | t_2 \rangle_{H_0} \exp\left[-i \int_{t_2}^{t_1} dt\, \lambda \langle H_1(t) \rangle \right] . \tag{17.82}$$

Now here the perturbing Hamiltonian is

$$H_1 = (y + y^\dagger)\mu\left(\frac{z + z^\dagger}{\sqrt{2}}\right)$$

$$= \left(e^{-i\omega(t - t_2)} y_2 + e^{-i\omega(t_1 - t)} y_1^\dagger \right) \mu \left(\frac{\left(1 + i\frac{t - t_2}{m}\right) z_1^\dagger + \left(1 + i\frac{t_1 - t}{m}\right) z_2}{\sqrt{2}\left(1 + \frac{i\tau}{2m}\right)} \right) , \tag{17.83}$$

in which we have, in particular, made use of the free particle solutions

$$z = \frac{\left(1 + i\frac{t_1 - t}{2m}\right) z_2 + \left(i\frac{t - t_2}{2m}\right) z_1^\dagger}{1 + \frac{i\tau}{2m}} , \tag{17.84a}$$

$$z^\dagger = \frac{i\frac{t_1 - t}{2m} z_2 + \left(1 + i\frac{t - t_2}{2m}\right) z_1^\dagger}{1 + \frac{i\tau}{2m}} . \tag{17.84b}$$

If we are constructing the transformation function $\langle y^{\dagger\prime} z^{\dagger\prime} t_1 | y'' z'' t_2 \rangle$, it is necessary to evaluate, in the operator function μ, the noncommuting operators z_2 and $z_1^{\dagger}$, using (17.64), or

$$[z_2, z_1^{\dagger}] = 1 + \frac{i\tau}{2m} . \tag{17.85}$$

We write

$$\mu(q) = \int_{-\infty}^{\infty} dk\, \mu(k) e^{ikq} , \qquad \mu(k)^* = \mu(-k) , \tag{17.86}$$

and remark that

$$e^{ikq} = \exp\left(ik[\alpha z_1^{\dagger} + \beta z_2] \right)$$

$$= \exp\left(ik\alpha z_1^{\dagger} \right) \exp\left(ik\beta z_2 \right) \exp\left(-\frac{k^2}{2}\alpha\beta[z_2, z_1^{\dagger}] \right) , \tag{17.87}$$

which implies that

$$\langle e^{ikq} \rangle = \exp\left[ik(\alpha z^{\dagger\prime} + \beta z'') \right] \exp\left[-\frac{k^2}{2}\alpha\beta\left(1 + i\frac{\tau}{2m} \right) \right]$$

$$= e^{ik\bar{q}} \exp\left[-\frac{k^2}{4}\frac{\left(1 + i\frac{t-t_2}{m}\right)\left(1 + i\frac{t_1-t}{m}\right)}{1 + \frac{i\tau}{2m}} \right] , \tag{17.88}$$

where we have used the abbreviation

$$\bar{q} = \frac{\left(1 + i\frac{t-t_2}{m}\right) z^{\dagger\prime} + \left(1 + i\frac{t_1-t}{m}\right) z''}{\sqrt{2}\left(1 + \frac{i\tau}{2m}\right)} = q'' + \frac{t - t_2}{m} p''$$

$$- \frac{i}{2}(p' - p'')\left(1 + i\frac{t - t_2}{m}\right) + \frac{1}{2}\left(q' - q'' - \frac{\tau}{m}\frac{p' + p''}{2} \right)\frac{1 + i\frac{t-t_2}{m}}{1 + \frac{i\tau}{2m}} . \tag{17.89}$$

Thus from (17.83)

$$\int_{t_2}^{t_1} dt\, \lambda \langle H_1 \rangle = \int_{t_2}^{t_1} dt\, \left(e^{-i\omega t_1} y^{\dagger\prime} e^{i\omega t} + e^{i\omega t_2} y'' e^{-i\omega t} \right) \int_{-\infty}^{\infty} dk\, \lambda \mu(k) e^{ik\bar{q}}$$

$$\times \exp\left\{ -\frac{k^2}{4}\left[1 + i\frac{\tau}{2m} + \frac{[(t - \frac{t_1+t_2}{2})/m]^2}{1 + \frac{i\tau}{2m}} \right] \right\}$$

$$= e^{-i\omega t_1} y^{\dagger\prime} i\gamma_1 - e^{i\omega t_2} y'' i\gamma_2^* . \tag{17.90}$$

With this last notation, our transformation function (17.82) is presented as

$$\langle y^{\dagger\prime} t_1 | y'' t_2 \rangle_0 \langle z^{\dagger\prime} t_1 | z'' t_2 \rangle_0 \exp\left(e^{-i\omega t_1} y^{\dagger\prime} \gamma_1 - e^{i\omega t_2} y'' \gamma_2^* \right) , \tag{17.91}$$

from which we derive the probability distribution from (17.50) and (17.70) (where we write $y^{\dagger\prime} = \frac{1}{\sqrt{2}}(q' - ip')$ and omit the factors $dq'' dp''/2\pi$, $dq' dp'/2\pi$)

$$\exp\left(-|y'e^{i\omega t_1} - y''e^{i\omega t_2}|^2\right)\left[1 + \left(\frac{\tau}{2m}\right)^2\right]^{-1/2}$$

$$\times \exp\left[-\frac{1}{2}\frac{\left(q' - q'' - \frac{\tau}{m}\frac{p'+p''}{2}\right)^2}{1 + (\tau/2m)^2}\right]\exp\left[-\frac{1}{2}(p'-p'')^2\right]$$

$$\times \exp\left(2\mathrm{Re}\left[y'e^{i\omega t_1}\gamma_1^* - y''e^{i\omega t_2}\gamma_2^*\right]\right)$$

$$= \exp\left(-|y'e^{i\omega t_1} - y''e^{i\omega t_2} - \gamma_1|^2\right)\left[1 + \left(\frac{\tau}{2m}\right)^2\right]^{-1/2}$$

$$\times \exp\left[-\frac{1}{2}\frac{\left(q' - q'' - \frac{\tau}{m}\frac{p'+p''}{2}\right)^2}{1 + (\tau/2m)^2}\right]\exp\left[-\frac{1}{2}(p'-p'')^2\right]$$

$$\times \exp\left(2\mathrm{Re}[y''e^{i\omega t_2}(\gamma_1^* - \gamma_2^*)]\right) , \tag{17.92}$$

to within the approximation considered ($|\gamma_1|^2$ being omitted in the exponent). Here from (17.90)

$$\gamma_1 - \gamma_2 = \int_{t_2}^{t_1} dt\, e^{i\omega t}\int_{-\infty}^{\infty} dk\, \lambda\mu(k)e^{ik\bar{q}_r}$$

$$\times \exp\left(-\frac{k^2}{4}\left[1 + \frac{([t-(t_1+t_2)/2]/m)^2}{1 + (\tau/2m)^2}\right]\right)$$

$$\times 2i\sinh\left(k\bar{q}_i + \frac{k^2}{4}\frac{i\tau}{2m}\left[1 - \frac{([t-(t_1+t_2)/2]/m)^2}{1 + (\tau/2m)^2}\right]\right) , \tag{17.93}$$

and from (17.89)

$$\bar{q}_r = q'' + \frac{t-t_2}{m}p'' + (p'-p'')\frac{t-t_2}{2m}$$

$$+ \frac{1}{2}\left(q'-q'' - \frac{\tau}{m}\frac{p'+p''}{2}\right)\frac{1 + \frac{t-t_2}{m}\frac{\tau}{2m}}{1 + (\tau/2m)^2} , \tag{17.94a}$$

$$\bar{q}_i = -\frac{1}{2}(p'-p'') + \left(q'-q'' - \frac{\tau}{m}\frac{p'+p''}{2}\right)\frac{(t - \frac{t_1+t_2}{2})/2m}{1 + (\tau/2m)^2} , \tag{17.94b}$$

are the real and imaginary parts of $\bar{q}$.

From the significance of k [$= \sqrt{2}k\Delta q$] as a ratio of Δq to a length of the order of the distance over which $\mu(q)$ differs from zero, it is apparent that a classical situation requires that $|k| \ll 1$. If we combine this with approximations valid for weak fields, $p' \approx p''$, $q' \approx q'' + (\tau/m)(p'+p'')/2$, we obtain an approximate computation of $\gamma_1 - \gamma_2$,

$$\gamma_1 - \gamma_2 \approx \int_{t_2}^{t_1} dt\, e^{i\omega t}\int dk\, \lambda\mu(k)\, e^{ik(q''+(t-t_2)p''/m)}ik$$

$$\times \left[-(p' - p'') + \left(q' - q'' - \frac{\tau}{m} \frac{p' + p''}{2} \right) \frac{\left(t - \frac{t_1 + t_2}{2} \right)/m}{1 + (\tau/2m)^2} \right]$$

$$= -(p' - p'') \int_{t_2}^{t_1} dt\, e^{i\omega t} \frac{\partial}{\partial q} \lambda \mu \left(q'' + \frac{t - t_2}{m} p'' \right)$$

$$+ \frac{\left(q' - q'' - \frac{\tau}{m} \frac{p' + p''}{2} \right)}{1 + (\tau/2m)^2} \int_{t_2}^{t_1} dt\, e^{i\omega t} \frac{t - \frac{t_1 + t_2}{2}}{m}$$

$$\times \frac{\partial}{\partial q} \lambda \mu \left(q'' + \frac{t - t_2}{m} p'' \right) , \tag{17.95}$$

where, in the same approximation,

$$\gamma_1 = -i \int_{t_2}^{t_1} dt\, e^{i\omega t} \lambda \mu \left(q'' + \frac{t - t_2}{2} p'' \right) . \tag{17.96}$$

The resulting probability distribution (17.92) can be written as

$$e^{-|y' - y_c'|} \left[1 + \left(\frac{\tau}{2m} \right)^2 \right]^{-1/2} \exp \left[-\frac{1}{2} \frac{(q' - q_c')^2}{1 + (\tau/2m)^2} \right] e^{-\frac{1}{2}(p' - p_c')^2} , \tag{17.97}$$

with

$$y_c' = y'' e^{-i\omega \tau} - i \int_{t_2}^{t_1} dt\, e^{-i\omega(t_1 - t)} \lambda \mu \left(q'' + \frac{t - t_2}{m} p'' \right) , \tag{17.98a}$$

$$q_c' = q'' + \frac{\tau}{m} p'' - \int_{t_2}^{t_1} dt \left(y'' e^{-i\omega(t - t_2)} + y''^* e^{i\omega(t - t_2)} \right) \frac{t_1 - t}{m}$$

$$\times \frac{\partial}{\partial q} \lambda \mu \left(q'' + \frac{t - t_2}{m} p'' \right) , \tag{17.98b}$$

$$p_c' = p'' - \int_{t_2}^{t_1} dt \left(y'' e^{-i\omega(t - t_2)} + y''^* e^{i\omega(t - t_2)} \right) \frac{\partial}{\partial q} \lambda \mu \left(q'' + \frac{t - t_2}{m} p'' \right) , \tag{17.98c}$$

being the classical solutions of the equations of motion in a weak-coupling approximation. The classical nature of the situation with $k\Delta q \ll 1$ is also evident from the following considerations. If the terminal times t_1 and t_2 are such that the particle has actually passed through the region where $\mu(q)$ differs sensibly from zero (the cavity), the same integrals are of the form

$$\int_{-\infty}^{\infty} dt\, e^{-i\omega t} \mu \left(q'' + \frac{t - t_2}{m} p'' \right) = \int_{-\infty}^{\infty} dk\, \mu(k) \int_{-\infty}^{\infty} dt\, e^{-i\omega t} e^{ik\left(q'' + \frac{t - t_2}{m} p'' \right)}$$

$$= 2\pi \int_{-\infty}^{\infty} dk\, \mu(k)\, e^{ik\left(q'' - \frac{t_2}{m} p'' \right)} \delta \left(\omega - k \frac{p''}{m} \right) , \tag{17.99}$$

which shows that only one value of k is significant,

$$k = \frac{\omega}{v} , \qquad p'' = mv .$$ (17.100)

The implicit dispersion of p is

$$\Delta p = \frac{\hbar}{\Delta q} \gg \hbar k$$ (17.101a)

$$v \Delta p = \Delta E \gg \hbar \omega .$$ (17.101b)

Hence under these conditions, the energy of the particle cannot be observed to the precision necessary to detect the absorption or emission of single quanta, and only effects arising from many quanta are significant.

17.5.1 Extreme Quantum Limit

The extreme quantum limit appears under conditions where all inequalities are reversed,

$$\Delta E \ll \hbar \omega , \qquad k \Delta q \gg 1 .$$ (17.102)

Since the initial particle momentum is precisely specified, a simple approach is provided by the $\langle q' t_1 | p' t_2 \rangle$ representation for the particle. In the absence of interactions, this transformation function is

$$\langle q' t_1 | p' t_2 \rangle = (2\pi)^{-1/2} e^{i \left(q' p' - p'^2 \tau / (2m) \right)} .$$ (17.103)

The first-order effect of the interaction requires the evaluation of

$$\langle \mu(q) \rangle = \left\langle \mu \left(q_1 - \frac{t_1 - t}{m} p_2 \right) \right\rangle = \int dk \, \mu(k) \, \langle e^{ik \left(q_1 - \frac{t_1 - t}{m} p_2 \right)} \rangle$$

$$= \int dk \, \mu(k) \, e^{ik(q' - (t_1 - t)p'/m)} e^{-i(t_1 - t)k^2/2m} ,$$ (17.104)

using the identity shown in (17.87), since

$$[q_1, p_2] = \left[q_2 + \frac{\tau}{m} p_2, p_2 \right] = i .$$ (17.105)

If the significant values of k are very small compared with p', we can neglect the last factor to obtain

$$\langle \mu(q) \rangle = \mu \left(q' - \frac{t_1 - t}{m} p' \right) ,$$ (17.106)

and the transformation function (17.82) is from (17.48), (17.103), and (17.83),

$$\langle y^{\dagger'} q' t_1 | y'' p' t_2 \rangle = e^{y^{\dagger'} y'' e^{-i\omega \tau}} \frac{1}{\sqrt{2\pi}} e^{i(q' p' - p'^2 \tau / 2m)}$$

$$\times \exp \left\{ -i \int_{t_2}^{t_1} dt \left(y'' e^{-i\omega(t - t_2)} + y^{\dagger'} e^{-i\omega(t_1 - t)} \right) \lambda \mu \left(q' - \frac{t_1 - t}{m} p' \right) \right\} .$$ (17.107)

If at the initial and final times

$$\mu(q') = \mu(q' - \tau v) = 0 , \tag{17.108}$$

that is, the interaction disappears before or after the particle is in the cavity, we have

$$\int_{t_2}^{t_1} dt\, e^{\mp i\omega t} \mu\left(q' - \frac{t_1 - t}{m}p'\right) = \int_{-\infty}^{\infty} dt\, e^{\mp i\omega t} \int_{-\infty}^{\infty} dk\, \mu(k) e^{ik(q' - (t_1 - t)\frac{p'}{m})}$$

$$= \int dk\, \mu(k)\, e^{ik(q' - t_1 p'/m)} 2\pi\delta(kv \mp \omega) = \frac{2\pi}{v} e^{\pm i\frac{\omega}{v}(q' - t_1 p'/m)} \mu(\pm\omega/v) ,$$

$$\tag{17.109}$$

which indicates that the neglect of the k^2 term is justified if $E \gg \hbar\omega$. This gives

$$\langle y^{\dagger'} q' t_1 | y'' p'' t_2 \rangle = e^{y^{\dagger'} e^{-i\omega t_1} y'' e^{i\omega t_2}} \frac{1}{\sqrt{2\pi}} e^{i(q'p' - p'^2 \tau/2m)}$$

$$\times \exp\left\{ -\frac{2\pi i}{v}\left[y'' e^{i\omega t_2} e^{i\frac{\omega}{v}q' - i\omega t_1} \lambda\mu\left(\frac{\omega}{v}\right) \right.\right.$$

$$\left.\left. + y^{\dagger'} e^{-i\omega t_1} e^{-i\frac{\omega}{v}q' + i\omega t_1} \lambda\mu^*\left(\frac{\omega}{v}\right) \right]\right\}$$

$$= e^{y^{\dagger'} e^{-i\omega t_1} y'' e^{i\omega t_2}} \frac{1}{\sqrt{2\pi}} \sum_{n_- n_+} e^{i(p' + (n_- - n_+)\omega/v)q'}$$

$$\times e^{-i(p' + (n_- - n_+)\omega/v)^2 t_1/2m} e^{ip'^2 t_2/2m} \frac{1}{n_-!} \frac{1}{n_+!}$$

$$\times \left(-\frac{2\pi i}{v}\lambda\mu\left(\frac{\omega}{v}\right) y'' e^{i\omega t_2}\right)^{n_-} \left(-\frac{2\pi i}{v}\lambda\mu^*\left(\frac{\omega}{v}\right) y^{\dagger'} e^{-i\omega t_1}\right)^{n_+}$$

$$= \sum_{\Delta n} \frac{1}{\sqrt{2\pi}} e^{i(p' - \Delta n\omega/v)q'} e^{-i(p' - \Delta n\omega/v)^2 t_1/2m} e^{ip'^2 t_2/2m}$$

$$\times e^{y^{\dagger'} e^{-i\omega t_1} y'' e^{i\omega t_2}} \left(\frac{\mu^*\left(\frac{\omega}{v}\right)}{\mu\left(\frac{\omega}{v}\right)} \frac{y^{\dagger'} e^{-i\omega t_1}}{y'' e^{i\omega t_2}}\right)^{\Delta n/2}$$

$$\times (-i)^{|\Delta n|} J_{|\Delta n|}\left(\frac{4\pi}{v}\lambda\left|\mu\left(\frac{\omega}{v}\right)\right| \sqrt{y^{\dagger'} e^{-i\omega t_1} y'' e^{i\omega t_2}}\right) ,$$

$$\tag{17.110}$$

using the series representation for the Bessel function (8.10), in which we recognize that $(n_- - n_+)\frac{\omega}{v} = -\Delta n\omega/v$, a net change in momentum, resulting in a corresponding small energy change for the particle. One can think of n_- and n_+ as the number of quanta absorbed and emitted, respectively, by the particle, with interference between all processes leading to a common energy change. The simplification of the situation is that the individual emission and absorption acts are uncorrelated.

From this transformation function, the transition probability referring to definite energy states for the oscillator can be derived, with the significance of Δn appearing again as the increase in quantum numbers for the oscillator. The result in the non-Hermitian representation can also be used directly, giving a qp probability distribution for the oscillator when the final momentum of the particle is known to be $p' - \Delta n\omega/v$,

$$
e^{-|y'e^{i\omega t_1} - y''e^{i\omega t_2}|^2} \left| \frac{y'e^{i\omega t_1}}{y''e^{i\omega t_2}} \right|^{\Delta n} \left| J_{\Delta n} \left(\frac{4\pi}{v} \lambda \left| \mu \left(\frac{\omega}{v} \right) \right| \sqrt{y^{\dagger\prime} e^{-i\omega t_1} y'' e^{i\omega t_2}} \right) \right|^2 .
$$
(17.111)

If the oscillator energy is large, initially, and changes by a small factor, we can write, approximately,

$$
\left| \frac{y'e^{i\omega t_1}}{y''e^{i\omega t_2}} \right|^{\Delta n} = \exp \left(\Delta n \operatorname{Re} \log \frac{y'e^{i\omega t_1}}{y''e^{i\omega t_2}} \right)
$$
$$
\equiv \exp \left[\Delta n \operatorname{Re} \frac{y'e^{i\omega t_1} - y''e^{i\omega t_2}}{y''e^{i\omega t_2}} \right] ,
$$
(17.112)

which gives the probability distribution

$$
\exp \left[- \left| y'e^{i\omega t_1} - y'' \sqrt{1 + \frac{\Delta n}{|y''|^2}} e^{i\omega t_2} \right|^2 \right] \left[J_{\Delta n} \left(\frac{4\pi}{v} \lambda |\mu(\omega/v)||y'| \right) \right]^2 .
$$
(17.113)

The latter shows that the oscillator energy is decreased by Δn, and that the probability for this momentum is given by the square of the Bessel function.

17.5.2 Correlations

We have discussed only the first approximation, in which correlations between successive emission and absorption acts are ignored. This is insufficient to describe amplification by the oscillator, which depends upon just such correlations. The completely formal description of a system with Hamiltonian $H = H_0 + \lambda H_1$ is given by

$$
\langle t_1 | t_2 \rangle = \langle t_1 | \left(e^{-i \int_{t_2}^{t_1} dt\, \lambda H_1(q)} \right)_+ | t_2 \rangle_{H_0} ,
$$
(17.114)

which is the matrix element of a time ordered product, with all operators and states varying in accordance with the Hamiltonian H_0. The general expansion in successive correlations is indicated by

$$
\langle t_1 | t_2 \rangle = \langle t_1 | t_2 \rangle_{H_0} \exp \left\{ - i \int_{t_2}^{t_1} dt\, \lambda \langle H_1(t) \rangle \right.
$$
$$
\left. - \frac{1}{2} \int_{t_2}^{t_1} dt\, dt'\, \lambda^2 \left[\langle (H_2(t) H_1(t'))_+ \rangle - \langle H_1(t) \rangle \langle H_1(t') \rangle \right] + \cdots \right\} .
$$
(17.115)

Alternatively, one can treat the two interacting systems asymmetrically. Thus, with $H_1 = (y^\dagger + y)\mu(q)$, an expansion in successive correlations for the particle only gives

$$\langle t_1|t_2\rangle = \langle t_1| \left(\exp\left\{ -i \int dt\, \lambda(y^\dagger + y)(t)\langle \mu(t)\rangle - \frac{1}{2} \int dt\, dt'\, \lambda^2(y^\dagger + y)(t) \right. \right.$$
$$\left. \left. \times (y^\dagger + y)(t')[\langle(\mu(t)\mu(t'))_+\rangle - \langle\mu(t)\rangle\langle\mu(t')\rangle] + \cdots \right\} \right)_+ |t_2\rangle_{H_0},$$

$$(17.116)$$

which is identical with the result obtained from the effective action operator referring only to the oscillator,

$$W = \int dt \left[iy^\dagger \cdot \frac{dy}{dt} - \omega y^\dagger \cdot y - \lambda(y + y^\dagger)\langle\mu(t)\rangle \right]$$
$$- \frac{i}{2} \int dt\, dt'\, \lambda^2 \left((y + y^\dagger)(t)(y + y^\dagger)(t') \right)_+$$
$$\times [\langle(\mu(t)\mu(t'))_+\rangle - \langle\mu(t)\rangle\langle\mu(t')\rangle] + \frac{1}{i} \log\langle t_1|t_2\rangle_{\text{part}} \cdot (17.117)$$

Thus, from the latter, we derive, approximately, the effective oscillator equation of motion

$$i\frac{dy}{dt} = \omega y + \lambda\langle\mu(t)\rangle + i\lambda^2 \int_{t_2}^{t_1} dt[\langle(\mu(t)\mu(t'))_+\rangle - \langle\mu(t)\rangle\langle\mu(t')\rangle](y + y^\dagger)(t)$$

$$(17.118)$$

that indicates the change in behavior produced by the presence of the particle. Further elaborations will be left to Harold.

17.6 Problems for Chap. 17

1. Derive the Gaussian distribution (17.59) as the large n limit of the Poisson distribution (17.57).
2. Verify the transformation function for a free particle, (17.69).
3. Verify the probability distribution for an electron interacting with an oscillator, (17.92), and then derive the approximate form (17.97).

Appendix

Electromagnetic Units

The question of electromagnetic units has been a vexing one for students of electromagnetic theory for generations, and is likely to remain so for the foreseeable future. It was thought by the reformers of the 1930s, Sommerfeld [32] and Stratton [33] in particular, that the rationalized system now encompassed in the standard Système International (SI) would supplant the older cgs systems, principally the Gaussian (G) and Heaviside–Lorentz (HL) systems. This has not occurred. This is largely because the latter are far more natural from a relativistic point of view; theoretical physicists, at least of the high-energy variety, use nearly exclusively rationalized or unrationalized cgs units. The advantage of the two mentioned cgs systems (there are other systems, which have completely fallen out of use) is that then all the electric and magnetic fields, **E**, **D**, **B**, **H**, have the same units, which is only natural since electric and magnetic fields transform into each other under Lorentz transformations. Electric permittivities and magnetic permeabilities correspondingly are dimensionless. The reason for the continued survival of two systems of cgs units lies in the question of "rationalization," that is, the presence or absence of 4πs in Maxwell's equations or in Coulomb's law. The rationalized Heaviside–Lorentz system is rather natural from a field theoretic point of view; but if one's interest is solely electromagnetism it is hard not to prefer Gaussian units.

In our previous book [9] we took a completely consistent approach of using Gaussian units throughout. However, such consistency is not present in any practitioner's work. Jackson's latest version of his classic text [13] changes horses midstream. Here we have adopted what may appear to be an even more schizophrenic approach: Where emphasis is on waveguide and transmission line descriptions, we use SI units, whereas more theoretical chapters are written in the HL system. This reflects the diverse audiences addressed by the materials upon which this book is based, engineers and physicists.

Thus we must live with disparate systems of electromagnetic units. The problem, however, is not so very complicated as it may first appear. Let us start by writing Maxwell's equations in an arbitrary system:

$$\nabla \cdot \mathbf{D} = k_1 \rho \,, \tag{A.1a}$$

$$\nabla \cdot \mathbf{B} = 0 \,, \tag{A.1b}$$

$$\nabla \times \mathbf{H} = k_2 \dot{\mathbf{D}} + k_1 k_2 \mathbf{J} \,, \tag{A.1c}$$

$$-\nabla \times \mathbf{E} = k_2 \dot{\mathbf{B}} \,, \tag{A.1d}$$

while the constitutive relations are

$$\mathbf{D} = k_3 \mathbf{E} + k_1 \mathbf{P} \,, \tag{A.2a}$$

$$\mathbf{H} = k_4 \mathbf{B} - k_1 \mathbf{M} \,. \tag{A.2b}$$

The Lorentz force law is

$$\mathbf{F} = e(\mathbf{E} + k_2 \mathbf{v} \times \mathbf{B}) \,. \tag{A.3}$$

The values of the four constants in the various systems of units are displayed in Table A.1. Here the constants appearing in the SI system have defined

Table A.1. Constants appearing in Maxwell's equations and the Lorentz force law in the different systems of units

constant	SI	HL	Gaussian
k_1	1	1	4π
k_2	1	$\frac{1}{c}$	$\frac{1}{c}$
k_3	ε_0	1	1
k_4	$\frac{1}{\mu_0}$	1	1

values:

$$\mu_0 = 4\pi \times 10^{-7} \, \mathrm{N\,A^{-2}} \,, \tag{A.4a}$$

$$\frac{1}{\sqrt{\epsilon_0 \mu_0}} = c \equiv 299\,792\,458 \, \mathrm{m/s} \,, \tag{A.4b}$$

where the value of the speed of light is defined to be exactly the value given. (It is the presence of the arbitrary additional constant μ_0 which seems objectionable on theoretical grounds.)

Now we can ask how the various electromagnetic quantities are rescaled when we pass from one system of units to another. Suppose we take the SI system as the base. Then, in another system the fields and charges are given by

$$\mathbf{D} = \kappa_D \mathbf{D}^{\mathrm{SI}} \,, \quad \mathbf{E} = \kappa_E \mathbf{E}^{\mathrm{SI}} \,, \tag{A.5a}$$

$$\mathbf{H} = \kappa_H \mathbf{H}^{\mathrm{SI}} \,, \quad \mathbf{B} = \kappa_B \mathbf{B}^{\mathrm{SI}} \,, \tag{A.5b}$$

$$\mathbf{P} = \kappa_P \mathbf{P}^{\mathrm{SI}} \,, \quad \mathbf{M} = \kappa_M \mathbf{M}^{\mathrm{SI}} \,, \tag{A.5c}$$

$$\rho = \kappa_\rho \rho^{\mathrm{SI}} \,, \quad \mathbf{J} = \kappa_J \mathbf{J}^{\mathrm{SI}} \,. \tag{A.5d}$$

We insert these into the Maxwell equations, and determine the κs from the constants in Table A.1. For the Gaussian system, the results are

$$\kappa_D = \sqrt{\frac{\varepsilon_0}{4\pi}} \, , \tag{A.6a}$$

$$\kappa_E = \frac{1}{\sqrt{4\pi\varepsilon_0}} \, , \tag{A.6b}$$

$$\kappa_H = \frac{1}{\sqrt{4\pi\mu_0}} \, , \tag{A.6c}$$

$$\kappa_B = \sqrt{\frac{\mu_0}{4\pi}} \, , \tag{A.6d}$$

$$\kappa_P = \kappa_\rho = \kappa_J = \sqrt{4\pi\varepsilon_0} \, , \tag{A.6e}$$

$$\kappa_M = \sqrt{\frac{4\pi}{\mu_0}} \, . \tag{A.6f}$$

The conversion factors for HL units are the same except the various 4πs are omitted. By multiplying by these factors any SI equation can be converted to an equation in another system.

Here is a simple example of converting a formula. In SI, the skin depth of an imperfect conductor is given by (13.118),

$$\delta = \sqrt{\frac{2}{\mu\omega\sigma}} \, . \tag{A.7}$$

Converting into Gaussian units, the conductivity becomes

$$\sigma = \frac{J}{E} \rightarrow 4\pi\varepsilon \frac{J}{E} = 4\pi\varepsilon\sigma \, . \tag{A.8}$$

Therefore, the skin depth becomes

$$\delta \rightarrow \sqrt{\frac{2}{4\pi\varepsilon\mu\sigma\omega}} = \frac{c}{\sqrt{2\pi\sigma\omega}} \, , \tag{A.9}$$

which is the familiar Gaussian expression.

Let us illustrate how evaluation works in another simple example. The so-called classical radius of the electron is given in terms of the mass and charge on the electron, m and e, respectively,

$$r_0 = \frac{e^2}{4\pi\varepsilon_0 mc^2}\bigg|_{SI} = \frac{e^2}{4\pi mc^2}\bigg|_{HL} = \frac{e^2}{mc^2}\bigg|_{G} \, . \tag{A.10}$$

where the charges are related by κ_ρ in (A.6e). Let us evaluate the formula in SI and G systems:

$$r_0 = \frac{(1.602 \times 10^{-19} \text{ C})^2 \times 10^{-7} \text{ N A}^{-2}}{9.109 \times 10^{-31} \text{ kg}} = 2.818 \times 10^{-15} \text{ m} , \qquad \text{(A.11a)}$$

$$r_0 = \frac{(4.803 \times 10^{-10} \text{ esu})^2}{9.109 \times 10^{-28} \text{ g} \times (2.998 \times 10^{10} \text{ cm/s})^2} = 2.818 \times 10^{-13} \text{ cm} .$$

$$\text{(A.11b)}$$

It is even easier to evaluate this in terms of dimensionless quantities, such as the fine structure constant

$$\alpha = \left.\frac{e^2}{\hbar c}\right|_{\text{G}} = \left.\frac{e^2}{4\pi\hbar c}\right|_{\text{HL}} = \left.\frac{e^2}{4\pi\varepsilon_0\hbar c}\right|_{\text{SI}} = \frac{1}{137.036} . \qquad \text{(A.12)}$$

The classical radius of the electron is then proportional to the Compton wavelength of the electron,

$$\lambda_c = \frac{\hbar c}{mc^2} = 3.8616 \times 10^{-13} \text{ m} , \qquad \text{(A.13)}$$

where a convenient conversion factor is $\hbar c = 1.97327 \times 10^{-5}$ eV cm. Thus

$$r_0 = \alpha\lambda_c = 2.818 \times 10^{-15} \text{ m} , \qquad \text{(A.14)}$$

which incidentally shows that the "classical radius" gives an unphysically small measure of the "size" of an electron.

More discussion of electromagnetic units can be found in the Appendix of [9]. For a rather complete discussion see [34].

References

1. J. Mehra, K.A. Milton: *Climbing the Mountain: The Scientific Biography of Julian Schwinger* (Oxford University Press, Oxford, 2000)
2. K.A. Milton, editor: *A Quantum Legacy: Seminal Papers of Julian Schwinger* (World Scientific, Singapore, 2000)
3. M. Flato, C. Fronsdal, K.A. Milton, editors: *Selected Papers (1937–1976) of Julian Schwinger* (Reidel, Dordrecht, 1979)
4. M. Kac: In: *Mark Kac: Probability, Number Theory, and Statistical Physics – Selected Papers*, ed by K. Baclawski, M.D. Dowsker (MIT Press, Cambridge, 1979)
5. J. Schwinger, D.S. Saxon: *Discontinuities in Waveguides: Notes on Lectures by Julian Schwinger* (Gordon and Breach, New York, 1968)
6. N. Marcuvitz, editor: *The Waveguide Handbook* (McGraw-Hill, New York, 1951)
7. J. Schwinger, J.R. Oppenheimer: Phys. Rev. **56**, 1066 (1939)
8. J. Schwinger: *Quantum Mechanics: Symbolism of Atomic Measurement* (Springer, Berlin, Heidelberg New York, 2001)
9. J. Schwinger, L.L. DeRaad Jr., K.A. Milton, Wu-yang Tsai: *Classical Electrodynamics* (Perseus/Westview, New York, 1998)
10. J. Schwinger: *Quantum Kinematics and Dynamics* (W. A. Benjamin, New York, 1970)
11. J. Schwinger, editor: *Selected Papers on Quantum Electrodynamics* (Dover, New York, 1958)
12. J. Schwinger: *Particles, Sources, and Fields*, vols. I–III (Addison-Wesley [Perseus Books], Reading, MA, 1970, 1973, 1988)
13. J.D. Jackson: *Classical Electrodynamics* (McGraw-Hill, New York, 1998)
14. J.A. Wheeler, R.P. Feynman: Rev. Mod. Phys. **17**, 157 (1945)
15. J. Schwinger: Found. Phys. **13**, 373 (1983)
16. L.R. Elias, et al.: Phys. Rev. Lett. **36**, 717 (1976); D.A.G. Deacon, et al.: Phys. Rev. Lett. **38**, 892 (1977); H. Boehmer, et al.: Phys. Rev. Lett. **48**, 141 (1982); M. Billardon, et al.: Phys. Rev. Lett. **51**, 1652 (1983)
17. E.T. Whittaker, G.N. Watson: *A Course in Modern Analysis* (Cambridge University Press, Cambridge, 1965)
18. S.Y. Lee: *Accelerator Physics*, 2nd edn. (World Scientific, Singapore, 2004)

19. Particle Data Group: Review of Particle Physics, Phys. Lett. B **592**, 1 (2004), http://pdg.lbl.gov/2004/reviews/collidersrpp.pdf
20. A.W. Chao, M. Tigner, editors: *Handbook of Accelerator Physics and Engineering* (World Scientific, Singapore, 2002)
21. A.W. Chao: *Physics of Collective Beam Instabilities in High Energy Accelerators* (Wiley, New York, 1993)
22. M. Reiser: *Theory and Design of Charged Particle Beams* (Wiley, New York, 1994)
23. P.A.M. Dirac: Proc. R. Soc. **167**, 148 (1938).
24. Julian Schwinger Papers (Collection 371), Department of Special Collections, University Research Library, University of California, Los Angeles
25. J. Schwinger: Phys. Rev. **75**, 1912 (1949)
26. D. Ivanenko, A.A. Sokolov: Dokl. Akad. Nauk SSSR [Sov. Phys. Dokl.] **59**, 1551 (1948)
27. A.A. Sokolov, I.M. Ternov: *Synchrotron Radiation* (Akademie-Verlag, Berlin; Pergamon Press, Oxford, 1968)
28. H. Wiedemann: *Synchrotron Radiation* (Springer, Berlin, Heidelberg, New York, 2003)
29. A. Hofmann: *The Physics of Synchrotron Radiation* (Cambridge University Press, Cambridge, 2004)
30. A. Sommerfeld: Math. Ann. **47**, 317 (1896); Zeits. f. Math. u. Physik **46**, 11 (1901)
31. P.M. Morse, P.J. Rubenstein: Phys. Rev. **54**, 895 (1938)
32. A. Sommerfeld: *Electrodynamics: Lectures in Theoretical Physics*, vol. 3 (Academic Press, New York, 1964)
33. J.A. Stratton: *Electromagnetic Theory* (Mc-Graw-Hill, New York, 1941)
34. F.B. Silsbee: *Systems of Electrical Units*, National Bureau of Standards Monograph 56 (U.S. Government Printing Office, Washington, 1962)

Index